Francis Stoessel

Thermal Safety of Chemical Processes

Further Reading

K. Bhagwati

Managing Safety

A Guide for Executives

2006

ISBN 978-3-527-31583-3

H. Schütz, P.M. Wiedemann,
W. Hennings, J. Mertens, M. Clauberg

Comparative Risk Assessment

Concepts, Problems and Applications

2006

ISBN 978-3-527-31667-0

R.L. Brauer

Safety and Health for Engineers

2005

ISBN 978-0-471-29189-3

C.A. Ericson

Hazard Analysis Techniques for System Safety

2005

ISBN 978-0-471-72019-5

F.P. Helmus

Process Plant Design

Project Management from Inquiry to Acceptance

2008

ISBN 978-3-527-31313-6

Francis Stoessel

Thermal Safety of Chemical Processes

Risk Assessment and Process Design

WILEY-VCH Verlag GmbH & Co. KGaA

The Author

Prof. Dr.-Ing. F. Stoessel
Swiss Institute for the
Promotion of Safety & Security
Schwarzwaldallee 215
4058 Basel
Switzerland

Library of Congress Card No.: applied for

British Library Cataloguing-in-Publication Data
A catalogue record for this book is available from the British Library.

Bibliographic information published by the Deutsche Nationalbibliothek
Die Deutsche Nationalbibliothek lists this publication in the Deutsche Nationalbibliografie; detailed bibliographic data are available on the Internet at <http://dnb.d-nb.de>.

Printed in the Federal Republic of Germany
Printed on acid-free paper

Typesetting SNP Best-set Typesetter Ltd., Hong Kong
Printing Strauss GmbH, Mörlenbach
Bookbinding Litges & Dopf GmbH, Heppenheim

ISBN: 978-3-527-31712-7

To Michèle,
Sébastien, Claire-Lise and Anne-Florence

Contents

Preface *XVII*

Part I General Aspects of Thermal Process Safety *1*

1 Introduction to Risk Analysis of Fine Chemical Processes *3*
1.1 Introduction *3*
1.2 Chemical Industry and Safety *4*
1.2.1 Chemical Industry and Society *4*
1.2.1.1 Product Safety *4*
1.2.1.2 Process Safety *5*
1.2.1.3 Accidents in Chemical Industry *5*
1.2.1.4 Risk Perception *5*
1.2.2 Responsibility *6*
1.2.3 Definitions and Concepts *7*
1.2.3.1 Hazard *7*
1.2.3.2 Risk *7*
1.2.3.3 Safety *8*
1.2.3.4 Security *8*
1.2.3.5 Accepted Risk *8*
1.3 Risk Analysis *8*
1.3.1 Steps of Risk Analysis *8*
1.3.1.1 Scope of Analysis *9*
1.3.1.2 Safety Data Collection *10*
1.3.1.3 Safe Conditions and Critical Limits *10*
1.3.1.4 Search for Deviations *10*
1.3.1.5 Risk Assessment *12*
1.3.1.6 Risk Profiles *14*
1.3.1.7 Risk Reducing Measures *14*
1.3.1.8 Residual Risk *16*
1.4 Safety Data *17*
1.4.1.1 Physical Properties *17*
1.4.1.2 Chemical Properties *17*

Thermal Safety of Chemical Processes: Risk Assessment and Process Design. Francis Stoessel

ISBN: 978-3-527-31712-7

1.4.1.3 Toxicity *18*
1.4.1.4 Ecotoxicity *19*
1.4.1.5 Fire and Explosion Data *19*
1.4.1.6 Interactions *20*
1.5 Systematic Search for Hazards *20*
1.5.1 Check List Method *21*
1.5.2 Failure Mode and Effect Analysis *22*
1.5.3 Hazard and Operability Study *23*
1.5.4 Decision Table *25*
1.5.5 Event Tree Analysis *25*
1.5.6 Fault Tree Analysis *26*
1.6 Key Factors for a Successful Risk Analysis *28*
References *29*

2 Fundamentals of Thermal Process Safety *31*
2.1 Introduction *33*
2.2 Energy Potential *34*
2.2.1 Thermal Energy *34*
2.2.1.1 Heat of Reaction *34*
2.2.1.2 Heat of Decomposition *35*
2.2.1.3 Heat Capacity *35*
2.2.1.4 Adiabatic Temperature Rise *37*
2.2.2 Pressure Effects *38*
2.2.2.1 Gas Release *39*
2.2.2.2 Vapor Pressure *39*
2.2.2.3 Amount of Solvent Evaporated *39*
2.3 Effect of Temperature on Reaction Rate *40*
2.3.1 Single Reaction *40*
2.3.2 Multiple Reactions *41*
2.4 Heat Balance *42*
2.4.1 Terms of the Heat Balance *42*
2.4.1.1 Heat Production *43*
2.4.1.2 Heat Removal *43*
2.4.1.3 Heat Accumulation *45*
2.4.1.4 Convective Heat Exchange Due to Mass Flow *46*
2.4.1.5 Sensible Heat Due to Feed *46*
2.4.1.6 Stirrer *46*
2.4.1.7 Heat Losses *47*
2.4.2 Simplified Expression of the Heat Balance *48*
2.4.3 Reaction Rate under Adiabatic Conditions *48*
2.5 Runaway Reactions *50*
2.5.1 Thermal Explosions *50*
2.5.2 Semenov Diagram *50*
2.5.3 Parametric Sensitivity *52*
2.5.4 Critical Temperature *52*

2.5.5 Time Frame of a Thermal Explosion, the TMR_{ad} Concept 54
2.6 Exercises 56
References 58

3 Assessment of Thermal Risks 59
3.1 Introduction 59
3.2 Thermal Risks 60
3.3 Systematic Assessment Procedure 61
3.3.1 Cooling Failure Scenario 61
3.3.2 Severity 64
3.3.3 Probability 66
3.3.4 Criticality of Chemical Processes 67
3.3.5 Assessment of the Criticality 67
3.3.6 Criticality Classes 68
3.3.6.1 Criticality Class 1 69
3.3.6.2 Criticality Class 2 69
3.3.6.3 Criticality Class 3 70
3.3.6.4 Criticality Class 4 70
3.3.6.5 Criticality Class 5 70
3.3.6.6 Remarks Concerning the Use of MTT as a Safety Barrier 71
3.4 Assessment Procedures 71
3.4.1 General Rules for Thermal Safety Assessment 71
3.4.2 Practical Procedure for the Assessment of Thermal Risks 72
3.5 Exercises 78
References 80

4 Experimental Techniques 81
4.1 Introduction 82
4.2 Calorimetric Measurement Principles 82
4.2.1 Classification of Calorimeters 82
4.2.2 Operating Modes of Calorimeters 83
4.2.3 Heat Balance in Calorimeters 84
4.2.3.1 Ideal Accumulation 85
4.2.3.2 Ideal Heat Flow 85
4.2.3.3 Isoperibolic Methods 85
4.3 Choice of Instruments Used in Safety Laboratories 85
4.3.1 Adiabatic Calorimeters 86
4.3.1.1 On the Evaluation of Adiabatic Experiments 86
4.3.1.2 Dewar Calorimeters 88
4.3.1.3 Accelerating Rate Calorimeter (ARC) 89
4.3.2 Micro Calorimeters 90
4.3.2.1 Differential Scanning Calorimetry (DSC) 90
4.3.2.2 Calvet Calorimeters 92
4.3.2.3 Thermal Activity Monitor 94
4.3.3 Reaction Calorimeters 95

4.4 Exercises 96
References 97

Part II Mastering Exothermal Reactions *101*

5 General Aspects of Reactor Safety *103*
5.1 Introduction *104*
5.2 Dynamic Stability of Reactors *105*
5.2.1 Parametric Sensitivity *105*
5.2.2 Sensitivity Towards Temperature: Reaction Number B *105*
5.2.3 Heat Balance *107*
5.2.3.1 The Semenov Criterion *107*
5.2.3.2 Stability Diagrams *107*
5.2.3.3 Heat Release Rate and Cooling Rate *107*
5.2.3.4 Using Dimensionless Criteria *109*
5.2.3.5 Chaos Theory and Lyapunov Exponents *110*
5.2.4 Reactor Safety After a Cooling Failure *111*
5.2.4.1 Potential of the Reaction, the Adiabatic Temperature Rise *111*
5.2.4.2 Temperature in Cases of Cooling Failure: The Concept of MTSR *112*
5.3 Example *112*
5.3.1 Example Reaction System *112*
References *116*

6 Batch Reactors *119*
6.1 Introduction *120*
6.2 Principles of Batch Reaction *121*
6.2.1 Introduction *121*
6.2.2 Mass Balance *121*
6.2.3 Heat Balance *122*
6.3 Strategies of Temperature Control *123*
6.4 Isothermal Reactions *123*
6.4.1 Principles *123*
6.4.2 Design of Safe Isothermal Reactors *123*
6.4.3 Safety Assessment *127*
6.5 Adiabatic Reaction *127*
6.5.1 Principles *127*
6.5.2 Design of a Safe Adiabatic Batch Reactor *128*
6.5.3 Safety Assessment *128*
6.6 Polytropic Reaction *128*
6.6.1 Principles *128*
6.6.2 Design of Polytropic Operation, Temperature Control *130*
6.6.3 Safety Assessment *133*
6.7 Isoperibolic Reaction *133*
6.7.1 Principles *133*

6.7.2 Design of Isoperibolic Operation, Temperature Control *134*
6.7.3 Safety Assessment *134*
6.8 Temperature Controlled Reaction *135*
6.8.1 Principles *135*
6.8.2 Design of Temperature Controlled Reaction *135*
6.8.3 Safety Assessment *136*
6.9 Key Factors for the Safe Design of Batch Reactors *138*
6.9.1 Determination of Safety Relevant Data *138*
6.9.2 Rules for Safe Operation of Batch Reactors *141*
6.10 Exercises *144*
References *146*

7 Semi-batch Reactors *147*
7.1 Introduction *148*
7.2 Principles of Semi-batch Reaction *149*
7.2.1 Definition of Semi-batch Operation *149*
7.2.2 Material Balance *149*
7.2.3 Heat Balance of Semi-batch Reactors *151*
7.2.3.1 Heat Production *151*
7.2.3.2 Thermal Effect of the Feed *151*
7.2.3.3 Heat Removal *151*
7.2.3.4 Heat Accumulation *152*
7.3 Reactant Accumulation in Semi-batch Reactors *153*
7.3.1 Fast Reactions *153*
7.3.2 Slow Reactions *156*
7.4 Design of Safe Semi-batch Reactors *158*
7.5 Isothermal Reaction *159*
7.5.1 Principles of Isothermal Semi-batch Operation *159*
7.5.2 Design of Isothermal Semi-batch Reactors *159*
7.6 Isoperibolic, Constant Cooling Medium Temperature *163*
7.7 Non-isothermal Reaction *166*
7.8 Strategies of Feed Control *167*
7.8.1 Addition by Portions *167*
7.8.2 Constant Feed Rate *167*
7.8.3 Interlock of Feed with Temperature *169*
7.8.4 Why to Reduce the Accumulation *170*
7.9 Choice of Temperature and Feed Rate *171*
7.10 Feed Control by Accumulation *173*
7.11 Exercises *176*
References *178*

8 Continuous Reactors *179*
8.1 Introduction *180*
8.2 Continuous Stirred Tank Reactors *180*
8.2.1 Mass Balance *181*

8.2.2 Heat Balance *182*
8.2.3 Cooled CSTR *182*
8.2.4 Adiabatic CSTR *183*
8.2.5 The Autothermal CSTR *185*
8.2.6 Safety Aspects *185*
8.2.6.1 Instabilities at Start-up or Shut Down *185*
8.2.6.2 Behavior in Case of Cooling Failure *186*
8.3 Tubular Reactors *189*
8.3.1 Mass Balance *189*
8.3.2 Heat Balance *190*
8.3.3 Safety Aspects *192*
8.3.3.1 Parametric Sensitivity *192*
8.3.3.2 Heat Exchange Capacities of Tubular Reactors *193*
8.3.3.3 Passive Safety Aspects of Tubular Reactors *193*
8.4 Other Continuous Reactor Types *198*
8.4.1.1 Cascade of CSTRs and Recycle Reactor *198*
8.4.1.2 Micro Reactors *199*
References *201*

9 Technical Aspects of Reactor Safety *203*
9.1 Introduction *204*
9.2 Temperature Control of Industrial Reactors *205*
9.2.1 Technical Heat Carriers *205*
9.2.1.1 Steam Heating *205*
9.2.1.2 Hot Water Heating *206*
9.2.1.3 Other Heating Media *207*
9.2.1.4 Electrical Heating *207*
9.2.1.5 Cooling with Ice *207*
9.2.1.6 Other Heat Carriers for Cooling *207*
9.2.2 Heating and Cooling Techniques *208*
9.2.2.1 Direct Heating and Cooling *208*
9.2.2.2 Indirect Heating and Cooling of Stirred Tank Reactors *208*
9.2.2.3 Single Heat Carrier Circulation Systems *209*
9.2.2.4 Secondary Circulation Loop Temperature Control Systems *211*
9.2.3 Temperature Control Strategies *212*
9.2.3.1 Isoperibolic Temperature Control *212*
9.2.3.2 Isothermal Control *212*
9.2.3.3 Isothermal Control at Reflux *214*
9.2.3.4 Non Isothermal Temperature Control *215*
9.2.4 Dynamic Aspects of Heat Exchange Systems *215*
9.2.4.1 Thermal Time Constant *215*
9.2.4.2 Heating and Cooling Time *217*
9.2.4.3 Cascade Controller *219*
9.3 Heat Exchange Across the Wall *219*
9.3.1 Two Film Theory *219*

9.3.2 The Internal Film Coefficient of a Stirred Tank 220
9.3.3 Determination of the Internal Film Coefficient 221
9.3.4 The Resistance of the Equipment to Heat Transfer 222
9.3.5 Practical Determination of Heat Transfer Coefficients 224
9.4 Evaporative Cooling 226
9.4.1 Amount of Solvent Evaporated 228
9.4.2 Vapor Flow Rate and Rate of Evaporation 228
9.4.3 Flooding of the Vapor Tube 229
9.4.4 Swelling of the Reaction Mass 230
9.4.5 Practical Procedure for the Assessment of Reactor Safety at the Boiling Point 231
9.5 Dynamics of the Temperature Control System and Process Design 233
9.5.1 Background 233
9.5.2 Modeling the Dynamic Behavior of Industrial Reactors 234
9.5.3 Experimental Simulation of Industrial Reactors 234
9.6 Exercises 236
References 240

10 Risk Reducing Measures 241
10.1 Introduction 243
10.2 Strategies of Choice 243
10.3 Eliminating Measures 244
10.4 Technical Preventing Measures 245
10.4.1 Control of Feed 245
10.4.2 Emergency Cooling 246
10.4.3 Quenching and Flooding 247
10.4.4 Dumping 248
10.4.5 Controlled Depressurization 248
10.4.6 Alarm Systems 251
10.4.7 Time Factor 252
10.5 Emergency Measures 253
10.5.1 Emergency Pressure Relief 253
10.5.1.1 Definition of the Relief Scenario 254
10.5.1.2 Design of the Relief Device 255
10.5.1.3 Design of Relief Devices for Multipurpose Reactors 255
10.5.1.4 Design of the Effluent Treatment System 256
10.5.2 Containment 256
10.6 Design of Technical Measures 257
10.6.1 Consequences of Runaway 257
10.6.1.1 Temperature 257
10.6.1.2 Pressure 258
10.6.1.3 Release 258
10.6.1.4 Closed Gassy Systems 258
10.6.1.5 Closed Vapor Systems 259

10.6.1.6 Open Gassy Systems 259
10.6.1.7 Open Vapor Systems 259
10.6.1.8 Extended Assessment Criteria for Severity 260
10.6.2 Controllability 260
10.6.2.1 Activity of the Main Reaction 261
10.6.2.2 Activity of Secondary Reactions 261
10.6.2.3 Gas Release Rate 262
10.6.2.4 Vapor Release Rate 262
10.6.2.5 Extended Assessment Criteria for the Controllability 263
10.6.3 Assessment of Severity and Probability for the Different Criticality Classes 264
10.6.3.1 Criticality Class 1 264
10.6.3.2 Criticality Class 2 264
10.6.3.3 Criticality Class 3 265
10.6.3.4 Criticality Class 4 266
10.6.3.5 Criticality Class 5 267
10.6.4 Protection System Based on Risk Assessment 273
10.6.4.1 Risk Assessment 273
10.6.4.2 Determination of the Required Reliability for Safety Instrumented Systems 273
10.7 Exercises 274
References 276

Part III Avoiding Secondary Reactions *279*

11 Thermal Stability *281*
11.1 Introduction 282
11.2 Thermal Stability and Secondary Decomposition Reactions 282
11.3 Consequences of Secondary Reactions 284
11.3.1 Stoichiometry of Decomposition Reactions 284
11.3.2 Estimation of Decomposition Energies 284
11.3.3 Decomposition Energy 284
11.4 Triggering Conditions 286
11.4.1 Onset: A Concept without Scientific Base 286
11.4.2 Decomposition Kinetics, the TMR_{ad} Concept 287
11.4.2.1 Determination of $q' = f(T)$ from Isothermal Experiments 288
11.4.2.2 Determination of T_{D24} 290
11.4.2.3 Estimation of T_{D24} from One Dynamic DSC Experiment 290
11.4.2.4 Empirical Rules for the Determination of a “Safe” Temperature 294
11.4.3 Complex Secondary Reactions 295
11.4.3.1 Determination of TMR_{ad} from Isothermal Experiments 296
11.4.3.2 Determination of $q' = f(T)$ from Dynamic Experiments 296
11.5 Experimental Characterization of Decomposition Reactions 298
11.5.1 Experimental Techniques 298
11.5.2 Choosing the Sample to be Analysed 299

11.5.2.1 Sample Purity *299*
11.5.2.2 Batch or Semi-batch Process *299*
11.5.2.3 Intermediates *301*
11.5.3 Process Deviations *302*
11.5.3.1 Effect of Charging Errors *302*
11.5.3.2 Effect of Solvents on Thermal Stability *303*
11.5.3.3 Catalytic Effects of Impurities *303*
11.6 Exercises *305*
References *307*

12 Autocatalytic Reactions *311*
12.1 Introduction *312*
12.2 Autocatalytic Decompositions *312*
12.2.1 Definitions *312*
12.2.1.1 Autocatalysis *312*
12.2.1.2 Induction Time *313*
12.2.2 Behavior of Autocatalytic Reactions *313*
12.2.3 Rate Equations of Autocatalytic Reactions *315*
12.2.3.1 The Prout–Tompkins Model *315*
12.2.3.2 The Benito–Perez Model *316*
12.2.3.3 The Berlin Model *317*
12.2.4 Phenomenological Aspects of Autocatalytic Reactions *318*
12.3 Characterization of Autocatalytic Reactions *319*
12.3.1 Chemical Characterization *319*
12.3.2 Characterization by Dynamic DSC *320*
12.3.2.1 Peak Aspect in Dynamic DSC *320*
12.3.2.2 Quantitative Characterization of the Peak Aspect *321*
12.3.2.3 Characterization by Isothermal DSC *322*
12.3.2.4 Characterization Using Zero-order Kinetics *323*
12.3.2.5 Characterization Using a Mechanistic Approach *324*
12.3.2.6 Characterization by Isoconversional Methods *324*
12.3.2.7 Characterization by Adiabatic Calorimetry *325*
12.4 Practical Safety Aspects for Autocatalytic Reactions *325*
12.4.1 Specific Safety Aspects of Autocatalytic Reactions *325*
12.4.2 Assessment Procedure for Autocatalytic Decompositions *331*
12.5 Exercises *332*
References *333*

13 Heat Confinement *335*
13.1 Introduction *335*
13.2 Heat Accumulation Situations *336*
13.3 Heat Balance *337*
13.3.1 Heat Balance Using Time Scale *338*
13.3.2 Forced Convection, Semenov Model *338*
13.3.3 Natural Convection *340*

13.3.4 High Viscosity Liquids and Solids *341*
13.4 Heat Balance with Reactive Material *343*
13.4.1 Conduction in a Reactive Solid with a Heat Source, Frank-Kamenetskii Model *344*
13.4.2 Conduction in a Reactive Solid with Temperature Gradient at the Wall, Thomas Model *348*
13.4.3 Conduction in a Reactive Solid with Formal Kinetics, Finite Elements Model *350*
13.5 Assessing Heat Accumulation Conditions *351*
13.6 Exercises *357*
References *359*

14 Symbols *361*

Index *367*

Preface

Often, chemical incidents are due to loss of control, resulting in runaway reactions. Many of these incidents can be foreseen and avoided, if an appropriate analysis of thermal process data is performed in the proper way and in due time. Chemical process safety is seldom part of university curricula and many professionals do not have the appropriate knowledge to interpret thermal data in terms of risks. As a result, even though responsible for the safety of the process, they do not have easy access to the knowledge. Process safety is often considered a specialist matter, thus most large companies employ specialists in their safety departments. However, this safety knowledge is also required at the front, where processes are developed or performed, that is in process development departments and production. To achieve this objective of providing professionals with the required knowledge on the thermal aspects of their processes, the methods must be made accessible to non-specialists. Such systematic and easy-to-use methods represent the backbone of this book, in which the methods used for the assessment of thermal risks are presented in a logical and understandable way, with a strong link to industrial practice.

The present book is rooted in a lecture on chemical process safety at graduate level (Masters) at the Swiss Federal Institute of Technology in Lausanne. It is also based on experience gained in numerous training courses for professionals held at the Swiss Institute for the Promotion of Safety & Security, as well as in a number of major chemical and pharmaceutical companies. Thus it has the character of a textbook and addresses students, but also addresses professional chemists, chemical engineers or engineers in process development and production of fine chemicals and pharmaceutical industries, as support for their practice of process safety.

The objective of the book is not to turn the reader into a specialist in thermal safety. It is to guide those who perform risk analysis of chemical processes, develop new processes, or are responsible for chemical production, to understand the thermal aspects of processes and to perform a scientifically founded – but practically oriented – assessment of chemical process safety. This assessment may serve as a basis for the optimization or the development of thermally safe processes. The methods presented are based on the author's long years of experience in the practice of safety assessment in industry and teaching students and professionals

Thermal Safety of Chemical Processes: Risk Assessment and Process Design. Francis Stoessel

ISBN: 978-3-527-31712-7

in this matter. It is also intended to develop a common and understandable language between specialists and non-specialists.

The book is structured in three parts:

Part I gives a general introduction and presents the theoretical, methodological and experimental aspects of thermal risk assessment. The first chapter gives a general introduction on the risks linked to the industrial practice of chemical reactions. The second chapter reviews the theoretical background required for a fundamental understanding of runaway reactions and reviews the thermodynamic and kinetic aspects of chemical reactions. An important part of Chapter 2 is dedicated to the heat balance of reactors. In Chapter 3, a systematic evaluation procedure developed for the evaluation of thermal risks is presented. Since such evaluations are based on data, Chapter 4 is devoted to the most common calorimetric methods used in safety laboratories.

Part II is dedicated to desired reactions and techniques allowing reactions to be mastered on an industrial scale. Chapter 5 introduces the dynamic stability of chemical reactors and criteria commonly used for the assessment of such stability. The behavior of reactors under normal operating conditions is a prerequisite for safe operation, but is not sufficient by itself. Therefore the different reactor types are reviewed with their specific safety problems, particularly in the case of deviations from normal operating conditions. This requires a specific approach for each reactor type, including a study of the heat balance, which is the basis of safe temperature control, and also includes a study of the behavior in cases where the temperature control system fails. The analysis of the different reactor types and the general principles used in their design and optimization is presented in Chapters 6 to 8. Chapter 6 presents the safety aspects of batch reactors with a strong emphasis on the temperature control strategies allowing safe processes. In Chapter 7, the semi-batch reactor is analysed with the different temperature control strategies, but also with the feed control strategies reducing the accumulation of non-converted reactants. In Chapter 8, the use of continuous reactors for mastering exothermal reactions is introduced. The temperature control requires technical means that may strongly influence operation safety. Therefore Chapter 9 is dedicated to the technical aspects of heat transfer, and the estimation of heat transfer coefficients. Since risk reducing measures are often required to maintain safe operation, such as in the failure of the process control system, Chapter 10 is specifically dedicated to the evaluation of the control of a runaway reaction and the definition and design of appropriate risk reducing measures.

Part III deals with secondary reactions, their characterization, and techniques to avoid triggering them. Chapter 11 reviews the general aspects of secondary reactions, determination of the consequences of loss of control and the risk assessment. Chapter 12 is dedicated to the important category of self-accelerating reactions, their characteristics, and techniques allowing their control. The problem of heat confinement, in situations where heat transfer is reduced, is studied in Chapter 13. The different industrial situations where heat confinement may occur are reviewed and a systematic procedure for their assessment is presented together with techniques that may be used for the design of safe processes.

Each chapter begins with a case history illustrating the topic of the chapter and presenting lessons learned from the incident. Within the chapters, numerous examples stemming from industrial practice are analysed. At the end of each chapter, a series of exercises or case studies are proposed, allowing the reader to check their understanding of the subject matter.

Acknowledgements

The methodology presented in this book is the result of long-term experience and concerns with the assessment of thermal risks in the chemical process industry gained in the Central Safety Research Laboratories of Ciba. Therefore the author would like to thank his colleagues: K. Eigenmann, F. Brogli, R. Gygax, H. Fierz, B. Urwyler, P. Lerena, and W. Regenass, who all participated in the development of the methodology and techniques covered in this book. He would also like to thank the management of the Swiss Institute for the Promotion of Safety & Security, M. Glor and H. Rüegg, who encouraged him to persevere in the project.

Many applications and methods were developed by students or young colleagues during diploma works, PhD-thesis or development projects. Among others, the author is grateful to J.M. Dien, O. Ubrich, M.A. Schneider, B. Zufferey, P. Reuse and B. Roduit.

Writing a book like this is a long-term project, which cannot be brought to its end without some sacrifices. Thus my last thoughts go to my family, especially my wife Michèle, who not only accepted neglect during the course of writing, but also encouraged and supported me and the work.

Part I
General Aspects of Thermal Process Safety

1
Introduction to Risk Analysis of Fine Chemical Processes

Case History

A multi-purpose reactor was protected against overpressure by a rupture disk, which lead directly to the outside through the roof of the plant. During a maintenance operation, it was discovered that this disk was corroded. Although it was decided to replace it, there was no spare part available. Since the next task to be carried out was a sulfonation reaction, it was decided to leave the relief pipe open without the rupture disk in place. In fact, a sulfonation reaction cannot lead to overpressure (sulfuric acid only starts to boil above 300 °C), so such a protection device should not be required. During the first batch a plug of sublimate formed in the relief line. This went unnoticed and production continued. After heavy rain, water entered the relief tube and accumulated above the sublimate plug. As the next batch began, the plug heated and suddenly ruptured, allowing the accumulated water to enter the reactor. This led to a sudden exothermal effect, due to the dilution of concentrated sulfuric acid. The increase in temperature triggered sudden decomposition of the reaction mass, causing the reactor to burst, resulting in huge damage.

Lessons drawn

This type of incident is difficult to predict. Nevertheless, by using a systematic approach to hazard identification it should become clear that any water entering the reactor could lead to an explosion. Therefore when changing some parts of the equipment, even if they are not directly involved in a given process, especially in multi-purpose plants, one should at least consider possible consequences on the safety parameters of the process.

1.1
Introduction

Systematic searches for hazard, assessment of risk, and identification of possible remediation are the basic steps of risk analysis methods reviewed in this chapter.

Thermal Safety of Chemical Processes: Risk Assessment and Process Design. Francis Stoessel

ISBN: 978-3-527-31712-7

After an introduction that considers the place of chemical industry in society, the basic concepts related to risk analysis are presented. The second section reviews the steps of the risk analysis of chemical processes discussed. Safety data are presented in the third section and the methods of hazard identification in the section after that. The chapter closes with a section devoted to the practice of risk analysis.

1.2 Chemical Industry and Safety

The chemical industry, more than any other industry, is perceived as a threat to humans, society, and the environment. Nevertheless, the benefits resulting from this activity cannot be negated: health, crop protection, new material, colors, textiles, and so on. This negative perception is more enhanced after major accidents, such as those at Seveso and Bhopal. Even though such catastrophic incidents are rare, they are spectacular and retain public attention. Thus, a fundamental question is raised: "What risk does society accept regarding the benefits of an activity, of a product?" Such a question assumes that one is able–a priori–to assess the corresponding risk.

In the present chapter, we focus on the methods of risk analysis as they are performed in the chemical industry, and especially in fine chemicals and pharmaceutical industries.

1.2.1 Chemical Industry and Society

The aim of the chemical industry is to provide industry and people in general with functional products, which have a precise use in different activities such as pharmaceuticals, mechanics, electricity, electronics, textile, food, and so on.

Thus, on one hand, safety in the chemical industry is concerned with product safety, that is, the risks linked with the use of a product. On the other hand, it is concerned with process safety, that is, the risks linked with manufacturing the product. In this book, the focus is on process safety.

1.2.1.1 Product Safety

Every product between its discovery and its elimination passes through many different steps throughout its history: conception, design, feasibility studies, market studies, manufacturing, distribution, use, and elimination, the ultimate step, where from functional product, it becomes a waste product [1].

During these steps, risks exist linked to handling or using the product. This enters the negative side of the balance between benefits and adverse effects of the product. Even if the public is essentially concerned with the product risks during its use, risks are also present during other stages, that is, manufacture, transportation, and storage. For pharmaceutical products, the major concerns are secondary effects. For other products, adverse effects are toxicity for people and/or for the environment, as well as fire and explosion. Whatever its form, once a product is no longer functional, it becomes a waste product and thus represents a potential source of harm.

Therefore, during product design, important decision have to be made in order to maximize the benefits that are expected from the product and to minimize the negative effects that it may induce. These decisions are crucial and often taken after a systematic evaluation of the risks. Commercialization is strictly regulated by law and each new product must be registered with the appropriate authorities. The aim of the registration is to ensure that the manufacturer knows of any properties of its product that may endanger people or the environment and is familiar with the conditions allowing its safe handling and use, and finally safe disposal at the end of the product's life. Thus products are accompanied by a Material Safety Data Sheet (MSDS) that summarizes the essential safety information as product identity, properties (toxicity, eco-toxicity, physical chemical properties), information concerning its life cycle (use, technology, exposure), specific risks, protection measures, classification (handling, storage, transportation), and labeling.

1.2.1.2 Process Safety

The chemical industry uses numerous and often complex equipment and processes. In the fine chemical industries (including pharmaceuticals), the plants often have a multi-purpose character, that is, a given plant may be used for different products. When we consider a chemical process, we must do it in an extensive way, including not only the production itself but also storage and transportation. This includes not only the product, but also the raw material.

Risks linked with chemical processes are diverse. As already discussed, product risks include toxicity, flammability, explosion, corrosion, etc. but also include additional risks due to chemical reactivity. A process often uses conditions (temperature, pressure) that by themselves may present a risk and may lead to deviations that can generate critical effects. The plant equipment, including its control equipment, may also fail. Finally, since fine chemical processes are work-intensive, they may be subject to human error. All of these elements, that is, chemistry, energy, equipment, and operators and their interactions, constitute what we call process safety.

1.2.1.3 Accidents in Chemical Industry

Despite some incidents, the chemical industry presents good accident statistics. A statistical survey of work accidents shows that chemistry is positioned close to the end of the list, classified by order of decreasing lost work days [2] (Table 1.1). Further, these accidents only constitute a minor part due to chemical accidents, the greatest part consisting of common accidents such as falls, cuts, and so on that can happen in any other activity.

1.2.1.4 Risk Perception

Another instructive comparison can be made by comparing fatalities in different activities. Here we use the Fatal Accident Rate index (FAR) that gives the number of fatalities for 10^8 hours of exposure to the hazard [3, 4]. Some activities are compared in Table 1.2. This shows that even with better statistics in terms of fatali-

Table 1.1 Accidents at work in different industries in Switzerland, from the statistics of the Swiss National Accident Insurance (2005).

Activity	Work accidents for 1000 insured
Construction	185
Wood	183
Mining	160
Metallurgy	147
Cement, glass, ceramics	130
Food	113
Rubber, plastics	95
Machinery	72
Transport	66
Energy	59
Textile, clothes	50
Offices, administration	46
Paper, graphics	45
Chemistry	**37**
Electricity, fine mechanics	33

Table 1.2 Some values of the FAR index for different activities.

Industrial activities	FAR	Non industrial activities	FAR
Coal mining	7.3	Alpinism	4000
Construction	5	Canoe	1000
Agriculture	3.7	Motor bike	660
Chemistry	**1.2**	Travel by air	240
Vehicle manufacturing	0.6	Travel by car	57
Clothing manufacturing	0.05	Travel by railway	5

ties, industrial activities are perceived as presenting higher risks. This may essentially be due to the risk perception. The difference in perception is that for traveling or sporting activities, the person has the choice as to whether to be exposed or not, whereas for industrial activities exposure to risk may be imposed. Industrial risks may also impinge on people who are not directly concerned with the activity. Moreover, the lack of information on these risks biases the perception [5].

1.2.2 Responsibility

In industrial countries, employers are responsible for the safety of their employees. On the other hand, legal texts often force the employees to apply the safety rules prepared by employers. In this sense, the responsibility is shared. Environment protection is also regulated by law. Authorities publish threshold limits for pollutants and impose penalties in cases where these limits are surpassed. In the European Union, the Seveso directive regulates the prevention of major accidents: if dangerous substances are used in amounts above prescribed limits, industries have to prepare a risk analysis that describes quantitatively possible emissions and their effect on the neighboring population. They also have to provide emergency plans in order to protect that population.

In what concerns process safety, the responsibility is shared within the company by the management at different levels. The Health Safety and Environment staff plays an essential role in this frame, thus during process design, safety should have priority.

1.2.3 Definitions and Concepts

1.2.3.1 Hazard

Definition of the European Federation of Chemical Engineering (EFCE) [6]:

> A situation that has the potential to cause harm to human, environment and property.

Thus, hazard is the antonym of safety. For the chemical industry, the hazard results from the simultaneous presence of three elements:

1. A threat stemming from the properties of processed substances, chemical reactions, uncontrolled energy release, or from equipment.
2. A failure that may be of technical origin or stem from human error, either during the operation or during process design. External events, such as weather conditions or natural catastrophe may also be at the origin of a failure.
3. An undetected failure in a system as non-identified hazards during risk analysis, or if insufficient measures are taken, or if an initially well-designed process gradually deviates from its design due to changes or lack of maintenance.

1.2.3.2 Risk

The EFCE defines risk as a measure of loss potential, and damage to the environment or persons in terms of probability and severity. An often-used definition is that risk is the product of severity time probability:

$$Risk = Severity \times Probability \tag{1.1}$$

In fact, considering risk as a product is somewhat restrictive: it is more general to consider it as a combination of the terms, severity and probability, that characterize the effects, that is, consequences and impact of a potential accident and its probability of occurrence. This also means that the risk is linked to a defined incident scenario. In other words, the risk analysis will be based on scenarios that must first be identified and described with the required accuracy, in order to be evaluated in terms of severity and probability of occurrence.

1.2.3.3 Safety

Safety is a quiet situation resulting from the real absence of any hazard [7].

Absolute safety (or zero risk) does not exist for several reasons: first, it is possible that several protection measures or safety elements can fail simultaneously; second, the human factor is a source of error and a person can misjudge a situation or have a wrong perception of indices, or may even make an error due to a moment's inattention.

1.2.3.4 Security

In common language, security is a synonym of safety. In the context of this book, security is devoted to the field of property protection against theft or incursion.

1.2.3.5 Accepted Risk

The accepted risk is a risk inferior to a level defined in advance either by law, technical, economical, or ethical considerations. The risk analysis, as it will be described in the following sections, has essentially a technical orientation. The minimal requirement is that the process fulfils requirements by the local laws and that the risk analysis is carried out by an experienced team using recognized methods and risk-reducing measures that conform to the state of the art. It is obvious that non-technical aspects may also be involved in the risk acceptation criteria. These aspects should also cover societal aspects, that is, a risk–benefit analysis should be performed

1.3 Risk Analysis

A risk analysis is not an objective by itself, but is one of the elements of the design of a technically and economically efficient chemical process [1]. In fact, risk analysis reveals the process inherent weaknesses and provides means to correct them. Thus, risk analysis should not be considered as a "police action," in the sense that, at the last minute, one wants to ensure that the process will work as intended. Risk analysis rather plays an important role during process design. Therefore, it is a key element in process development, especially in the definition of process control strategies to be implemented. A well-driven risk analysis not only leads to a safe process, but also to an economic process, since the process will be more reliable and give rise to less productivity loss.

1.3.1 Steps of Risk Analysis

There are many risk analysis methods, but all have three steps in common:

1. search for hazards,
2. risk assessment, and
3. definition of risk-reducing measures.

Table 1.3 Causes of incidents and their remediation.

Causes	Remediation
Lack of knowledge concerning the properties of material and equipment, the reactivity, the thermal data, etc.	Collection and evaluation of process data, physical properties, safety data, thermal data. Definition of safe process conditions and critical limits
No-identified deviation or failure	Systematic search for deviations from normal operating conditions
Wrong risk assessment (misjudged)	Interpretation of data, clearly defined assessment criteria, professional experience
No adequate measures provided	Process improvement, technical measures
Measures neglected	Plant management, management of change

If these three steps are at the heart of the risk analysis, it is also true that performing these steps requires preliminary work and other steps that should not be bypassed [1, 8].

By systematically studying past incidents in the chemical industry, several causes can be identified. These are summarized in Table 1.3.

Thus, the risk analysis must be well prepared, meaning that the scope of the analysis must be clearly defined; data must be available and evaluated, to define the safe process conditions and the critical limits. Then, and only then, the systematic search for process deviations from the safe conditions can be started. The identified deviations lead to the definition of scenarios, which can be assessed in terms of severity and probability of occurrence. This work can advantageously be summarized in a risk profile, enhancing the major risks that are beyond the accepted limits. For these risks, reduction measures can then be defined. The residual risk, that is, the risk remaining after implementation of the measures, can be assessed as before and documented in a residual risk profile showing the progress of the analysis and the risk improvement. These steps are reviewed in the next sections.

1.3.1.1 Scope of Analysis

The scope of the analysis aims to identify the process under consideration, in which plant it will take place, and with which chemicals it will be performed. The chemical reactions and unit operations must be clearly characterized. In this step, it is also important to check for interface problems with other plant units. As an example, when considering raw material delivery, it can be assumed that the correct raw material of the intended quantity and quality is delivered from a tank farm. Thus, it can be referred to the tank farm risk analysis, or the tank farm is to be included in the scope of the analysis. Similar considerations can be made for energy supply, to ensure that the appropriate energy is delivered. Nevertheless, loss of energy must be considered in the analysis, but it will be assumed that if nitrogen is required, nitrogen will be delivered. This allows checking for non-analysed items in a whole plant, completing the analysis.

1.3.1.2 Safety Data Collection

The required data must be collected prior to the risk analysis. This can be done gradually during process development as the knowledge on the process increases. The data can be summarized on data sheets devoted to different aspects of the process. They typically should encompass the following:

- involved chemical compounds,
- chemical reactions,
- technical equipment,
- utilities,
- operators.

The required data are reviewed in detail in Section 1.4. In order to be economic and efficient, the data collection is accompanied by their interpretation in terms of risks. This allows adapting the amount and accuracy of the data to the risk. This procedure is illustrated in the example of thermal data in Section 3.4.

1.3.1.3 Safe Conditions and Critical Limits

Once the safety data have been collected and documented, they must be evaluated with regard to the process conditions in terms of their significance for process safety. With the interpretation of the safety data, the process conditions that provide safe operation and the limits that should not be surpassed become clear. This defines the critical limits of the process, which are at the root of the search for deviations in the next step of the risk analysis.

This task should be performed by professionals having the required skills. Practice has shown that it is advantageous to perform, or at least to review, the interpretation with the risk analysis team. This ensures that the whole team has the same degree of knowledge and understanding of the process features.

1.3.1.4 Search for Deviations

During this step, the process is considered in its future technological environment, that is, the plant equipment, the control systems including the operators, and the delivery of raw material. The utilities are included in the critical examination of deviations from normal operating conditions. Here the following fields may be distinguished:

- deviations from operating mode, which are a central part in batch processes,
- technical failures of equipment, such as valves, pumps, control elements, and so on, which represent the central part of the equipment-oriented risk analysis,
- deviations due to external causes, such as climatic impacts (frost, flooding, storms),
- failure of utilities, especially electrical power or cooling water.

With continuous processes, different stages must be considered: steady state, start up and shut down, emergency stops, and so on.

The methods for search of hazards can be classified into three categories:

1. Intuitive methods, such as brainstorming.
2. Inductive methods, such as check lists, Failure Mode and Effect Analysis (FMEA), event trees, decision tables, Analysis of Potential Problems (APP). These methods proceed from an initial cause of the deviation and construct a scenario ending with the final event. They are based on questions of the type: "What if?"
3. Deductive methods, such as the Fault Tree Analysis (FTA) that proceeds by starting from the top event and looking for failures that may cause it to happen. These methods are based on questions of the type: "How can it happen?"

Some examples of those methods, commonly used for hazard search in chemical processes, are presented in Section 1.5.

The triggering mechanism to make a real threat out of a potential threat is called the cause. Each potential threat can have several potential causes, which should be listed. The possible consequences of a triggered event are referred to as the effects. This description of hazard causes and effects build an event scenario. The listing of the hazards in a table with an identifier, a short description a list of possible causes and the consequences, makes up the hazard catalog. The table may also contain risk assessment, a description of risk-reducing measures, assessment of residual risk, and who is responsible for the action decided on. This is of great help for the follow-up of the project. An example of such a hazard catalog is presented in Figure 1.1.

Company: **Id.-Nr.:**
Product: **Unit:** **Process dated:**
Author: **Date:**

#	Hazard	Causes	Effects	S1	P1	Action	S2	P2	Resp.	Status	Remarks

Figure 1.1 Example of Hazards Catalogue with deviation causes effects and actions decided by the team as well as their status.

1.3.1.5 Risk Assessment

The deviation scenarios found in the previous step of the risk analysis must be assessed in terms of risk, which consists of assigning a level of severity and probability of occurrence to each scenario. This assessment is qualitative or semi-quantitative, but rarely quantitative, since a quantitative assessment requires a statistical database on failure frequency, which is difficult to obtain for the fine chemicals industry with such a huge diversity of processes. The severity is clearly linked to the consequences of the scenario or to the extent of possible damage. It may be assessed using different points of view, such as the impact on humans, the environment, property, the business continuity, or the company's reputation. Table 1.4 gives an example of such a set of criteria. In order to allow for a correct assessment, it is essential to describe the scenarios with all their consequences. This is often a demanding task for the team, which must interpret the available data in order to work out the consequences of a scenario, together with its chain of events.

The probability of occurrence (P) is linked to the causes of the deviations. It is often expressed as frequency (f), referring to an observation period (T) often of one year:

$$P = f \cdot T \Rightarrow f = \frac{P}{T} \tag{1.2}$$

Table 1.4 Example assessment criteria for the severity.

Category	1. Negligible	2. Marginal	3. Critical	4. Catastrophic
Life/health in company	Injury, ambulant treatment	Injury requiring hospitalization	Injury with long-term disability	Fatality
Life/health outside company	No effect	No effect	First aid cases	Severe injury
Environment	No effect	Only on-site effects, effect on water treatment plant	Pollution outside site, recovery within 1 month	Long-term pollution of water, soil
Property	Not significant	Production line to be repaired	Loss of production line	Loss of plant
Business continuity	Not affected	Production stopped over 1 week	Delivery to customers must be interrupted several weeks	Business interruption more than 1 month
Image	No report outside company	Report in local media	Report in national media	Report in international media

Table 1.5 Example assessment criteria for the probability.

Category	Frequency	Definition/Examples
Frequent	Several times in a week	Hazards occurring at each batch if no measures are taken, e.g. charging powders in flammable solvent, exposure during handling of liquid or solid chemicals, ignition effective electrostatic discharge (if nothing is done against charging)
Moderate	Once or twice a month	Pump failure, failure of data acquisition, weighing error, wrong set point setting
Occasional	Several times a year	Imprecise communication between production, e.g. tank farm, failure of utilities, failure of a motor, explosive mixture after a failure
Remote	Once a year	Wrong piping connection after repair, mix-up of chemicals, programming error of control system, leakage at reactor or tank jacket, total power failure in the site
Unlikely	Once in 10 years	Simultaneous failure of redundant level control, e.g. LAH and LAHH, leak at flange
Almost impossible	Once in 100 years or more	Undiscovered failure of self controlling data acquisition, simultaneous failure of multiple technical safety measures, heavy earthquake, aircraft impact

A probability of 0.01 is equivalent to an occurrence of 1 incident in 100 years. An example of evaluation criteria for the probability is given in Table 1.5. There are two approaches for the assessment of probability: one is the qualitative approach, based on experience and using analogies to similar situations. The other is the quantitative approach, based on statistical data obtained from equipment failure databases [4]. These data were mainly gathered from the petrochemicals industry and bulk chemical industry, working essentially with dedicated plant units. For the fine chemicals and pharmaceutical industries, where the processes are carried out in multi-purpose plants, this approach is more difficult to use. This is because the equipment may work under very different conditions from process to process, which obviously has an impact on its reliability. The quantitative analysis must be based on a method, to allow identification of the interaction between different failures. Such a method, such as the fault tree analysis, is presented in Section 1.5.4. To get a better idea of the probability, a semi-quantitative approach consists of listing the logical relationships between the different causes. This allows identifying if the simultaneous failure of several elements is required to obtain the deviation and gives access to a semi-quantitative assessment.

The criteria mentioned in Tables 1.4 and 1.5 are given as an example of a possible practice, but as a part of the company's risk policy, they must be defined for each company with respect to its actual situation. Severity and probability of occurrence of an event form the two coordinates of the risk profile.

frequent				
moderate				
occasional				
remote				
unlikely				
almost impossible				
	negligible	marginal	critical	catastrophic

Figure 1.2 Example risk diagram with the accepted risk in white, non-accepted risk in dark gray, and conditionally accepted risks in light gray.

1.3.1.6 Risk Profiles

Risk assessment is not an objective by itself, but represents the required step for the risk evaluation. This is the step whereby it is decided if a risk is acceptable, or if it should be reduced by appropriate measures. This is usually done by comparing the risk to acceptance criteria defined in advance. This can be done graphically by using a risk diagram or risk matrix, as the example presented in Figure 1.2. The numbers characterizing the different scenarios can be placed into the matrix, thus allowing a visual risk evaluation. Such a risk diagram must comprise two zones corresponding to the clearly accepted (white in Figure 1.2) and clearly rejected risks (dark gray in Figure 1.2). Often a third zone (light grey in Figure 1.2) is also used. This third zone corresponds to risks that should be reduced, as far as reasonably applicable measures can be defined, the decision being based on technical and economical considerations. This practice corresponds to the As Low As Reasonably Practicable (ALARP) principle [9]. The borderline separating the white zone from the others is called the protection level: this is the limit of accepted risks and represents an important decision for the risk policy of a company.

The risk matrix presented in Figure 1.2 is based on Tables 1.4 and 1.5 and defines a 4×6 matrix. Experience has shown that choosing too narrow a matrix, for example, a 3×3 matrix, with the levels Low, Medium, and High, has the drawback of being too rough. It is unable to show the improvement of a risk situation especially with high severities, since such a situation often remains with high severity and low probability, even if additional measures are defined. On the other hand, too precise a matrix is not useful for risk evaluation and may lead to tedious discussions during its assessment.

1.3.1.7 Risk Reducing Measures

If the risk linked to a scenario falls into the non-acceptable zone, it must be reduced by appropriate risk-reducing measures. These are usually classified following two viewpoints, the action level and the action mode. The action level can

Table 1.6 Example of measures classified following their action level and their action mode.

	Elimination	Prevention	Mitigation
Technical	Alternative synthesis route	Alarm system with automatic interlock	Emergency pressure relief system
Organizational	No operator in hazardous field	Control by operators	Emergency services
Procedural	Access control	Instruction for behavior in abnormal situations	Instruction for emergency response

be elimination of the hazard, risk prevention, or mitigation of the consequences. For the action mode, different means can be used: technical measures that do not require any human intervention, or organizational measures that require human intervention and are accompanied by procedural measures defining the operating mode of the measure. Some examples are given in Table 1.6.

Eliminating measures are the most powerful since they avoid the risk, meaning that the incident can simply not occur or at least they strongly reduce the severity of the consequences of an eventual incident. This type of measures was especially promoted by Trevor Kletz in the frame of the development of inherently safer processes [10–12]. For a chemical process, eliminating the risks can mean that the synthesis route must be changed avoiding instable intermediates, strongly exothermal reactions, or highly toxic material. The choice of the solvent may also be important in this frame, the objective being to avoid flammable, toxic, or environmentally critical solvents. Concerning runaway risks, an eliminating measure aims to reduce the energy in such a way that no runaway can take place.

Preventive measures provide conditions where the incident is unlikely to happen, but its occurrence cannot be totally avoided. In this category, we find measures such as inventory reduction for critical substances, the choice of a continuous rather than a batch process leading to smaller reactor volumes, and a semi-batch rather than a full batch process providing additional means of reaction control. Process automation, safety maintenance plans, etc. are also preventative measures. The aim of these measures is to avoid triggering the incident and thus reducing its consequences. In the frame of runaway risks, a runaway remains theoretically possible, but due to process control, its severity is limited and the probability of occurrence reduced, such that it can be controlled before it leads to a critical situation.

Mitigation measures have no effect on triggering the incident, but avoid it leading to severe consequences. Examples of such measures are emergency plans, organization of emergency response, and explosion suppression. In the frame of runaway risks, such a risk may be triggered but its impact is limited, for example, by a blow down system that avoids toxic or flammable material escaping to the environment.

Technical measures are designed in such a way that they require no intervention, nor need to be triggered or executed. They are designed to avoid human error (in their action, but not in their design!). Technical measures are often built as automated control systems, such as interlocks or safety trips. In certain instances, they must be able to work under any circumstances, even in the case of utility failure. Therefore, great care is required in their design, which should be simple and robust. Here the simplification principle of inherent safety, the KISS principle (Keep It Simple and Stupid), should be followed. Depending on the risk level, they must also present a certified high degree of reliability. This is described in the international standard IEC 61511 [9] that advises on the different Safety Integrity Levels (SIL) with the required reliability as a function of the risk.

Organizational measures are based on human action for their performance. In the fine chemicals and pharmaceutical industries, reactor-charging operations are often manual operations and the product identification relies on the operator. In this context, quality systems act as support to safety, since they require a high degree of traceability and reliability. Examples of such measures are labeling, double visual checks, response to acoustic or optical alarms, in process control, and so on. The efficiency of theses measures is entirely based on the discipline and instruction of the operators. Therefore, they must be accompanied by programs of instructions, where the adequate procedures are learned in training.

During the risk analysis, the measures must be accurately described to establish terms of reference, but no detailed engineering must be done during the analysis. It is also advisable to define a responsible person for the design and establishment of these measures.

1.3.1.8 **Residual Risk**

This is the last step of risk analysis. After having completed the risk analysis and defined the measures to reduce risks, a further risk assessment must be carried out to ensure risks are reduced to an accepted level. The risks cannot be completely eliminated: risk zero does not exist, thus a residual risk remains. This is also because only identified risks were reduced by the planned measures. Thus, the residual risk has three components:

1. the consciously accepted risk,
2. the identified, but misjudged risk, and
3. the unidentified risk.

Thus, a rigorous and consciously performed risk analysis should reduce both of the last components. This is the responsibility of the risk analysis team. Hence, it becomes obvious that risk analysis is a creative task that must anticipate events, which may occur in the future and has the objective of defining means for their avoidance. This may also be seen in opposition to laws that react on events from the past. Therefore, it is a demanding task oriented to the future, which requires excellent engineering skills.

At this stage, a second risk profile can be constructed, in a similar way to that shown in Section 1.3.1.6. This allows the identification of the risks that are now strongly reduced and thus the measures, which require special care in their design, should perhaps be submitted to a reliability analysis, as described in Section 1.3.1.7.

1.4 Safety Data

In this section, a safety dataset, resulting from over 20 years of practical experience with risk analysis of chemical processes, is presented. These data build the base of risk analysis in the fine chemicals and pharmaceutical industries, essentially in multi-purpose plants. Therefore, the dataset introduces plant considerations only at its end. This allows exchanging them without any need for recollecting the whole dataset, in cases where the process is transferred from one plant unit to another. Moreover, this dataset may be used in the frame of different risk analysis methods.

There are many different sources for safety data, such as Material Safety Data Sheet (MSDS), databases [13, 14], company databases, and reports. Great care is required, when using MSDS, since experience has shown that they are not always reliable.

The safety data used in risk analysis can be grouped into different categories, described in the following sections. The data should be provided for raw material, intermediates, and products, as well as for reaction mixtures or wastes as they are to be handled in the process. Missing data, important in risk analysis, may be marked with a letter "I," to indicate that this information is missing or as a default by a letter "C," if its value is unknown but judged to be critical.

1.4.1.1 Physical Properties

Physical properties such as melting point, boiling point, and vapor pressure, as well as densities and solubility in water, are especially important in case of a release, but also give important restrictions to the process conditions. For instance, the melting point may indicate that the contents of a stirred vessel solidify below this temperature. This gives a lower limit to the heating or cooling system temperature, which would forbid using an emergency cooling system. In a similar way, the vapor pressure may define an upper temperature limit if a certain pressure level is not to be surpassed. Densities may also indicate what the upper and lower phase in a mixture is. Solubility in water is important in case of spillage.

1.4.1.2 Chemical Properties

The chemical properties allow summarizing observations or experiences made during process development or previous production campaigns. The following characteristic chemical properties should be identified during the risk analysis: acidity, auto-ignition temperature, pyrophoric properties, reaction with water, light sensitivity, air sensitivity, and storage stability. Further, impurities in the product may affect the toxic and ecotoxic properties of substances or mixtures.

1.4.1.3 Toxicity

The odor limit compared to other limits may indicate an early warning of a leak. The maximum allowed work place concentration (MAC), is the maximum allowed average concentration expressed in $mg\,m^{-3}$ of a gas, vapor, or dust in air in a workplace, which has no adverse effects on health for an exposure of 8 hours per day or 42 hours per week for the majority of a population. Since it is an average, maintaining the concentration below this value does not guarantee no effects, since the sensitivity may differ within a population. On the other hand, a short-term exposure to a concentration above MAC does not imply consequences on health.

A distinction is made between acute toxicity and chronic toxicity. For acute toxicity, the following indicators may be used:

- Lethal dose LD_{50}: gives the concentration that caused 50% of fatalities within 5 days in an animal population exposed once to the concentration. It may be an oral or dermal exposure and is expressed in $mg\,kg^{-1}$ of organism with a specification of the test animal used.
- Lethal concentration LC_{50}: is the concentration in air that caused 50% of fatalities within 5 days in a test in an animal population exposed to this concentration. It is through inhalation and is expressed in $mg\,kg^{-1}$ of organism with a specification of the test animal used.

The LD_{50} and TC_{50} for humans would be more directly applicable but, for obvious reasons, only very sparse data are available:

- The toxic dose lowest (TDL_0 oral) is the lowest dose that induced diseases in humans by oral absorption.
- The toxic concentration lowest (TCL_0 oral) is the lowest concentration in the air that induced diseases in humans by inhalation.

More qualitative indicators are also useful: absorption through healthy skin, irritation to skin, eyes, and respiratory system, together with sensitization with the following indicators: carcinogenic, mutagenic, teratogenic, reprotoxic, and so on. These properties can be summarized by indication of a toxicity class.

To judge the effect of short-term exposure, such as during a spillage, the short-term exposure limit (e.g. IDLH), must be known. The different levels given by the Emergency Response Planning Guidelines (EPRG), issued by the American Department of Energy and the Department of Transport, may also be used in this frame.

The use of carcinogenic material should be avoided as far as possible, by replacement with non-toxic or at least less toxic substances. If their use cannot be avoided, appropriate technical and medicinal measures should be applied in order to protect the workers from their effects. Among such measures, the reduction of the exposure in terms of concentration and duration as well as a medical follow-up may be required. The exposure can be limited by using closed systems, avoiding any direct contact with the substance, or personal protection equipment. Moreover, the number of exposed operators should be limited.

1.4.1.4 Ecotoxicity

In instances of spillage or release, not only humans may be concerned, but the damage may also affect the environment. The following data are required:

- biological degradability, bacteria toxicity (IC_{50}),
- algae toxicity (EC_{50}),
- daphnia toxicity (EC_{50}),
- fish toxicity (LC_{50}).

The *Po*/w, that is, the distribution coefficient between octanol and water, indicates a possible accumulation in fat. Malodorous or odor intense compounds should also be indicated.

The symbol LC_{50} means lethal concentration for 50% of a test population. The symbol EC_{50} means efficiency concentration for mobility suppression of 50% a test population. The symbol IC_{50} means inhibition concentration for 50% of a population in a test for respiratory suppression.

1.4.1.5 Fire and Explosion Data

The most common property in the assessment of fire hazards is the flashpoint that is applicable to liquids or melts, and is the lowest temperature at which the vapor above the substance may be ignited and continue to burn. The reference pressure for the flashpoint is 1013 mbar.

The combustion index is applicable to solids and gives a qualitative indication about combustibility, ranging from one to six. Index 1 corresponds to no combustion and Index 6 to a violent combustion with fast propagation. From Index 4, the combustion propagates through to the solid.

The self-sustaining decomposition is a phenomenon whereby the decomposition is initiated by a hot spot, and then propagates through to the solid with a velocity of some millimeters to centimeters per second. The decomposition does not require oxygen, so it cannot be avoided by using an inert atmosphere.

Electrostatic charges may provide an ignition source for the explosion of a gas, vapor, or dust cloud. Electrostatic charges can accumulate only if a separation process is involved. Since this is an often-occurring phenomenon as soon as a product is in motion, separation processes are common in chemical processes, during pumping, agitation, pneumatic transport, and so on. Charge accumulation occurs when the conductivity is too low to allow charge relaxation. This may lead to an electrostatic discharge that may ignite an explosion if present at the same time as explosive atmosphere. For this to occur the concentration of combustible must be in a given range and oxygen must be present. In order to assess such situations, the explosion characteristics are required.

Explosion limits indicate in which concentration range a mixture of combustible substance can be ignited. There are two limits, the lower explosion limit (LEL), below which the concentration is too low to produce an explosion and the upper explosion limit (UEL), above which the oxygen is in default and no explosion occurs. Further, the explosion is characterized by the maximum explosion pressure and its violence by the maximum pressure increase rate. In order

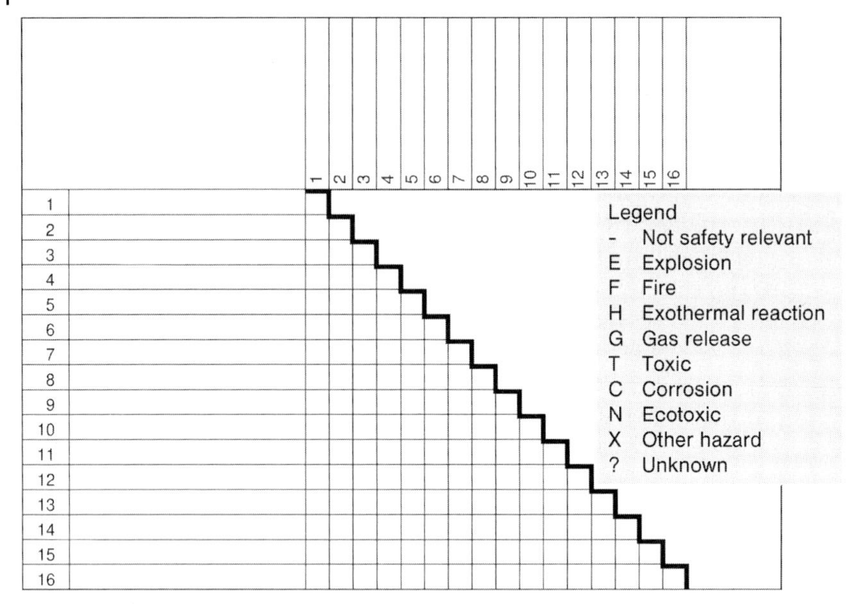

Figure 1.3 Interaction matrix, also called hazard matrix, summarizing the safety data of chemicals involved in a process.

to decide if an explosion can be ignited, the minimum ignition energy (MIE) is required.

The shock and friction sensitivity of a solid is also an important parameter, especially when it is to be submitted to mechanical stress during processing.

1.4.1.6 **Interactions**

The reactivity of chemicals used in a process must be assessed, since these chemicals may become in contact in a desired way or accidentally during the process. These interactions are usually analysed in a triangular matrix where the desired and undesired reactions are marked at the intersection of each row and column. Beside chemicals or mixtures, the different fluids (i.e. heat carrier), waste streams, and construction materials must also be considered. An example of such a matrix, summarizing the safety data and the interactions, is represented in Figure 1.3.

1.5 Systematic Search for Hazards

In this section, a selection of commonly used hazard identification techniques is presented. These techniques can be used in the fine chemicals and pharmaceutical industries. The methods presented here are designed to provide a systematic search for hazards with the final objective of providing a comprehensive analysis.

1.5.1
Check List Method

The check list method is based on past experience. The process description, the operating mode, is screened using a list of possible failures or deviations from this particular operating mode. Thus, it is obvious that the quality and comprehensiveness of the check list directly govern its efficiency. Indeed, the experience of the authors confirms that the check list is essential. This method is well adapted to discontinuous processes as practised in the fine chemicals and pharmaceutical industries, where processes are often performed in multi-purpose plants. The basic document for the hazard identification is the process description, also called operating mode. Each step of the process is analysed with the check list.

The check list presented here is constructed as a matrix with a row for each keyword of the check list and a column for each process step. The list itself is in two parts: the first (Figure 1.4) is devoted to the utilities and the corresponding question is: "May the failure of the considered utility lead to a hazard in a given process step?" In the second part (Figure 1.5), the operating mode is analysed using the check list, by questioning if a deviation from these conditions may lead to a hazard. This also allows checking the thoroughness of the process description, to see if the process conditions are given with sufficient precision and to avoid any misunderstandings.

The check list presents some intended redundancies, for example, equipment cleaning and impurities, or flow rate and feed rate, that are intended to ensure the comprehensiveness of the analysis. If a critical situation is identified, the corresponding box is marked with a cross, and the corresponding hazard identified by the coordinates of the box (e.g. F6: referring to the effect of failure of compressed air in sequence F), as described in the hazard catalog (Figure 1.1) in terms of possible causes, effects, risk assessment, measures, and residual risk. For an efficient analysis, it is advisable to group the process steps into sequences in order to avoid getting lost in useless detail. As an example, the preparation of a reactor may comprise a sequence of steps, such as the check for cleanness, proper connections, valve positions, inerting, heating to a given temperature, and so on.

Deviation	Process step:	A	B	C	D	E	F	G	H
1	Electrical power								
2	Water								
3	Steam								
4	Brine								
5	Nitrogen								
6	Compressed air								
7	Vacuum								
8	Ventilation								
9	Absorption								

Figure 1.4 Check list for utilities.
Question: "May the failure of a utility lead to a hazard?"

Deviation	Process step:	A	B	C	D	E	F	G	H
10	Cleaning								
11	Equipment check								
12	Emptying								
13	Equipment ventilation								
14	Charging, feeding								
15	Amount, flow rate								
16	Feed rate								
17	Order of charging								
18	Mixup of chemicals								
19	Electrostatic hazards								
20	Temperature								
21	Pressure								
22	pH								
23	Heating / cooling								
24	Agitation								
25	Reaction with heat carrier								
26	Catalyst, inhibitor								
27	Impurities								
28	Separation, settling								
29	Connections								
30	Pumping								
31	Waste elimination								
32	Process interruption								
33	Sampling								

Figure 1.5 Check list for the operating mode.
Question: "May a deviation from these conditions lead to a hazard?"

1.5.2
Failure Mode and Effect Analysis

The Failure Mode and Effect Analysis (FMEA) is based on the systematic analysis of failure modes for each element of a system, by defining the failure mode and the consequences of this failure on the integrity of that system. It was first used in the 1960s in the field of aeronautics for the analysis of the safety of aircraft [15]. It is required by regulations in the USA and France for aircraft safety. It allows assessing the effects of each failure mode of a system's components and identifying the failure modes that may have a critical impact on the operability safety and maintenance of the system. It proceeds in four steps:

1. the system is to be defined with the function of each of its components,
2. the failure modes of the components and their causes are established,
3. the effects of the failure are studied, and
4. conclusions and recommendations are derived.

One important point in this type of analysis is to define clearly the different states of the working system, to ensure that it is in normal operation, in a waiting state, in emergency operation, in testing, in maintenance, and so on. The depth of decomposition of the system into its components is crucial for the efficiency of the analysis.

In order to illustrate the method, we can take the example of a pump as a component. It may fail to start or to stop when requested, provide too low a flow rate or too low a pressure, or present an external leak. The internal causes for pump failure may be mechanical blockage, mechanical damage, or vibrations. The external causes may be power failure, human error, cavitation, or too high a head loss. Then the effect on the operation of the system and external systems must be identified. It is also useful to describe the ways for detecting the failure. This allows establishing the corrective actions and the desired frequency of checks and maintenance operations.

As it can be seen from this example, the AMDE may rapidly become very work-intensive and tedious. Therefore, a special adaptation has been made for the chemical process industry: the Hazard and Operability study.

1.5.3 Hazard and Operability Study

The Hazard and Operability Study (HAZOP) was developed in the early 1970s by ICI [16], after the Flixborough incident [17]. It is derived from the Failure Mode and Effect Analysis, but specially adapted for the process industry in general, and in the chemical industry in particular. It is essentially oriented towards the identification of risks stemming from the process equipment. It is particularly well suited for the analysis of continuous processes in the steady state, but can also be used for batch processes. The first steps of the risk analysis, of scope definition, data collection, safe conditions definition, are the same as for other methods. Using the process and instruments design (PID) and the Process Flow Diagram (PFD) as basic documents, the plant is divided into nodes and lines. For each of these divisions, a design intention is written that precisely summarizes its function. For example, a feed line could be defined as: "the line A129 is designed to feed 100 kg $hour^{-1}$ of product A from Tank B101 to reactor R205."

Then in a kind of guided brainstorming approach, using predefined guidewords applied to different parameters of the design intention, the process is systematically analysed. These guidewords are listed in Table 1.7, together with examples. As can be seen, there is some redundancy in the guidewords, for example, a temperature may be too high due to over-heating. This, again, is intentional and allows ensuring a comprehensive analysis. In cases where batch processes are to be analysed by the HAZOP technique, additional guidewords concerning time and sequencing, for example, too early, too late, too often, too few, too long, or too short may also be added. It is then verified that the deviation generated by applying the guideword to a parameter is meaningful. For example, "reverse flow" may be meaningful, but it would hardly be the case for "reverse temperature." If the generated deviation has no sense, it is skipped and the next deviation is generated with

Table 1.7 HAZOP guidewords with definitions and examples.

Guideword	Definition	Example
No/not	Negation of the design intention. No part of the design intention is realized	No flow, no pressure, no agitation
Less	Quantitative decrease, deviation from the specified value towards lower value. This may refer to state variables as temperature, quantities, as well as to actions such as heating	Flow rate too low, temperature too low, reaction time too short
More	Quantitative increase: deviation from the specified value towards higher value. This may refer to state variables as temperature, quantities, as well as to actions such as heating	Flow rate too high, temperature too high, too much product
Part of	Qualitative decrease: only part of the design intention is realized	Charging only a part of a predefined amount, omission of a compound at charging, reactor partly emptied
As well as	Qualitative increase: the design intention is realized, but at the same time something else happens	Heating and feeding at the same time, raw material contaminated by impurity with catalytic effect
Reverse	The design intention is reversed, logical opposite of design intention	Reversed flow, back flow, heating instead of cooling
Other/else	Total substitution: The design intention is not realized, but something else happens instead	Heating instead of dosing, charging A instead of B, mix-up of chemicals

the next guideword. For traceability of the thoroughness of the analysis, it may be marked as not applicable, "n.a."

For the meaningful deviations identified by the procedure described above, the possible causes for triggering the deviation are systematically searched. As an example, possible causes for "no flow" may be an empty feed tank, a closed valve, an inadvertently open valve to another direction, a pump failure, a leak, and so on. In this context, it may be useful to indicate the logical relationship between the causes, such as where simultaneous failure of several elements is required in order to trigger the deviation. This is of great help for the assessment of the probability of occurrence.

The effects are searched in order to allow the assessment of the severity. These results are documented together with the risk evaluation and, where required, with risk-reducing measures in a hazard catalog, as presented in Figure 1.1.

The analysis is performed on the totality of the nodes and lines defined by the division of the plant. This allows checking the comprehensiveness of the analysis. The HAZOP technique, as its name indicates, is not only devoted to the identification of hazards, but also to the identification of operability issues. In this frame, the hazard catalog also provides a list of possible symptoms for the early identification of abnormal situations and remediation. Then it becomes an efficient tool for process design, especially for the design of automation systems and interlocks.

1.5.4
Decision Table

The decision table method consists of logically combining all possible states of each element of a system and outlining the consequences on the entire system. It can be applied to a part of a system or to an operating mode. The combinations are analysed by Boole's algebra that gives the analysis a strong logical backbone. A part of such a decision table is shown by the example of the collision of a car with a deer (Figure 1.6). It is the most powerful method for analysing combinations of failures, exhaustive in this respect. Nevertheless, the combinations rapidly become so numerous that it is difficult to retain an overview of the system by this method. Thus, it has a more academic character.

Deer on the road ?	No	Yes	Yes	Yes	Yes	Yes	Yes	...
Driver sees it in time ?	No	No	No	No	No	No	No	...
Brake in time ?	No	No	No	No	No	Yes	Yes	...
Brakes fails ?	No	No	No	Yes	Yes	No	No	...
Deer stays on road ?	No	No	Yes	No	Yes	No	Yes	...
Collision ?	No	No	Yes	No	Yes	No	No	...

Figure 1.6 Decision table for the collision of a car with a deer [8].

1.5.5
Event Tree Analysis

The event tree analysis (ETA) is an inductive method that starts from an initial event and searches for the different possible effects. It is especially useful for studying the scenario of what may happen after the initial event when developing emergency plans. Starting from the initial event, one searches for consecutive events, until the system reaches a final state. These different generations of events are represented as a tree. An example, again based on the collision of a car with a deer, is represented in Figure 1.7. The vertical lines leading from one event to the next are related in a logical "AND" relationship and the corresponding probabilities must be multiplied. Horizontal lines indicate a logical "OR" relationship and the corresponding probabilities must be added. Thus, the tree can be quantified for the probability of entering one or the other branch after an event is known. Thus, it allows assessing quantitatively the effects of different possible chains of events and focuses the measures on the avoidance of the most critical chains.

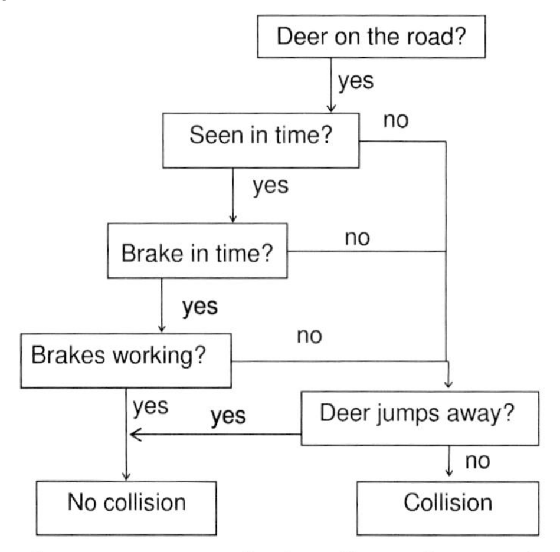

Figure 1.7 Event tree for the collision of a car with a deer.

1.5.6
Fault Tree Analysis

The fault tree analysis (FTA) is a deductive method, whereby the top event is given and the analysis focuses on the search of the causes that may trigger it. The principle is to start from the top event and identify the immediate causes or failures. Then each of these failures is again considered as an event and is analysed to identify the next generation of causes or failures. In this way, a hierarchy of the causes is built up, where each cause stems from parent causes as in a generation tree (Figure 1.8). Such a tree may be developed to infinity; nevertheless, the depth of the analysis can easily be adjusted to function as the objectives of the analysis. In most cases, the depth of the analysis is adjusted to allow the design of risk-reducing measures. For example, in the analysis of a chemical process, when a pump failure is found, it is not useful to find out what caused the pump failure. For the process safety, it may be more appropriate to provide a back-up pump or to increase the maintenance frequency of the pump. Thus, in general the analysis is stopped at the failure of elementary devices as valves, pumps, control instruments, and so on.

A special feature of the FTA is that different events are linked by logical relationships. One possibility is the logical "AND", meaning that two parent events must be realized simultaneously in order to generate the child event. The other possibility is the logical "OR" meaning, whereby only the realization of one parent event is sufficient to generate the child event. It becomes clear that the realization of an event behind an "AND" gate is less likely to occur than events behind an "OR" gate. This allows for a quantification of the fault tree.

The probability of occurrence of an event C depending on the simultaneous realization of two events A and B, that is, behind a logical gate "AND", is the conditional probability of A AND B:

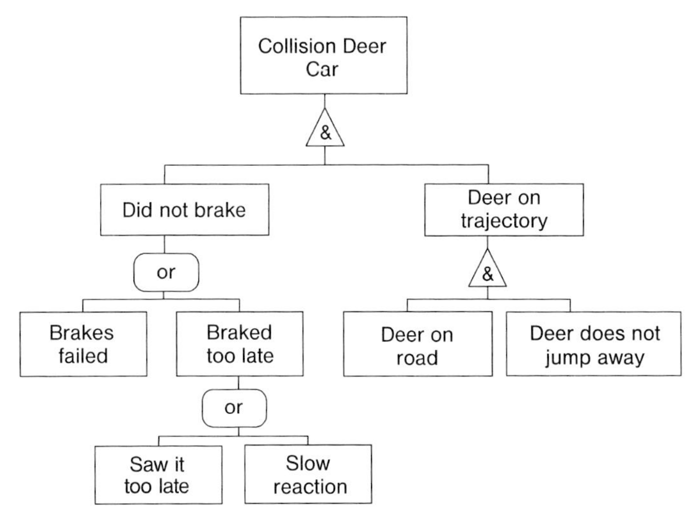

Figure 1.8 Example fault tree analysis for the collision of a car with a deer.

$$P_C = P_A \cdot P_B \tag{1.3}$$

Since probabilities are comprised between zero and one and should be low figures, the conditional probability usually becomes extremely small. In other terms, an "AND" gate strongly reduces the probability of the occurrence of an event and it is advisable to design a safety system in order to provide such "AND" relationships before the top event.

The probability of occurrence of an event C, where only the realization of one parent event from A or B is required (behind an "OR" gate), the probability is the sum of probabilities of all parent events:

$$P_C = P_A + P_B - P_A \cdot P_B \tag{1.4}$$

In this expression, the subtraction of the product of probabilities takes into account the fact that the simultaneous realization of both events is still taken into account in the realization of individual events. This correction is usually very small, since individual probabilities are small.

In this way, the fault tree can be quantified, which makes this technique very powerful for the reliability analysis of protection systems. The prerequisite is the availability of statistical reliability data of the different devices and instruments that is often difficult to obtain for multi-purpose plants, where devices can be exposed to very different conditions when changing from one process to another. Nevertheless, if the objective is to compare different designs, semi-quantitative data are sufficient.

1.6
Key Factors for a Successful Risk Analysis

The quality of a risk analysis depends essentially on three factors:

1. the systematic and comprehensive hazard identification,
2. the experience of the risk analysis team members,
3. the quality and comprehensiveness of the data used during the analysis.

The hazard identification methods presented in Sections 1.5.1 to 1.5.6 above are all based on strongly systematic procedures. In the check list method, the systematic is provided by the check list itself. The comprehensiveness can be verified in the matrix (see Figures 1.4 and 1.5). With the FMEA, the systematic is provided by the division of the system into elements and the failure modes considered. In the HAZOP study, the systematic stems from the division of the plant into nodes and lines, then the systematic application of the keywords. With the decision table method, the systematic is inherent to the table. For the FTA and ETA, the systematic is given by the tree and the logical ports. Nevertheless, the work of the team must be traceable, even by persons who did not participate to the analysis. Thus, it is recommended to also document the hazards that were not considered as critical.

Obviously, the composition of the risk analysis team is of primary importance for the quality of the work. Here the professional experience of the participants plays a key role, since the objective of the analysis is to identify events that have not yet occurred. It is a creative task to identify the hazards, but also to define risk-reducing measures. Thus, different professions must be represented in the team, including chemists, chemical engineers, engineers, automation engineers, and operators. When a new process is to be analysed, the experience gained during process development should be available to the team, hence members of the process development team must be represented in the risk analysis. The plant manager, who is the risk owner, takes a determining part in the analysis.

The team leader or moderator is responsible for the quality of the analysis; caring for its thoroughness, for discipline in the team, and for the time management. In the choice of risk-reducing measures, the moderator drives the group towards efficient solutions. More generally, the group dynamics is important, so the participants should also be creative and open-minded. The moderator ensures that all opinions can be expressed, leading the team towards consensual solutions. It is advantageous that the moderator has a sound industrial experience and, if possible, some experience in dealing with risks or in incident analysis.

The risk analysis represents an important part of the process know how and therefore the hazards catalog (see Figure 1.1) cannot be a static document, but a part of the process documentation at the same level as the operating mode and mass balances. It may be useful to describe the risk-reducing measures together with the status, such as new, accepted, rejected, implemented, and so on. The hazard catalog then becomes a management tool and a living document, which must regularly be updated and accompany the process throughout its life. The list of measures is a significant part of the documentation, since it also describes the function of all safety relevant elements.

References

1 Hungerbühler, K. and Ranke, J. and Mettier, T. (1998) *Chemische Produkte und Prozesse; Grundkonzepte zum Umweltorientierten Design*, Springer, Berlin.
2 SUVA (2006) *Unfallstatistik UVG 2005*, SUVA, Lucerne.
3 Laurent, A. (2003) *Sécurité des procédés chimiques, connaissances de base et méthodes d'analyse de risques*. Tec&Doc. Lavoisier, Paris.
4 Lees, F.P. (1996) *Loss Prevention in the Process Industries Hazard Identification Assessment and Control*, 2nd edn, Vol 1–3, Butterworth-Heinemann, Oxford.
5 Stoessel, F. (2002) On risk acceptance in the industrial society. *Chimia*, **56**, 132–6.
6 Jones, D. (1992) *Nomenclature for Hazard and Risk Assessment in the Process Industries*, 2nd edn., Institution of Chemical Engineers, Rugby.
7 Rey, A., ed. (1992) *Le Robert dictionnaire d'aujourd'hui*, Dictionnaires Le Robert, Paris.
8 Schmalz, F. (1996) *Lecture script: Sicherheit und Industriehygiene*, Zürich.
9 IEC (2004) *Funktionale Sicherheit – Sicherheitstechnische Systeme für die Prozessindustrie IEC 61511*, DIN VDE.
10 Hendershot, D.C. (1997) Inherently safer chemical process design. *Journal of Loss Prevention in the Process Industries*, **10** (3), 151–7.
11 Kletz, T.A. (1996) Inherently safer design: the growth of an idea. *Process Safety Progress*, **15** (1), 5–8.
12 Crowl, D.A., ed. (1996) *Inherently Safer Chemical Processes. A Life Cycle Approach*, CCPS Concept book, Center for Chemical Process Safety, New York, p. 154.
13 Sorbe, S. (2005) *Kenndaten chemischer Stoffe*, Ecomed Sicherheit, Landsberg.
14 Lewis, R.J. (2005) *Sax's Dangerous Properties of Industrial Materials*, 11th edn, Reinhold, New York.
15 Villemeur, A. (1988) *Sûreté de fonctionnement des systèmes industriels*, Eyrolles, Paris.
16 Kletz, T. (1992) *Hazop and Hazan: Identifying and Assessing Process Industry Hazards*, 3rd edn, Institution of Chemical Engineers, Rugby.
17 Kletz, T. (1988) *Learning from Accidents in Industry*, Butterworths, London.

2 Fundamentals of Thermal Process Safety

Case History "Storage During Weekend"

After the synthesis in batches of 2600 kg, an intermediate was obtained in the form of a melt. This product was kept in an unstirred storage vessel at 90 °C. The vessel was heated by hot water circulation; the heating system being open at ambient pressure, the temperature physically limited to 100 °C. Under normal circumstances, the melt was immediately flacked and transferred to smaller containers flacking. On a Friday evening, this transfer operation could not be carried out for technical reasons so the melt was left in the vessel over the weekend. Since it was known that the product was prone to decomposition, the plant manager studied the available information on the stability of the product. Quality tests indicated that the melt would degrade at a rate of 1% per day, if left at 90 °C. Having no other choice under the circumstances, this quality loss would have been tolerated by the plant management. Additional information was a DSC-thermogram showing an exothermal decomposition with an energy of 800 kJ kg^{-1}, detected from a temperature of 200 °C. Considering that during 3 days the decomposition would be 3%, he estimated that the energy released by the decomposition would be 24 kJ kg^{-1}, corresponding to an approximate temperature rise of 12 °C. Thus, he decided to maintain this reactive mass in the vessel during the weekend. During Sunday night to Monday morning, the storage vessel exploded, causing significant material damage.

A correct assessment of the situation would have predicted the explosion. The main error was considering the storage isothermal. In fact, such large vessels, when they are not agitated, behave quasi adiabatically. The correct estimation of the initial heat release rate allows calculation of the temperature increase rate under adiabatic conditions. By taking into account the acceleration of the reaction with increasing temperature, the approximate time of the explosion would have been predictable. This is left as an exercise for the reader (see Worked Example 2.1).

Thermal Safety of Chemical Processes: Risk Assessment and Process Design. Francis Stoessel

ISBN: 978-3-527-31712-7

Lessons drawn

- Runaway reactions may have serious consequences.
- The correct assessment of thermal phenomena requires a special knowledge.
- Rigorous methods are required for the thermodynamic analysis of such situations.

Worked Example 2.1: Storage Over the Weekend

After the synthesis in batches of 2600 kg an intermediate is obtained in the form of a melt. This product is kept in an unstirred vessel, which is heated by an open hot water heating system (ambient pressure). Under normal circumstances, the melt is immediately flacked transferred to smaller containers. On a Friday evening, the transfer operation cannot be performed for technical reasons so the melt must be left in the vessel over the weekend.

Data:
- From quality tests it is known that the melt will degrade at a rate of 1% per day if left at 90 °C. Having no other choice, under the present circumstances, this loss would be tolerated by the plant management.
- A DSC thermogram measured in a closed pressure resistant crucible at a scanning rate of $4\,\text{K}\,\text{min}^{-1}$ of the intermediate shows an exothermal peak starting from 200 °C with energy of $800\,\text{kJ}\,\text{kg}^{-1}$.

Question:
What considerations would you use to judge the management's plan for storage over the weekend?

The plant manager's error:
The plant manager implicitly assumed that the decomposition observed at 200 °C in the DSC-Thermogram and the loss of quality measured at 90 °C were due to the same reaction. In fact, this is a worst case assumption and is true. The loss of quality is due to the decomposition, and is 1% per day at 90 °C, that is 3% during a weekend. Thus the energy released during the storage is

$$800\,\text{kJ}\,\text{kg}^{-1} \times 0.03 = 24\,\text{kJ}\,\text{kg}^{-1}$$

corresponding with a specific heat capacity of $2\,\text{kJ}\,\text{kg}^{-1}\,\text{K}^{-1}$, to an adiabatic temperature rise of

$$\Delta T_{ad} = \frac{24\,\mathrm{kJ\,kg^{-1}}}{2\,\mathrm{kJ\,kg^{-1}\,K^{-1}}} = 12\,\mathrm{K}$$

This conclusion is wrong because it implicitly assumes that the temperature is constant at 90 °C, which is not true for an unstirred mass of 2600 kg close to the melting point.

The correct solution of the problem:
At such a scale and without stirring, the heat transfer with the surroundings, that is the jacket with hot water at 90 °C is very poor. Thus, adiabatic conditions should be assumed as a worst case approach. Under these circumstances, the heat released in the reaction mass serves to increase its temperature. Thus, we have to calculate the heat release rate. For a conversion of 1% per day, the heat release rate is

$$q' = \frac{Q'}{t} = \frac{800000\,\mathrm{J\,kg^{-1}} \times 0.01}{24\,\mathrm{h} \times 3600\,\mathrm{s\,h^{-1}}} = 0.01\,\mathrm{J\,s^{-1}\,kg^{-1}} = 100\,\mathrm{mW\,kg^{-1}}$$

This power may be converted to adiabatic temperature increase rate:

$$\frac{dT}{dt} = \frac{q'}{c'_P} = \frac{0.1\,\mathrm{W\,kg^{-1}}}{2000\,\mathrm{J\,kg^{-1}\,K^{-1}}} = 5 \cdot 10^{-5}\,\mathrm{K\,s^{-1}} = 0.18\,^{\circ}\mathrm{C\,h^{-1}} \approx 0.2\,^{\circ}\mathrm{C\,h^{-1}}$$

Applying van't Hoff rule, whereby the reaction rate doubles for a temperature increase of 10 K, the rate would be 0.4 $^{\circ}\mathrm{C\,h^{-1}}$ at 100 °C. Assuming an average rate of 0.3 $^{\circ}\mathrm{C\,h^{-1}}$ in the temperature range from 90 to 100 °C, the time required to reach 100 °C is 33 hours, that is about 32 hours. The next 10 K increase to 110 °C would take 16 hours, then 8 hours to 120 °C and so on. This is a geometric progression and the sum of its terms is 2 × 32 hours = 64 hours. Thus, an explosion during the weekend is predictable.

The right decision is to transfer the melt into a stirred vessel, where the temperature can be actively controlled and monitored. A heat release rate of 0.1 $\mathrm{W\,kg^{-1}}$ can easily be removed from the vessel.

2.1 Introduction

The introduction of this knowledge and a presentation of these methods are the objective of this book. In the present chapter, the essential theoretical aspects of thermal process safety are reviewed. Often-used fundamental concepts of thermodynamics are presented in the first section with a strong focus on process safety. In the second section, important aspects of chemical kinetics are briefly reviewed. The third section is devoted to the heat balance, which also governs chemical reactors and physical unit operations such as calorimetric measurement techniques. In the last section, a theoretical background on runaway reactions is given.

2.2
Energy Potential

2.2.1
Thermal Energy

2.2.1.1 Heat of Reaction

Most of the chemical reactions performed in the fine chemicals industry are exothermal, meaning that thermal energy is released during the reaction. It is obvious that the amount of energy released is directly linked to the potential damage in the case of an incident. For this reason, the heat of reaction is one of the key data, which allow assessment of the risks linked to a chemical reaction at the industrial scale.

The unit of energy (J) is related to other units as follows:

- $1\,\text{J} = 1\,\text{N}\cdot\text{m} = 1\,\text{W}\cdot\text{s} = 1\,\text{kg}\,\text{m}^2\,\text{s}^{-2}$
- $1\,\text{J} = 0.239\,\text{cal}$ or $1\,\text{cal} = 4.18\,\text{J}$

The units used for heat of reaction are:

- Molar enthalpy of reaction: ΔH_r: $\text{kJ}\,\text{mol}^{-1}$
- Specific heat of reaction: Q'_r: $\text{kJ}\,\text{kg}^{-1}$

The latter, the specific heat of reaction, is practical for safety purposes, because most of the calorimeters directly deliver the specific heat of reaction in $\text{kJ}\,\text{kg}^{-1}$. Further, since it is a specific entity, it can easily be scaled to the intended process conditions. Both heat of reaction and molar enthalpy are related by

$$Q'_r = \rho^{-1} C(-\Delta H_r) \tag{2.1}$$

Obviously, the heat of the reaction depends on the concentration of reactant (C). By convention, exothermal reactions have negative enthalpies, whereas endothermic reactions have positive enthalpies.[1)] Some typical values of reaction enthalpies are given in Table 2.1 [1].

Reaction enthalpies also may be obtained from enthalpies of formation (ΔH_f), given in tables of thermodynamic properties [2, 3]:

$$\Delta H_r^{298} = \sum_{\text{products}} \Delta H_{f,i}^{298} - \sum_{\text{reactants}} \Delta H_{f,i}^{298} \tag{2.2}$$

1) *Pro memoria* in this book, in opposition to the thermodynamic convention, we consider all effects, which increase the temperature of the system as positive. Thus, the enthalpy has a minus sign.

Table 2.1 Typical values of reaction enthalpies [1].

Reaction	ΔH_R kJ mol^{-1}
Neutralization (HCl)	−55
Neutralization (H_2SO_4)	−105
Diazotization	−65
Sulfonation	−150
Amination	−120
Epoxydation	−100
Polymerization (Styrene)	−60
Hydrogenation (Alkene)	−200
Hydrogenation (Nitro)	−560
Nitration	−130

Other sources of enthalpies of formation are the Benson-group increments [4]. These values consider molecules in the gas phase, thus for liquid phase reactions they must be corrected by the latent enthalpy of condensation. Hence, these values can be used as a first and rough approximation. However, one must be aware that reaction enthalpies may vary over a great range, depending on the operating conditions. As an example, the enthalpy of sulfonation reactions may vary from $-60\,\text{kJ mol}^{-1}$ to $-150\,\text{kJ mol}^{-1}$, depending on the sulfonation reactant and its concentration. In addition to the heat of reaction, heat of crystallization and heat of mixing may also affect the values measured in practice. For this reason, it is recommended to measure the heat of reaction under practical conditions, whenever possible.

2.2.1.2 Heat of Decomposition

A large part of the compounds used in the chemical industry is in a so-called meta-stable state. The consequence is that an additional energy input, for example, a temperature increase, may bring the compound into a more energetic and instable intermediate state that relaxes to a more stable state by an energy release that may be difficult to control. Such a reaction path is shown in Figure 2.1. Along the reaction path, the energy first increases and then decreases to a lower level. The energy of decomposition (ΔH_d) is released along the reaction path. It is often higher than common reaction energies, but remains well below combustion energies. The decomposition products are often unknown or not well defined. This means the estimation of decomposition energies by standard enthalpies of formation is difficult. Decomposition energies are treated in detail in Chapter 11.

2.2.1.3 Heat Capacity

By definition, the heat capacity of a system is the amount of energy required to raise its temperature by 1 K. The unit is J K^{-1}. To allow calculations and comparisons, the specific heat capacity is more commonly used:

- Heat capacity: c_P J K^{-1}
- Specific heat capacity: c'_P $\text{kJ kg}^{-1}\,\text{K}^{-1}$

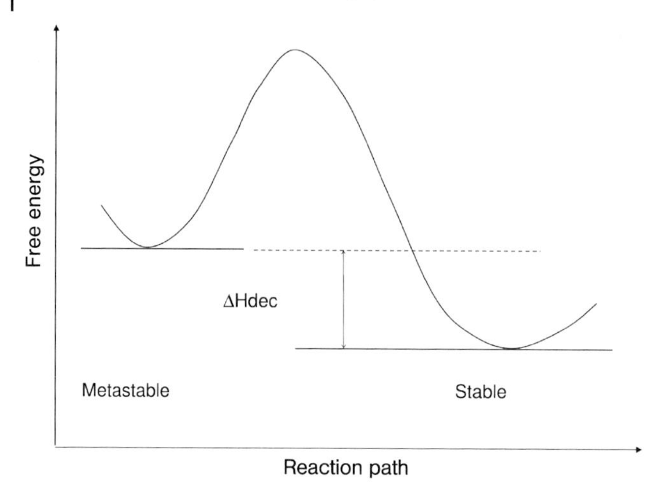

Figure 2.1 Variation of the free energy along a reaction path.

Table 2.2 Typical values of specific heat capacities.

Compound	c_P' kJ·kg^{-1}K^{-1}
Water	4.2
Methanol	2.55
Ethanol	2.45
2-Propanol	2.58
Acetone	2.18
Aniline	2.08
n-Hexane	2.26
Benzene	1.74
Toluene	1.69
p-Xylene	1.72
Chlorobenzene	1.3
Tetrachloromethane	0.86
Chloroform	0.97
NaOH 10 mol% in water	1.4
Sulfuric acid 100%	1.4
NaCl	4.0

Water has a relatively high specific heat capacity, whereas inorganic compounds have lower heat capacities. Organic compounds are in the medium range (Table 2.2).

The specific heat capacity of a mixture can be estimated from the specific heat capacities of the different compounds by a mixing rule:

$$c_P' = \frac{\sum_i M_i c_{P_i}'}{\sum_i M_i} \tag{2.3}$$

The heat capacity increases with temperature, for example, for liquid water at 20 °C the specific heat capacity is 4.182 kJ kg^{-1} K^{-1} and at 100 °C is 4.216 kJ kg^{-1} K^{-1} [2]. Its variation is frequently described by the polynomial expression (virial equation):

$$c'_P(T) = c'_{P0}[1 + aT + bT^2 + \cdots] \tag{2.4}$$

In order to obtain accurate results, this function should be accounted for when the temperature of a reaction mass tends to vary over a wider range. However, in the condensed phase the variation of heat capacity with temperature is small. Moreover, in case of doubt and for safety purposes, the specific heat capacity should be approximated by lower values. Thus, the effect of temperature can be ignored and generally the heat capacity determined at a (lower) process temperature is used for the calculation of the adiabatic temperature rise.

2.2.1.4 **Adiabatic Temperature Rise**

The energy of a reaction or of decomposition is directly linked with the severity, that is, the potential of destruction of a runaway. Where a reactive system cannot exchange energy with its surroundings, adiabatic conditions prevail. In such a case, the whole energy released by the reaction is used to increase the system's temperature. Thus, the temperature rise is proportional to the energy released. For most people, the order of magnitude of energies is often difficult to value. Thus, the adiabatic temperature rise is a more convenient way, and therefore more commonly used criterion, to assess the severity of a runaway reaction. It can be calculated by dividing the energy of reaction by the specific heat capacity:

$$\Delta T_{ad} = \frac{(-\Delta H_r)C_{A0}}{\rho \cdot c'_P} = \frac{Q'_r}{c'_P} \tag{2.5}$$

The central term in Equation 2.5 enhances the fact that the adiabatic temperature rise is a function of reactant concentration and molar enthalpy. Therefore, it is dependant on the process conditions, especially on feed and charge concentrations. The right-hand term in Equation 2.5, showing the specific heat of reaction, is especially useful in the interpretation of calorimetric results, which are often expressed in terms of the specific heat of the reaction. Thus, the interpretation of calorimetric results must always be performed in connection with the process conditions, especially concentrations. This must be accounted for when results of calorimetric experiments are used for assessing different process conditions.

The higher the adiabatic temperature rise, the higher the final temperature will be if the cooling system fails. This criterion is static in the sense that it gives only an indication of the excursion potential of a reaction, but no information about the dynamics of the runaway.

As an example, for the assessment of the potential severity of the loss of control of a reaction, Table 2.3 shows the effect of typical energies of a desired synthesis reaction and decomposition and their equivalents in the form of adiabatic

Table 2.3 Energy equivalents for a typical reaction and decomposition.

Reaction	Desired	Decomposition
Specific energy	100 kJ kg^{-1}	2000 kJ kg^{-1}
Adiabatic temperature rise	50 K	1000 K
Evaporation of methanol per kg of reaction mixture	0.1 kg	1.8 kg
Mechanical potential energy height at which 1 kg is raised	10 km	200 km
Mechanical kinetic energy velocity at which 1 kg is accelerated	0.45 km s^{-1} (mach 1.5)	2 km s^{-1} (mach 6.7)

temperature rise and mechanical energy. The equivalent mechanical energies are calculated for 1 kg of reaction mass.

It becomes obvious that while desired reactions may not by themselves be inherently dangerous, decomposition reactions may lead to dramatic effects. In order to illustrate this, the amount of solvent, such as methanol, which may be evaporated when the boiling point is reached during a runaway, is calculated. In the example given in Table 2.3, it is unlikely that the energy stemming from the desired reaction alone would cause an effect in a properly designed industrial reactor. However, this will certainly not be true for the decomposition reaction, where even the amount of methanol that could be evaporated (1.8 kg) cannot be available in 1 kg of reaction mass. Thus, a possible secondary effect of the evaporation of solvents is a pressure increase in the reactor, followed by a rupture and formation of an explosive vapor cloud, which in turn may lead to a severe room explosion if ignited. The risk of such an occurrence must be assessed.

2.2.2 Pressure Effects

The destructive effect of a runaway reaction is always due to pressure. Besides the temperature increase, secondary decomposition reactions often result in the production of small molecules (fragments), which are gaseous or present a high vapor pressure and thus cause a pressure increase. As high energies are often involved in decomposition reactions, the temperature increase results in pyrolysis of the reaction mixture. In such cases, the thermal runaway is always accompanied by a pressure increase and no specific study of the pressure effect is required. Nevertheless, in the first stages of a runaway, the pressure increase may cause the rupture of the reactor before the runaway starts to show acceleration. In such cases, a study of the pressure effects may be required. If the temperature increase occurs in a reaction mixture with volatile compounds, their vapor pressure may also cause a pressure increase. Some simple ways for assessing the resulting pressure are described in the next subsections.

2.2.2.1 Gas Release

Gases often form during decomposition reactions. Depending on the operating conditions, the effects of a gas release are different. In a closed vessel, the pressure increase may lead to the rupture with evolving gas or aerosol or even to the explosion of the vessel. A first approximation can estimate the pressure using the ideal gas law:

$$P = \frac{NRT}{V} \tag{2.6}$$

The universal gas constant used in this equation is $83.15 \cdot 10^{-6}\,\mathrm{m^3\,bar\,kmol^{-1}\,K^{-1}}$. In an open vessel, the gas production may result in evolving gas, liquid, or aerosols, which may also have secondary effects such as toxicity, burns, fire, and ecological, and even a secondary unconfined vapor or dust explosion. The volume of resulting gas can be estimated, using the same ideal gas law:

$$V = \frac{NRT}{P} \tag{2.7}$$

Thus, the amount of gas being released during a reaction or decomposition is an important element for the assessment of the severity of a potential incident.

2.2.2.2 Vapor Pressure

By increasing the temperature, the vapor pressure of the reaction mass may increase. The resulting pressure can be estimated by the Clausius–Clapeyron law, which links the pressure to the temperature and the latent enthalpy of evaporation ΔH_v:

$$\ln\frac{P}{P_0} = \frac{-\Delta H_v}{R}\left(\frac{1}{T} - \frac{1}{T_0}\right) \tag{2.8}$$

The universal gas constant to be used in this equation is $8.314\,\mathrm{J\,mol^{-1}\,K^{-1}}$ and the molar enthalpy of vaporization is expressed in $\mathrm{J\,mol^{-1}}$. Since the vapor pressure increases exponentially with temperature, the effects of a temperature increase, for example due to an uncontrolled reaction, may be considerable. As a rule of thumb, the vapor pressure doubles for a temperature increase of about 20 K.

2.2.2.3 Amount of Solvent Evaporated

If the boiling point is attained during runaway, a possible secondary effect of the evaporation of a solvent is the formation of an explosive vapor cloud, which in turn can lead to a severe explosion if ignited. In some cases, there is enough solvent present in the reaction mixture to compensate the energy release, allowing the temperature to stabilize at the boiling point. This is only possible if the solvent can be safely refluxed or distilled off into a catch pot or a scrubber. Moreover, the equipment must be designed for the resulting vapor flow rate. Also, the thermal stability of the resulting concentrated reaction mixture must be verified.

The amount of solvent evaporated can easily be calculated by using the energy of the reaction and/or of the decomposition as follows:

$$M_v = \frac{Q_r}{\Delta H'_v} = \frac{M_r \cdot Q'_r}{\Delta H'_v} \tag{2.9}$$

After a cooling failure, when boiling point is reached, a fraction of the energy released is used to heat the reaction mass to the boiling point and the remaining fraction of the energy results in evaporation. The amount of evaporated solvent can be calculated from the "distance" to the boiling point:

$$M_v = \left(1 - \frac{T_b - T_0}{\Delta T_{ad}}\right) \frac{Q_r}{\Delta H'_v} \tag{2.10}$$

In Equations 2.9 and 2.10, the enthalpy of evaporation used is the specific enthalpy of evaporation, expressed in $\mathrm{kJ\,kg^{-1}}$. These expressions only give the amount of solvent evaporated, which is a static parameter. They give no information about the vapor flow rate, which is related to the dynamics of the process, that is, the reaction rate (see Section 9.4). This aspect is discussed in the chapter on technical aspects of reactor safety.

2.3 Effect of Temperature on Reaction Rate

When considering thermal process safety, the key of mastering the reaction course lays in governing the reaction rate, which is the driving force of a runaway reaction. This is because the heat release rate of a reaction is proportional to the reaction rate. Thus, reaction kinetics plays a fundamental role in the thermal behavior of a reacting system. In the present section, some specific considerations on reaction kinetics with regard to process safety consider the dynamic aspects of reactions.

2.3.1 Single Reaction

For a single reaction $A \rightarrow P$ following an nth-order kinetic law, the reaction rate is given by

$$-r_A = kC_{A0}^n(1 - X_A)^n \tag{2.11}$$

which shows that the reaction rate decreases as conversion progresses. By following Arrhenius model, the rate constant k is an exponential function of temperature:

$$k = k_0 e^{-E/RT} \tag{2.12}$$

In this equation, k_0 is the frequency factor, also called the pre-exponential factor, and E the activation energy of the reaction in $J\,mol^{-1}$. Since the reaction rate is always expressed in $mol \cdot volume^{-1} time^{-1}$, the rate constant and the pre-exponential factor have dimensions depending on the order of the reaction $volume^{(n-1)} mol^{-(n-1)} time^{-1}$. The universal gas constant used in this equation is $8.314\,J\,mol^{-1}\,K^{-1}$. The van't Hoff rule, can be used as a rough approximation of the temperature effect on reaction rate:

- Reaction rate doubles for a temperature increase of 10 K.

The activation energy, an important factor in reaction kinetics, may be interpreted in two ways: The first is an energy barrier to be overcome by the reaction, such as that depicted in Figure 2.1. The second is the sensitivity of the reaction rate towards changes in temperature. For synthesis reactions, the activation energy usually varies between 50 and $100\,kJ\,mol^{-1}$. In decomposition reactions, it may reach $160\,kJ\,mol^{-1}$. For values above these, an autocatalytic behavior should be suspected. Low activation energy ($<40\,kJ\,mol^{-1}$) may indicate a mass transfer controlled reaction. Higher activation energies give a higher sensitivity towards temperature. A very slow reaction at low temperatures may become fast and therefore dangerous at higher temperatures.

2.3.2 Multiple Reactions

Reaction mixtures encountered in industrial practice often show complex behavior and the overall reaction rate comprises several individual reactions, forming a multiple reaction scheme. There are two basic reaction schemes allowing the construction of more complex ones [5]. The consecutive reactions are also called reaction in series:

$$A \xrightarrow{k_1} P \xrightarrow{k_2} S \text{ with } \begin{cases} r_A = -k_1 C_A \\ r_P = k_1 C_A - k_2 C_P \\ r_S = k_2 C_P \end{cases} \tag{2.13}$$

The second basic reaction scheme is competitive reactions, also called reaction in parallel:

$$\begin{cases} A \xrightarrow{k_1} P \\ A \xrightarrow{k_2} S \end{cases} \text{ with } \begin{cases} r_A = -(k_1 + k_2) C_A \\ r_P = k_1 C_A \\ r_S = k_2 C_A \end{cases} \tag{2.14}$$

In Equations 2.13 and 2.14, the reactions are supposed to be first-order in each compound, but different reaction orders may be encountered. With multiple

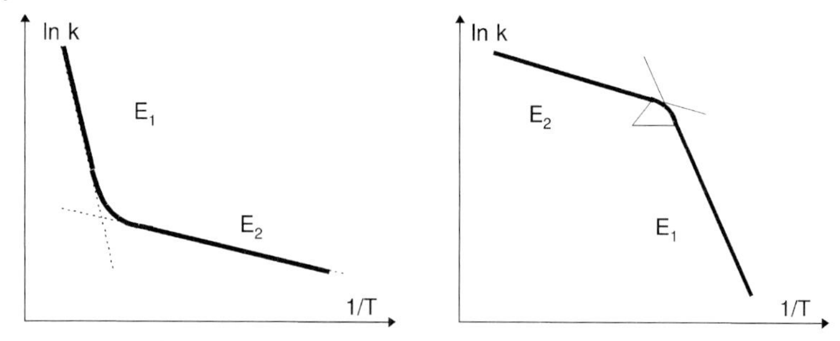

Figure 2.2 With multiple reactions, the apparent activation energy may vary with temperature, depending on which reaction is dominant.

reactions, the activation energies of the different steps are generally different, thus various reactions show different sensitivity towards temperature changes. The consequence is that depending on the temperature, one of the other reactions (or reaction mechanisms) become dominant, thus one must be extremely careful when kinetic data is extrapolated over a large temperature range. In the left-hand example in Figure 2.2, if calorimetric measurements are performed at a higher temperature, for example, in order to obtain a measurable signal, the activation energy will be E_1 and the extrapolation towards a lower temperature will predict too low reaction rates: this is unsafe. In the right-hand example, the measured activation energy is E_2 and the extrapolation towards a lower temperature is too conservative. For this reason, it is helpful to measure the thermal behavior close to the operating or storage temperature. High sensitive calorimeters, such as the Thermal Activity Monitor (Thermometrics™), are very useful for this purpose (see Section 4.3.2).

2.4 Heat Balance

Understanding the heat balance is essential when considering thermal process safety. This also applies to the industrial scale for reactors or storage units, as well as at laboratory scale for understanding the results of calorimetric experiments. In fact, the same heat balance terms will serve in both situations. For this reason, we first present the different terms of the heat balance of a reactor with a reacting system. This is followed by an often-used and simplified heat balance and finally we will study how reaction rate is affected by adiabatic conditions.

2.4.1 Terms of the Heat Balance

In chemical thermodynamics, the convention is that exothermal effects are negative and endothermal positive. Here, since we consider the heat balance for practical and safety reasons, all effects that increase the temperature, such as exothermal reactions, are positive. The different most common terms used in the heat balance are:

2.4.1.1 Heat Production

The heat production corresponds to the heat release rate by the reaction. Therefore, it is proportional to the reaction enthalpy and to the reaction rate:

$$q_{rx} = (-r_A)V(-\Delta H_r) \tag{2.15}$$

This term is of primary importance with respect to reactor safety: mastering the heat release by the reaction is the key of reactor safety. For a single *n*th-order reaction, the reaction rate can be expressed as

$$-r_A = k_0 \cdot e^{-E/RT} \cdot C_{A0}^n \cdot (1-X)^n \tag{2.16}$$

This expression enhances the fact that the heat release rate is a function of the conversion and will therefore vary with time in discontinuous reactors or during storage. In a batch reaction, there is no steady state. It is constant in the Continuous Stirred Tank Reactor (CSTR) and is a function of the location in the tubular reactor (see Chapter 8). The heat release rate is

$$q_{rx} = k_0 \cdot e^{-E/RT} \cdot C_{A0}^n \cdot (1-X)^n \cdot V \cdot (-\Delta H_r) \tag{2.17}$$

Two features of this expression are important for safety purposes. First, the heat release rate of a reaction is an exponential function of temperature and second, since it is proportional to the volume, it will vary with the cube of the linear dimension of the vessel (L^3) containing the reacting mass.

2.4.1.2 Heat Removal

There are several possible mechanisms for the heat exchange between a reacting medium and a heat carrier: radiation, conduction and forced or natural convection. Here we shall consider convection only. Other mechanisms are considered in the chapter on heat accumulation. The heat exchanged with a heat carrier (q_{ex}) across the reactor wall by forced convection is proportional to the heat exchange area (A) and to the driving force, that is, the temperature difference between the reaction medium and the heat carrier. The proportionality coefficient is the overall heat transfer coefficient (U):

$$q_{ex} = U \cdot A \cdot (T_c - T_r) \tag{2.18}$$

In the case of significant change of the physical chemical properties of the reaction mixture, the overall heat exchange coefficient (U) will also be a function of time. The heat transfer properties are usually a function of temperature, where changes in the viscosity of the reaction mass play a dominant role.

Table 2.4 Specific heat exchange areas for different reactor scales.

Scale	Reactor volume m^3	Heat exchange area m^2	Specific area m^{-1}
Research lab.	0.0001	0.01	100
Bench scale	0.001	0.03	30
Pilot plant	0.1	1	10
Production	1	3	3
Production	10	13.5	1.35

With respect to safety, two important features must be considered here. The heat removal is a linear function of temperature and since it is proportional to the heat exchange area, it varies as the square of the linear dimension of the equipment (L^2). This means that when the dimensions of a reactor have to be changed, as for scale-up, the heat removal capacity increases more slowly than the heat production rate does. Therefore, the heat balance becomes more critical for larger reactors. Some typical dimensions are shown in Table 2.4. The heat exchange area varies within certain limits by design of the vessel geometry, but for stirred tank reactors these limits represent hard constraints. The cylindrical geometry with a height to diameter ratio of approximately 1 : 1 is imposed.

Thus, the specific cooling capacity of reaction vessels varies by approximately two orders of magnitude, when scaling up from laboratory scale to production scale. This has a great practical importance, because if an exothermal effect is not detected at laboratory scale, this does not mean that the reaction is safe at a larger scale. At laboratory scale, the cooling capacity may be as high as $1000\,W\,kg^{-1}$, whereas at plant scale it is only in the order of $20–50\,W\,kg^{-1}$ (Table 2.5). This also means that the heat of reaction can be measured only in calorimetric devices and cannot be deduced from the measurement of a temperature difference between the reaction medium and the coolant.

In Equation 2.18, the heat transfer coefficient U plays an important role. Therefore, some methods to estimate or even to measure this coefficient are presented

Table 2.5 Typical specific cooling capacity[a] for different reactor scales.

Scale	Reactor volume m^3	Specific cooling capacity $W\,kg^{-1}\,K^{-1}$	Typical cooling capacity $W\,kg^{-1}$
Research lab.	0.0001	30	1500
Bench scale	0.001	9	450
Pilot plant	0.1	3	150
Production	1	0.9	45
Production	10	0.4	20

a) The specific cooling capacity has been calculated for the vessel filled to its nominal volume with an overall heat transfer coefficient of $300\,W\,m^{-2}\,K^{-1}$, a density of $1000\,kg\,m^{-3}$ and a temperature difference of 50 K between the reactor contents and the cooling medium.

in the chapter on technical aspects of reactor safety (Section 9.3). For a given composition of the reactor contents, the Reynolds number strongly influences the heat transfer coefficient. This means that for stirred tank vessels, the type and shape of the stirrer and its revolution velocity affect the heat transfer coefficient. The temperature gradient across the reactor wall, the driving force of heat exchange, must sometimes be limited to avoid crystallization or fouling of the reactor wall. This can be achieved by limiting the minimum heat carrier temperature above the melting point of the reaction mass. In other cases, it can be that the limitation is due to a constraint on the temperature or on the flow rate of the cooling medium.

2.4.1.3 **Heat Accumulation**

Heat accumulation represents the variation of the energy contents of a system with temperature:

$$q_{ac} = \frac{d\sum_i (M_i \cdot c'_{P_i} \cdot T_i)}{dt} = \sum_i \left(\frac{dM_i}{dt} \cdot c'_{P_i} \cdot T_i\right) + \sum_i \left(M_i \cdot c'_{P_i} \cdot \frac{dT_i}{dt}\right) \tag{2.19}$$

The sum is calculated by taking into account every system component, that is, the reaction mass and the equipment. Hence, the heat capacity of the reactor or of the vessel – at least the parts directly in contact with the reacting system – must be considered. For a discontinuous reactor, the heat accumulation can be written with mass or volume units:

$$q_{ac} = M_r c'_P \frac{dT_r}{dt} = \rho V c'_P \frac{dT_r}{dt} \tag{2.20}$$

Since heat accumulation is the consequence of the difference between heat production rate and cooling rate, it results in a variation of the temperature of the reactor contents. Hence, if the heat exchange does not compensate exactly the heat release rate of the reaction, the temperature will vary as

$$\frac{dT_r}{dt} = \frac{q_{rx} - q_{ex}}{\sum_i M_i \cdot c'_{P_i}} \tag{2.21}$$

In Equation 2.21, the index i refers to all compounds of the reaction mass and to the reactor itself. However, in practice, for stirred tank reactors, the heat capacity of the reactor is often negligible compared to that of the reaction mass. In order to simplify the expressions, the heat capacity of the equipment can be ignored. This is justified by the following example. For a $10\,m^3$ reactor, the heat capacity of the reaction mass is in the order of magnitude of $20\,000\,kJ\,K^{-1}$; whereas the metal mass in contact with the reaction medium is about 400 kg, representing a heat capacity of about $200\,kJ\,K^{-1}$, that is, ca. 1% of the overall heat capacity. Further, the error leads to a more critical assessment of the situation, which is a good practice in matters of safety assessment. Nevertheless, in certain specific applications, it will be introduced as required as, for example, in continuous reactors and especially tubular reactors, where the heat capacity of the reactor itself may intentionally be used to increase the overall heat capacity and therefore the reactor safety by design. This point will be examined in detail in Chapter 8.

2.4.1.4 Convective Heat Exchange Due to Mass Flow

In continuous systems, the feed stream is not always at the same temperature as the reactor outlet. This temperature difference between the reactor feed (T_0) and exit streams (T_f) results in a convective heat exchange with the surroundings. The heat flow is proportional to the heat capacity and the volume flow rate ($\dot{v}$):

$$q_{cx} = \rho \cdot \dot{v} \cdot c'_P \cdot \Delta T = \rho \cdot \dot{v} \cdot c'_P \cdot (T_f - T_0) \tag{2.22}$$

This is an overall heat balance of a continuous reactor: more detailed heat balances are introduced in Chapter 8.

2.4.1.5 Sensible Heat Due to Feed

If a feed stream to a reactor is at a different temperature (T_{fd}) than the reactor's contents (T_r), the thermal effect of the feed stream must be accounted for in the heat balance. This effect is also called "sensible heat":

$$q_{fd} = \dot{m}_{fd} \cdot c'_{P_{fd}} \cdot (T_{fd} - T_r) \tag{2.23}$$

This effect is especially important in the semi-batch reactor. If the temperature difference between reactor and feed is important and/or the feed rate is high, this term may play a dominant role, the sensible heat significantly contributing to reactor cooling. In such cases, when the feed is stopped, it may result in an abrupt increase of the reactor temperature. This term is also important in calorimetric measurements, where the appropriate correction must be performed.

2.4.1.6 Stirrer

The mechanical energy dissipated by the agitator is converted into viscous friction energy and finally altered into thermal energy. In most cases, this may be negligible when compared to the heat released by a chemical reaction. However, with viscous reaction masses, for example, polymerization reactions, this term must be integrated into the heat balance. The energy dissipated by the stirrer may also play an important role, when holding a reaction mass in a stirred tank. It can be estimated from

$$q_s = Ne \cdot \rho \cdot n^3 \cdot d_s^5 \tag{2.24}$$

The computation of the thermal energy dissipated by a stirrer requires knowledge of the power number (Ne) and of the geometry of the stirrer. Some examples of power numbers for common stirrers are given in Table 2.6.

Table 2.6 Newton number and stirrer geometric characteristics of some common stirrer types.

Stirrer type	Turbulent Ne	Flow type
Propeller	0.35	Axial
Impeller	0.20	Radial with axial component at the bottom
Anchor	0.35	Tangential close to the wall
Flat blade disk turbine	4.6	Radial with high shear effect
Pitched blade turbine	0.6–2.0	Axial with strong radial component
Mig (2 stages)	0.55	Complex with axial, radial and tangential flows
Intermig (2 stages)	0.65	Complex with radial component and high turbulence at wall

2.4.1.7 Heat Losses

Industrial reactors are thermally insulated for safety reasons (hot surfaces) and for economical reasons (heat losses). Nevertheless, at higher temperatures, heat losses may become important. Their calculation may become tedious, since heat losses are often due to a combination of losses by radiation and by natural convection. If an estimation is required, a simplified expression using a global overall heat transfer coefficient (α) may be useful:

$$q_{loss} = \alpha \cdot (T_{amb} - T_r) \quad (2.25)$$

Some values of the heat loss coefficient, α, are given in Table 2.7, which also compares them with laboratory equipment [6]. These values were measured by applying natural cooling to the vessel and determining the half-life of the cooling (Newtonian cooling, see Section 13.3.2). The heat losses may change by two orders of magnitude between industrial reactors and laboratory equipment. This explains why an exothermal reaction may remain undetected in small-scale tests, whereas it may become critical in large-scale equipment. A 1-liter glass Dewar flask has heat losses equivalent to those of a $10\,m^3$ industrial reactor. The simplest way of

Table 2.7 Typical heat losses from industrial vessels and laboratory equipment.

Vessel volume	Specific heat loss $W\,kg\,K^{-1}$	$t_{1/2}$ h
$2.5\,m^3$ reactor	0.054	14.7
$5\,m^3$ reactor	0.027	30.1
$12.7\,m^3$ reactor	0.020	40.8
$25\,m^3$ reactor	0.005	161.2
10 ml test tube	5.91	0.117
100 ml glass beaker	3.68	0.188
DSC-DTA	0.5–5	–
1 l glass Dewar flask	0.018	43.3

estimating this overall heat transfer coefficient is by direct measurement at plant scale.

2.4.2
Simplified Expression of the Heat Balance

A heat balance, taking into account all the terms explained above, can be established:

$$q_{ac} = q_r + q_{ex} + q_{fd} + q_s + q_{loss} \tag{2.26}$$

However, in most cases a simplified heat balance, which comprises the two first terms on the right-hand side of Equation 2.26, is sufficient for safety purposes. Let us consider a simplified heat balance, neglecting terms such as the heat input by the stirrer or heat losses. Then, the heat balance for a batch reactor can be written as

$$q_{ac} = q_r - q_{ex} \Leftrightarrow \rho V c'_P \frac{dT_r}{dt} = (-r_A)V(-\Delta H_r) - UA(T_r - T_c) \tag{2.27}$$

For a reaction of order n rearranged to enhance the variation of temperature with time:

$$\frac{dT_r}{dt} = \Delta T_{ad} \frac{-r_A}{C_{A0}^{n-1}} - \frac{UA}{\rho V c'_P}(T_r - T_c) \tag{2.28}$$

where the adiabatic temperature rise corresponding to the conversion is defined as

$$\Delta T_{ad} = \frac{(-\Delta H_r) C_{A0} X_A}{\rho c'_P} \tag{2.29}$$

The term $\dfrac{UA}{\rho V c'_P}$ is the inverse of the thermal time constant of a reactor (Section 9.2.4.1).

2.4.3
Reaction Rate under Adiabatic Conditions

Running an exothermal reaction under adiabatic conditions leads to a temperature increase, and therefore to acceleration of the reaction, but at the same time, the reactant depletion leads to a decreasing reaction rate. Hence, these two effects act in an opposite way: the temperature increase leads to an exponential increase of the rate constant and therefore of the reaction rate. The reactant depletion slows down the reaction. The resultant effect of these two antagonist variations depends on the relative importance of both terms.

For a first-order reaction performed under adiabatic conditions, the rate varies with temperature as

$$-r_A = \underbrace{k_0 e^{-E/RT}}_{\nearrow} \underbrace{C_{A0}(1 - X_A)}_{\searrow} \tag{2.30}$$

Since under adiabatic conditions, there is a linear relationship between temperature and conversion following Equation 2.29, depending on the heat of reaction, the temperature increase produced for a given conversion may or may not dominate the balance. In order to illustrate this effect, the rate was calculated as a function of temperature for two reactions: the first is a weakly exothermal reaction with an adiabatic temperature increase of only 20 K, whereas the second is a more exothermal reaction with an adiabatic temperature increase of 200 K. The results are presented in Table 2.8. For the first reaction, with a low adiabatic temperature rise of 20 K, the reaction rate only slowly increases during the first 4 degrees. Afterwards, the reactant depletion dominates and the reaction rate decreases. This cannot be considered a thermal explosion, being a self-heating phenomenon. In the case of the second reaction, with an adiabatic temperature rise of 200 K, the reaction rate increases sharply over a large temperature range. The reactant depletion becomes visible only at higher temperatures.

Table 2.8 Reaction rate under adiabatic conditions with different reaction enthalpies, corresponding to an adiabatic temperature increase of 20 and 200 K.

Temperature	100	104	108	112	116	120	–	200
Rate constant s^{-1}	1.00	1.27	1.61	2.02	2.53	3.15	–	118
Rate (ΔT_{ad} = 20 °C)	1.00	1.02	0.96	0.81	0.51	0.00	–	
Rate (ΔT_{ad} = 200 °C)	1.00	1.25	1.54	1.90	2.33	2.84	–	59

This type of behavior is called thermal explosion. In Figure 2.3, the evolution of temperature under adiabatic conditions is showed for a series of reactions with different reaction energies, but with the initial heat release rate and activation energy. For the lower energies, that is, ΔT_{ad} < 200 K, the reactant depletion leads to an S-shaped curve and the curve does not present the character of a thermal explosion, but rather that of a self-heating. This effect disappears with the more exothermal reactions, meaning that the reactant depletion has practically no influence on the reaction rate. In fact, the rate decrease only appears at very high conversion. For a total energy corresponding to an adiabatic temperature rise above 200 K, even a conversion of about 5% leads to a temperature increase of 10 K and more. Thus, the acceleration due to temperature increase dominates far over the effect of reactant depletion. This is equivalent to considering a zero-order reaction. For this reason, in the frame of thermal explosions, the kinetics are often

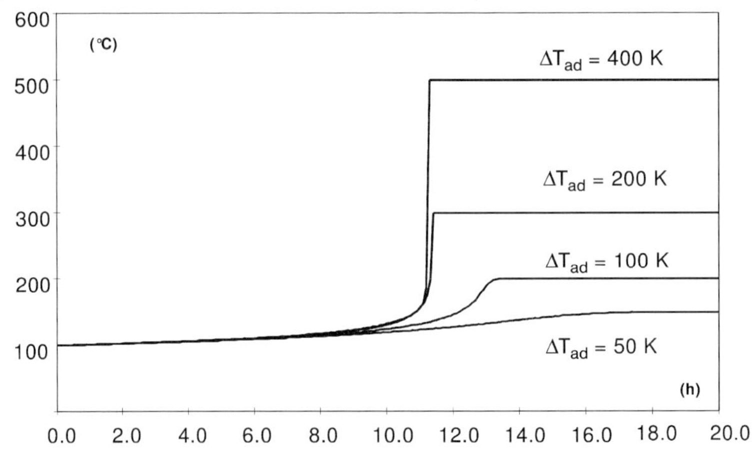

Figure 2.3 Adiabatic temperature course as a function of time for reactions with different energies. The S-shape is only visible for lower energies.

simplified to zero-order. This also represents a conservative approximation: zero-order reaction results in shorter time to explosion than higher reaction orders.

2.5 Runaway Reactions

2.5.1 Thermal Explosions

If the power of the cooling system is lower than the heat production rate of a reaction, the temperature increases. The higher temperature results in a higher reaction rate, which in turn causes a further increase in heat production rate. Because the heat production of the reaction can increase exponentially, while the cooling capacity of the reactor increases only linearly with the temperature, the cooling capacity becomes insufficient and the temperature increases. A runaway reaction or thermal explosion develops.

2.5.2 Semenov Diagram

Let us consider a simplified heat balance involving an exothermal reaction with zero-order kinetics. The heat release rate of the reaction $q_{rx} = f(T)$ varies as an exponential function of temperature. The second term of the heat balance, the heat removal by a cooling system $q_{ex} = f(T)$, with Newtonian cooling (Equation 2.18), varies linearly with temperature. The slope of this straight line is $U \cdot A$ and the intersection with the abscissa is the temperature of the cooling system T_c. This

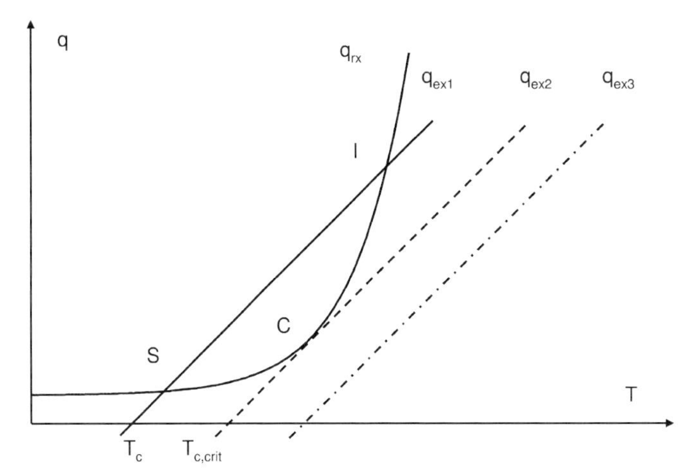

Figure 2.4 Semenov diagram: the intersections S and I between the heat release rate of a reaction and the heat removal by a cooling system represent an equilibrated heat balance. Intersection S is a stable operating point, whereas I represent an instable operating point. Point C corresponds to the critical heat balance.

heat balance can be represented in a diagram called the Semenov Diagram (Figure 2.4). The heat balance is in equilibrium when the heat production is equal to heat removal ($q_{rx} = q_{ex}$). This happens at the two intersections of the exponential heat release rate curve q_{rx} with the straight line of the heat removal curve q_{ex} in the Semenov Diagram. The intersection at lower temperature (S) corresponds to a stable equilibrium point.

When the temperature deviates from S to a higher value, the heat removal dominates and the temperature decreases until production and removal become equal. The system recovers its stable equilibrium. Inversely, for a deviation to lower temperatures, the heat production dominates, resulting in a temperature increase until equilibrium is reached again. Therefore, the intersection at lower temperature corresponds to a stable operating point. The same consideration for the intersection at higher temperature (I) shows that the system becomes instable. For a small deviation to a lower temperature, cooling dominates and the temperature decreases until point (S) is reached again, whereas a small deviation to a higher temperature results in excess heat production, thus a runaway condition develops.

The intersection of the cooling line q_{ex1} (solid line) with the temperature axis represents the temperature of the cooling system, T_c. Thus, for a higher cooling system temperature, the straight line corresponding to the power of the cooling system is shifted to the right parallel to itself (dashed line in Figure 2.4). Both intersection points become closer to each other until they merge at one point. This point corresponds to a tangent and is an instable operating point. The corresponding temperature of the cooling system is called the critical temperature ($T_{c,,crit}$). For a cooling medium temperature above $T_{c,,crit}$, the cooling line q_{ex3} (dash-dot line) has

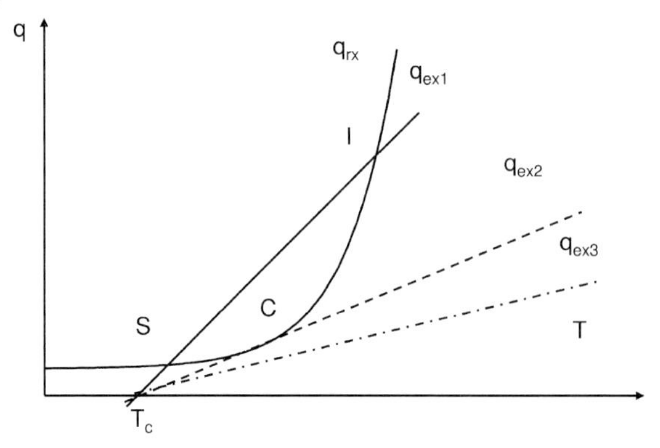

Figure 2.5 Semenov diagram: effect of a change in the heat transfer characteristics of the reactor *UA*.

no intersection with the heat release curve q_{rx}, meaning that the heat balance equation has no solution and runaway is inevitable.

2.5.3
Parametric Sensitivity

When a reactor is operated with a critical cooling medium temperature, an infinitely small increase of the cooling medium temperature leads to a runaway situation. This is known as parametric sensitivity, which is a small change in one of the operating parameters leading from a controlled situation to runaway. Moreover, a similar effect can be observed if, instead of changing the temperature of the cooling system, the heat transfer coefficient is changed. Since the slope of the heat removal line is equal to UA (Equation 2.18), a decrease of the overall heat transfer coefficient (U) results in a decrease in the slope of q_{ex}, from (q_{ex1}) to (q_{ex3}), which may also lead to a critical situation (point C in Figure 2.5). This may happen when fouling occurs in the heat exchange system, or when crusts or solid deposits form on the inner reactor wall of a reactor. The same effect is observed for a change in the heat transfer area (A), as during scale-up. This "switch" from a stable to an instable situation may occur even with very small changes in the operating parameters, such as U, A, and T_c. The consequence is a potentially high sensitivity of the reactor stability towards these parameters, rendering the control of the reactor difficult in practice. Therefore, the assessment of the stability of a chemical reactor requires the knowledge of the heat balance of a chemical reactor. The concept of critical temperature is useful for this purpose.

2.5.4
Critical Temperature

As stated above, if a reactor is operated with a cooling medium temperature close to the critical cooling medium temperature, a small variation of the coolant

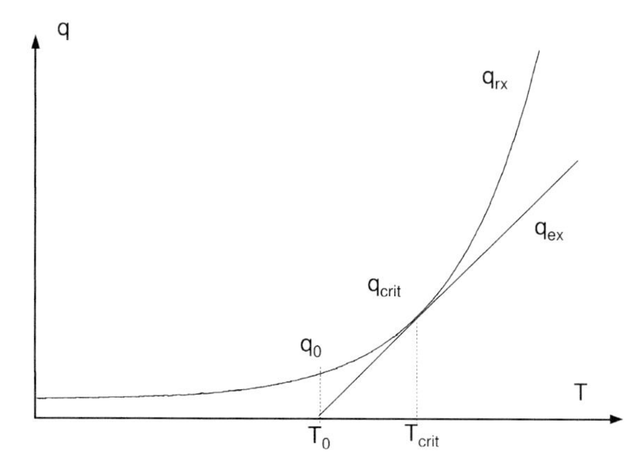

Figure 2.6 Semenov-diagram: calculation of the critical temperature.

temperature may result in an over-critical heat balance, and a runaway situation develops. Thus, in order to assess the stability of the operating conditions, it is important to know if the reactor is operated far away from or close to the critical temperature.[2)] This situation can be assessed using the Semenov diagram (Figure 2.6). We consider a zero-order reaction with a heat release rate expressed as a function of temperature as

$$q_r = k_0 \cdot e^{-E/RT_{crit}} \cdot Q_r \tag{2.31}$$

where the heat of reaction is in absolute units (J). If we consider the critical situation, the heat release rate of the reaction is equal to the cooling capacity of the reactor:

$$q_r = q_{ex} \Leftrightarrow k_0 \cdot e^{-E/RT_{crit}} \cdot Q_r = UA(T_{crit} - T_0) \tag{2.32}$$

Since at this point both lines are at a tangent to each other, their derivatives are also equal:

$$\frac{dq_r}{dT} = \frac{dq_{es}}{dT} \Leftrightarrow k_0 \cdot e^{-E/RT_{crit}} \cdot Q_r \cdot \frac{E_a}{RT_{crit}^2} = UA \tag{2.33}$$

Both equations are verified simultaneously for the critical temperature difference:

$$\Delta T_{crit} = T_{crit} - T_0 = \frac{E}{RT_{crit}^2} \tag{2.34}$$

2) The critical temperature used in this context has nothing to do with the thermodynamic critical temperature.

The critical temperature (T_{crit}) can be evaluated from

$$T_{crit} = \frac{E}{2R}\left(1 \pm \sqrt{1 - \frac{4RT_0}{E}}\right) \tag{2.35}$$

or

$$\Delta T_{crit} = \frac{R(T_0 + \Delta T_{crit})^2}{E} = \frac{RT_0^2\left(1 + \frac{\Delta T_{crit}}{T_0}\right)^2}{E} \tag{2.36}$$

which can be written as

$$\Delta T_{crit} = \frac{RT_0^2}{E}\left(1 + \frac{2\Delta T_{crit}}{T_0} + \frac{\Delta T_{crit}^2}{T_0}\right) \tag{2.37}$$

This means that for a given reaction characterized by its thermo-kinetic constants (k_0, E, Q_r) processed in a given reactor, and characterized by its heat exchange parameters (U, A, T_0), there is a minimum temperature difference required for stable performance of the reactor:

$$\Delta T_{crit} = T - T_0 \geq \frac{RT_{crit}^2}{E} \tag{2.38}$$

Hence, the assessment of the situation requires both the thermo-kinetic parameter of the reaction and the heat exchange parameter of the cooling system of the reactor. The same principle can be applied to a storage situation with the thermo-kinetic parameter of the decomposition reaction and the heat exchange characteristics of the storage vessel.

2.5.5 Time Frame of a Thermal Explosion, the TMR_{ad} Concept

Another important characteristic of a runaway reaction is the time a thermal explosion takes to develop under adiabatic conditions, or Time to Maximum Rate under adiabatic conditions (TMR_{ad}). To calculate this time, we consider the heat balance under adiabatic conditions for a zero-order reaction:

$$\frac{dT}{dt} = \frac{q}{\rho V c_P'} \tag{2.39}$$

with

$$q = q_0 \exp\left[\frac{-E}{R}\left(\frac{1}{T_0} - \frac{1}{T}\right)\right] \tag{2.40}$$

Here T_0 is the initial temperature from which the thermal explosion develops. If T_c is close to T_0, that is $(T_0 + 5\,K) \leq T_{crit} \leq (T_0 + 30\,K)$, an approximation can be made, $T_0 \cdot T \approx T_0^2$ then

$$\frac{1}{T_0} - \frac{1}{T} = \frac{T - T_0}{T_0 T} \approx \frac{T - T_0}{T_0^2} \tag{2.41}$$

and

$$\left.\begin{aligned} T_0 - T &= \Delta T \\ \frac{RT_{crit}^2}{E} &= \Delta T_{crit} \end{aligned}\right\} \Rightarrow q = q_0 e^{\Delta T/\Delta T_{crit}} \tag{2.42}$$

the variables are changed in

$$\frac{\Delta T}{\Delta T_{crit}} = \theta \Rightarrow T = \Delta T_{crit} \cdot \theta \tag{2.43}$$

and Equation 2.42 becomes

$$\theta = \theta_0 e^{\theta} \tag{2.44}$$

$$\frac{d\theta}{dt} = \theta_0 e^{\theta} \Rightarrow \int_0^t dt = \frac{1}{\theta_0} \int_0^1 e^{-\theta} d\theta \tag{2.45}$$

By integration, the time required to reach the temperature T_{crit} from T_0, that is $\theta = 0 \rightarrow \theta = 1$ is

$$t = \frac{1}{\theta_0} \int_0^1 e^{-\theta} d\theta = \left[\frac{1}{\theta_0 e^{-\theta}} \right]_0^1 = \frac{1}{\theta_0} (1 - e^{-1}) \tag{2.46}$$

$$t = (1 - e^{-1}) \frac{\Delta T}{T_0} = 0{\cdot}632 \frac{c_P' R T_{crit}^2}{q_0 E} \approx 0{\cdot}632 \frac{c_P' R T_0^2}{q_0 E} \tag{2.47}$$

This time is also called the Time of No Return (TNR): after this time has elapsed under adiabatic conditions and even if the cooling system has been recovered, it is impossible to cool the reactor, since its heat balance becomes super-critical:

$$TNR = 0{\cdot}632 \frac{c_P' R T_0^2}{q_0 E} \tag{2.48}$$

The TNR is an important feature if an emergency cooling system has to cope with an imminent runaway reaction: it must become efficient in a time shorter than *TNR*.

A further time of interest is that required by a thermal explosion to run to completion. To calculate this, the integration is performed between T_0 and $T_0 + \Delta T_{ad}$ or $\theta \to \infty$

$$t = \frac{1}{\theta_0}\int_0^{\infty} e^{-\theta} d\theta = \left[\frac{1}{\theta_0} e^{-\theta}\right]_0^{\infty} = \frac{1}{\theta_0} \tag{2.49}$$

$$t = \frac{\Delta T}{T_0} = \frac{c'_P R T_{crit}^2}{q_0 E} \approx \frac{c'_P R T_0^2}{q_0 E} \tag{2.50}$$

The time to maximum rate under adiabatic conditions is

$$TMR_{ad} = \frac{c'_P R T_0^2}{q_0 E} \tag{2.51}$$

TMR_{ad} is a function of the reaction kinetics. It can be evaluated based on the heat release rate of the reaction q_0 at the initial conditions T_0 by knowing the heat specific capacity of the reaction mass c'_P and the activation energy of the reaction E. Since q_0 is an exponential function of temperature, TMR_{ad} decreases exponentially with temperature and decreases with increasing activation energy:

$$TMR_{ad}(T) = \frac{c'_P R T_0^2}{q_0 e^{-E/RT_0} E} \tag{2.52}$$

This concept of TMR_{ad} was initially developed by Semenov [7] and was reintroduced by Townsend and Tou [8] as they developed the accelerating rate calorimeter (see Section 4.3.1.3). It is used to characterize decomposition reactions, as described in Chapters 3 and 11.

2.6
Exercises

▶ Exercise 2.1

A thermally instable insecticide has to be transported in a 200-liter drum. The degree of filling is 90% and the product has a specific weight of $1000\,kg\,m^{-3}$. The complete decomposition is accompanied by the evolution of $0.1\,m^3$ of gas/kg insecticide at 30 °C.

Question:
What percentage of the contents of the drum is allowed to decompose if the drum resists an overpressure of max. 0.45 bar? A storage temperature of 30 °C is assumed.

Table 2.9 Thermal data and physical properties of cyclohexane.

Thermal data		Cyclohexane	
Heat of reaction	80 kJ kg^{-1}	Molecular weight	84 g mol^{-1}
Heat of decomposition	140 kJ kg^{-1}	Boiling point	81 °C
Specific heat capacity	2.0 kJ kg^{-1} K^{-1}	Heat of evaporation	30 kJ mol^{-1}
		Lower explosion limit	1.3% v/v
		Molecular volume of vapor at 25 °C	25 liters

▶ Exercise 2.2

A chemical reaction is performed at 10 °C in cyclohexane. A secondary decomposition reaction becomes dominant above 30 °C. In case of cooling failure, the reaction mass will reach boiling point and cyclohexane will evaporate. Thermal data and physical properties of cyclohexane are summarized in Table 2.9.

Question:
Calculate the volume of the largest flammable vapor cloud at 25 °C, corresponding to the runaway of 1 kg of reaction mass.

▶ Exercise 2.3

An intermediate for a dyestuff is prepared by sulfonation and nitration of an aromatic compound at 40 °C. The intermediate product has to be precipitated by dilution of the sulfuric acid with water to a final concentration of 60%. This dilution is performed under adiabatic conditions (no cooling) and the final temperature is 80 °C. This temperature of 80 °C is important for the crystallization and for the following filtration. After the temperature has reached 80 °C, the mixture is immediately cooled down to 20 °C by applying the full cooling capacity of the reactor.

The thermal study of the reaction mixture shows a heat release rate of 10 W kg^{-1} at 80 °C and a total heat of decomposition of 800 kJ kg^{-1}. The specific heat is 2 kJ kg^{-1} K^{-1}.

Questions:
1. Assess the severity linked to this operation in case of a cooling failure.
2. How much time is left to organize emergency measures?
3. Assess the probability of the incident.

Hint:

The induction time of the thermal explosion can be estimated using the van't Hoff rule: the reaction rate doubles when the temperature is increased by 10 K. The temperature increase rate can be approximated by

$$\frac{\Delta T}{\Delta t} \approx \frac{dT}{dt} = \frac{q'(T)}{c'_P}$$

References

1 Gygax, R. (1993) *Thermal Process Safety, Data Assessment, Criteria, Measures,* Vol. 8 (ed. ESCIS), ESCIS, Lucerne.

2 Perry, R. and Green, D. (eds) (1998) *Perry's Chemical Engineer's Handbook,* 7th edn, McGraw-Hill, New York.

3 Weast, R.C. (ed.) (1974) *Handbook of Chemistry and Physics,* 55th edn, CRC Press, Cleveland.

4 Poling, B.E., Prausnitz, J.M., and O'Connell, J.P. (2001) *The Properties of Gases and Liquids,* 5th edn, McGraw-Hill, New York.

5 Levenspiel, O. (1972) *Chemical Reaction Engineering,* John Wiley & Sons, New York.

6 Barton, A. and Rogers, R. (1997) *Chemical Reaction Hazards,* Institution of Chemical Engineers, Rugby.

7 Semenov, N.N. (1928) Zur theorie des Verbrennungsprozesses. *Zeitschrift für Physik,* **48**, 571–82.

8 Townsend, D.I. (ed.) (1981) *Accelerating Rate Calorimetry,* Vol. 68, *Industrial Chemical Engineering Series,* IchemE.

3
Assessment of Thermal Risks

Case History "Sulfonation"

2-Chloro-5-nitrobenzene sulfonic acid is synthesized by addition of p-chloronitrobenzene as a melt to 20% Oleum (20% SO_3 in H_2SO_4) at 100 °C [1]. This is added over 20 minutes to Oleum heated at 50 °C. The temperature then rises to 120–125 °C due to the heat of reaction. The conversion is achieved by maintaining this temperature during several hours with 2 bar of steam.

This operation was performed as usual, but a further temperature increase above 125 °C went unnoticed. This led to an explosion, in which the reactor disintegrated, its lid crashing through the roof of the building. In fact, the abnormally high temperature triggered a secondary decomposition reaction, which caused the heavy damage. No heat exchange can take place with the steam in the jacket once the reactor temperature is above the condensation temperature of the steam (120 °C at 2 bar). Thus, it was not possible to control the temperature at this stage.

Before the incident, neither the reaction and decomposition energy potentials nor the triggering conditions of the decomposition were known. Thus, a potentially severe process was entirely under manual control, without provision for an alarm system and emergency measures. A correct assessment of the energies and triggering conditions of the decomposition predicts such an incident, giving the opportunity to design a process that will avoid such incidents.

3.1
Introduction

In this chapter, after introducing some definitions, a systematic assessment procedure, based on the cooling failure scenario, is outlined. This scenario formulates six key questions that comprise the database for the assessment. Relying on the characteristic temperature levels arising from the scenario, criticality classes are defined. They provide a selection of the required risk-reducing

Thermal Safety of Chemical Processes: Risk Assessment and Process Design. Francis Stoessel

ISBN: 978-3-527-31712-7

measures. The chapter is closed with a practical assessment procedure, which represents the thread followed throughout this book. Section 2 is devoted to the design of safe reactors, or safe operating procedures for industrial reactors. Section 3 is devoted to the characterization and avoidance of secondary reactions.

3.2
Thermal Risks

Traditionally, "risk" is defined as the product of the severity of a potential incident by its probability of occurrence. Hence, risk assessment requires the evaluation of both the severity and the probability. Obviously, the results of such an analysis aid in designing measures for the reduction of the risk (Figure 3.1). The question that arises now is: "What do severity and probability mean in the case of thermal risks inherent to a particular chemical reaction or process?"

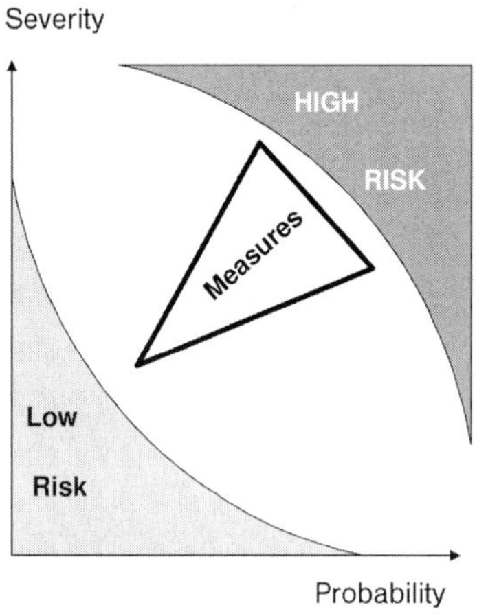

Figure 3.1 Risk diagram.

The thermal risk linked to a chemical reaction is the risk of loss of control of the reaction and associated consequences (e.g. triggering a runaway reaction). Therefore, it is necessary to understand how a reaction can "switch" from its normal course to a runaway condition. In order to make this assessment, the theory of thermal explosion (see Chapter 2) needs to be understood, along with the concepts of risk assessment. This implies that an incident scenario was identified and described, with its triggering conditions and the resulting consequences, in order to assess the severity and probability of occurrence. For thermal risks, the worst case will be to lose the cooling of a reactor or in general to consider that the reaction mass or the substance to be assessed is submitted to adiabatic conditions. Hence, we consider a cooling failure scenario.

3.3 Systematic Assessment Procedure

3.3.1 Cooling Failure Scenario

The behavior of a chemical system during a runaway will be demonstrated by using the example of an exothermal batch reaction. A classical procedure is as follows: reactants are charged into the reactor at room temperature and heated under stirring to the reaction temperature. They are then held at this level, where cycle time and yield are optimized. After the reaction is complete, the reactor is cooled and emptied (dashed line in Figure 3.2).

The scenario presented here was developed by R. Gygax [1, 2]. Let us assume that, while the reactor is at the reaction temperature (T_P), a cooling failure occurs (point 4 in Figure 3.2). The scenario consists of the description of the temperature evolution after the cooling failure. If, at the instant of failure, unconverted material is still present in the reactor, the temperature increases due to the completion of the reaction. This temperature increase depends on the amount of non-reacted material, thus on the process conditions. It reaches a level called the Maximum Temperature of the Synthesis Reaction (MTSR). At this temperature, a secondary decomposition reaction may be initiated. The heat produced by this reaction may

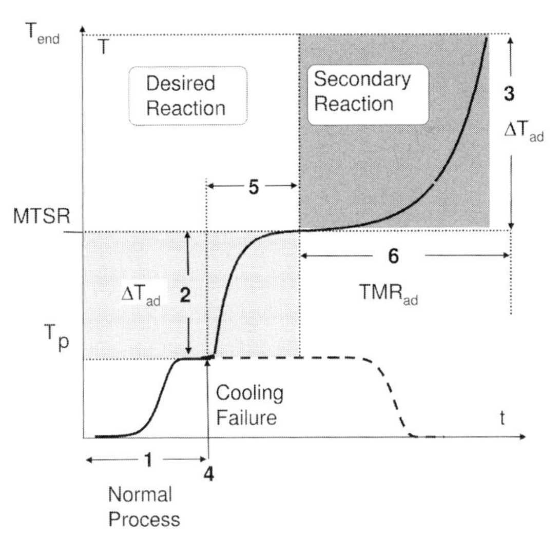

Figure 3.2 Cooling Failure Scenario: After a cooling failure, the temperature rises from process temperature to the maximum temperature of synthesis reaction. At this temperature, a secondary decomposition reaction may be triggered. The left-hand part of the scheme is devoted to the desired reaction and the temperature increase to the *MTSR* in case of a failure. In the right-hand part, the temperature increase due to a secondary exothermal reaction is shown, with its characteristic time to maximum rate. The numbers represent the six key questions.

lead to a further increase in temperature (period 6 in Figure 3.2), reaching the final temperature (T_{end}).

Here we see that by losing control of the main reaction (or synthesis reaction), we may trigger a secondary reaction. This distinction between main and secondary reactions simplifies the assessment, since both reactions are virtually separated, allowing them to be studied separately, but may still be connected by the temperature level $MTSR$.

The following questions represent six key points that help to develop the runaway scenario and provide guidance for the determination of data required for the risk assessment:

Question 1: Can the process temperature be controlled by the cooling system?

During normal operation, it is essential to ensure sufficient cooling in order to control the temperature of the reactor, hence to control the reaction course. This typical question should be addressed during process development. To ensure the thermal control of the reaction, the power of the cooling system must be sufficient to remove the heat released in the reactor. Special attention must be devoted to possible changes in the viscosity of the reaction mass as for polymerizations, and to possible fouling at the reactor wall (see Chapter 9). An additional condition, which must be fulfilled, is that the reactor is operated in the dynamic stability region, as described in Chapter 5.

The required data are the heat release rate of the reaction (q_{rx}) and the cooling capacity of the reactor (q_{ex}). These can best be obtained from reaction calorimetry.

Question 2: What temperature can be attained after runaway of the desired reaction?

If after the cooling failure unconverted reactants are still present in the reaction mixture, they will react in an uncontrolled way and lead to an adiabatic temperature increase. The remaining unconverted reactants are referred to as accumulated reactants. The available energy is proportional to the accumulated fraction. Thus, the answer to this question necessitates the study of the reactant conversion as a function of time, in order to determine the degree of accumulation of unconverted reactants (X_{ac}). The concept of Maximal Temperature of the Synthesis Reaction (MTSR) was developed for this purpose:

$$MTSR = T_p + X_{ac} \cdot \Delta T_{ad,rx} \tag{3.1}$$

These data can be obtained from reaction calorimetry, which delivers the heat of reaction required for the determination of the adiabatic temperature rise (ΔT_{ad}). The integration of the heat release rate can be used to determine the thermal conversion and the thermal accumulation (X_{ac}). The accumulation may also be obtained from analytical data.

Question 3: What temperature can be attained after runaway of the secondary reaction?

Since the temperature of the MTSR is higher than the intended process temperature, secondary reactions may be triggered. This will lead to further runaway due to the uncontrolled secondary reaction, which may be decomposition. The thermal data of the secondary reaction allows us to calculate the adiabatic temperature rise and determine the final temperature starting from the temperature level MTSR:

$$T_{end} = MTSR + \Delta T_{ad,d} \tag{3.2}$$

This temperature (T_{end}) gives an indication of the possible consequences of a runaway, as will be shown below.

The data may be obtained from calorimetric methods usually employed for the study of secondary reaction and thermal stability as DSC, Calvet calorimetry, and adiabatic calorimetry.

Question 4: At which moment does the cooling failure have the worst consequences?

Since the time of the cooling failure is unknown, it must be assumed that it occurs at the worse instant, that is, at the time where the accumulation is at a maximum and/or the thermal stability of the reaction mixture is critical. The amount of unconverted reactants and the thermal stability of the reaction mass may vary with time. Thus, it is important to know at which instant the accumulation, and therefore the thermal potential, is highest. The thermal stability of the reaction mass may also vary with time. This is often the case when a reaction proceeds over intermediate steps. Hence, both the synthesis reaction and secondary reactions must be known in order to answer this question. The combination of a maximum accumulation with the "minimum" thermal stability defines the worst case. Obviously, the safety measures have to account for it.

The data required to answer this question may be obtained from reaction calorimetry for the accumulation in combination with DSC, Calvet calorimetry, or adiabatic calorimetry for the thermal stability.

Question 5: How fast is the runaway of the desired reaction?

Starting from the process temperature, reaching the MTSR will take some time. However, industrial reactors are usually operated at temperatures where the desired reaction is fast. Hence, a temperature increase above the normal process temperature will cause a significant acceleration of the reaction. In most cases, this period is very short (see period 5 in Figure 3.2).

The duration of the main reaction runaway may be estimated using the initial heat release rate of the reaction and the concept of TMR_{ad}:

$$TMR_{ad} = \frac{c'_P \cdot R \cdot T_P^2}{q_{(T_P)} \cdot E} \tag{3.3}$$

Question 6: How fast is the runaway of the decomposition starting at MTSR?

Since the temperature of the MTSR is higher than the intended process temperature, secondary reactions may be triggered. This will lead to further runaway due to the uncontrolled secondary reaction, which may be through decomposition. The dynamics of the secondary reaction plays an important role in the determination of the probability of an incident. The concept of Time to Maximum Rate under adiabatic conditions (TMR_{ad}) [3] was used for that purpose (see Section 2.5.5):

$$TMR_{ad} = \frac{c'_P \cdot R \cdot T_{MTSR}^2}{q_{(MTSR)} \cdot E} \tag{3.4}$$

The six key questions presented above ensure that the essential knowledge about the thermal safety of a process is addressed. In this sense, they represent a systematic way of analysing the thermal safety of a process and building the cooling failure scenario. Once the scenario is defined, the next step is the actual assessment of the thermal risks, which requires assessment criteria. The criteria used for the assessment of severity and probability are presented below.

3.3.2 Severity

Since most reactions in the fine chemicals industry are exothermal, the consequences of loss of control of a reaction are linked to the energy released. As shown in Section 2.2.1.4, the adiabatic temperature rise, which is proportional to the reaction energy, represents an easy to use criterion for the evaluation of the severity, that is, the potential of destruction of an uncontrolled energy release as a runaway reaction. The adiabatic temperature rise can be calculated by dividing the energy of the reaction by the specific heat capacity:

$$\Delta T_{ad} = \frac{Q'}{c'_P} \tag{3.5}$$

In this expression, Q' represents the specific energy of the reaction or of the decomposition. The specific heat capacities given in Table 2.2 can be used. However, as a first approximation, the following specific heat capacities may be useful:

- water $4.2\,kJ\,kg^{-1}\,K^{-1}$
- organic liquids $1.8\,kJ\,kg^{-1}\,K^{-1}$
- inorganic acids $1.3\,kJ\,kg^{-1}\,K^{-1}$
- easy to remember $2.0\,kJ\,kg^{-1}\,K^{-1}$

The higher the final temperature, the worse are the consequences of the runaway. In the case of a large temperature increase, some components of the reaction mixture may be vaporized or the decomposition may produce gaseous compounds. Therefore, the pressure of the system will increase. This may result in the rupture of the vessel and heavy damage. As an example, a final temperature of 200 °C in a solvent such as acetone may be critical, whereas a final temperature of 80 °C in water will not be.

The adiabatic temperature rise is not only important in the determination of the temperature levels, but also has important consequences on the dynamic behavior of a runaway reaction. As a general rule, high energies result in fast runaway or thermal explosion, while low energies (adiabatic rise below 100 K) result in slower temperature increase rates (Figure 2.3), given the same activation energy, the same initial heat release rate, and starting temperature (see Section 2.4.3).

The **severity** of the runaway can be evaluated using the **temperature** levels attained, if the desired reaction (Question 2, Section 3.3.1) and the decomposition reaction (Question 3, Section 3.3.1) proceed under adiabatic conditions.

A proposal for assessment criteria on a four levels scale is presented in Table 3.1. This four levels scale for the severity is commonly used in the fine chemicals industry [4] and has its roots in the Zurich Hazard Analysis (ZHA) developed by the Zurich Insurance Company [5]. If the assessment is performed on a three levels scale, the upper levels "critical" and "catastrophic" may be merged into one level, "high."

The assessment of the severity is based on the fact that for a temperature increase of 200 K and above the temperature increase under adiabatic conditions as a function of time becomes very sharp (Figure 2.3). This results in a violent reaction and consequently heavy consequences. At the opposite end of the scale, for a temperature increase of 50 K and below, the behavior of the reaction mass under adiabatic conditions cannot be a thermal explosion. The temperature curve is smooth and corresponds to self-heating rather than to an explosion. Thus, if there is no risk of pressurization, for example, due to dissolved gases, the severity is low.

Table 3.1 Assessment criteria for the severity of a runaway reaction.

Simplified	Extended	ΔT_{ad} (K)	Order of magnitude of Q' kJ kg^{-1}
High	Catastrophic	>400	>800
	Critical	200–400	400–800
Medium	Medium	50–100	100–400
Low	Negligible	<50 and no pressure	<100

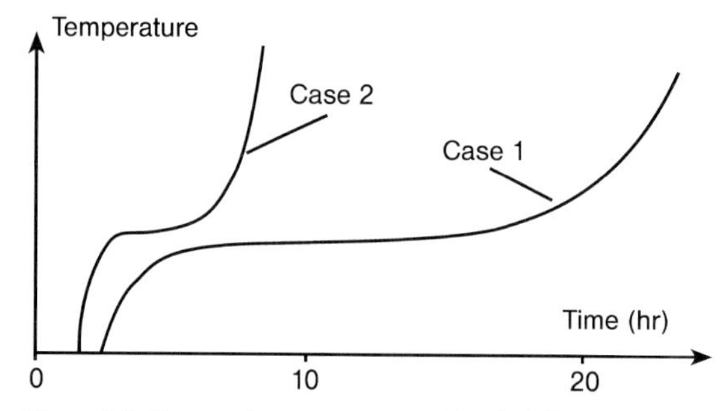

Figure 3.3 Time-scale as a measure of probability.

3.3.3 Probability

Presently there is no direct quantitative measure of the probability of occurrence of an incident, or in the case of thermal process safety, of the occurrence of a runaway reaction. Nevertheless, if we consider the runaway curves presented in Figure 3.3, the two cases presented are very different. In case 1, after the temperature increase due to the main reaction, there is enough time left to take measures to regain control or recover a safe situation. If we compare the probability of runaway in both cases, it becomes clear that the probability of triggering the runaway is higher in case 2 than in case 1. Thus, while we cannot easily quantify probabilities, we can at least compare them on a semi-quantitative scale.

> The **probability** can be evaluated using the **time-scale**: If, after the cooling failure (Question 4), there is enough time left (Questions 5 and 6) to take emergency measures before the runaway becomes too fast, the probability of the runaway will remain low.

For the assessment of probabilities, a six levels scale is often used [4], as proposed by the ZHA method [5]. Assessment criteria based on such a scale is presented in Table 3.2.

If a simple three levels scale is to be used, the levels "frequent" and "probable" are merged in one level "high" and the levels "seldom," "remote," and "almost impossible" are merged in one level "low." The intermediate level "occasional" then becomes "medium." For chemical reactions on an industrial scale (not for storage or transportation), we can consider a probability to be low if the time to maximum rate of a runaway reaction under adiabatic conditions is longer than 1 day. The probability becomes high if the time to maximum rate becomes less than 8 hours (1 shift). These time-scales are only orders of magnitude and are dependent on many factors, among them the degree of automation, the training

Table 3.2 Assessment criteria for the severity of a runaway reaction.

Simplified	Extended	TMR_{ad} (h)
High	Frequent	<1
	Probable	1–8
Medium	Occasional	8–24
Low	Seldom	24–50
	Remote	50–100
	Almost impossible	>100

of the operators, the frequency of electrical power failures, size of the reactor, and so on. This scaling of probabilities is only valid if something is being done to cope with the known severity and applies to reactions, not to storage.

3.3.4 Criticality of Chemical Processes

3.3.5 Assessment of the Criticality

The cooling failure scenario presented above uses the temperature scale for the assessment of severity and the time-scale for the probability assessment. Starting from the process temperature (T_P), in the case of a failure, the temperature first increases to the maximum temperature of the synthesis reaction ($MTSR$). At this point, a check must be made to see if a further increase due to secondary reactions could occur. For this purpose, the concept of TMR_{ad} is very useful. Since TMR_{ad} is a function of temperature (see Section 2.5.5) it may also be represented on the temperature scale. For this, we can consider the variation of TMR_{ad} with temperature and look for the temperature at which TMR_{ad} reaches a certain value (Figure 3.4), for example, 24 hours or 8 hours, which are the levels in the assessment criteria presented in Sections 3.3.2 and 3.3.3.

In addition to the three temperature levels (T_P, $MTSR$, and T_{D24}), there is another important temperature: that at which technical limits of the equipment arc rcached. This may be due to the resistance of construction materials, or to the reactor design parameter as pressure or temperature, and so on. In an open reacting system, operated at atmospheric pressure, the boiling point is often used. In a closed system, operated under pressure, it may be the temperature on reaching the set pressure of the pressure relief system.

Thus, by considering the temperature scale, and for reactions presenting a thermal potential, we consider the relative position of four temperature levels:

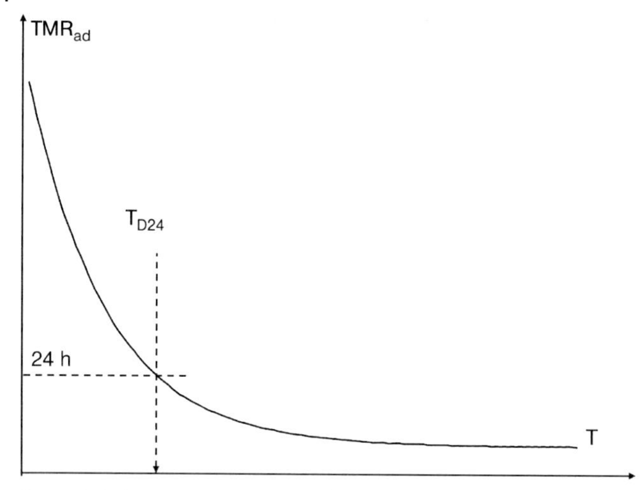

Figure 3.4 Variation of TMR_{ad} as a function of temperature. The maximum allowed temperature with respect to thermal stability (T_{D24}) is given at the point where TMR_{ad} is equal to 24 hours.

1. The process temperature (T_P): the initial temperature in the cooling failure scenario. In case of non-isothermal processes, the initial temperature will be taken at the instant when the cooling failure has the heaviest consequences (worst case).
2. Maximum temperature of synthesis reaction (MTSR): this temperature depends essentially on the degree of accumulation of unconverted reactants and so is strongly dependant on process design.
3. Temperature at which TMR_{ad} is 24 hours (T_{D24}): this temperature is defined by the thermal stability of the reaction mixture (see Chapter 11). It is the highest temperature at which the thermal stability of the reaction mass is unproblematic.
4. Maximum temperature for technical reasons (MTT): is the boiling point in an open system. For a closed system, it is the temperature at the maximum permissible pressure, that is, the set pressure of a safety valve or bursting disk.

These four temperature levels classify the scenarios into five different classes, ranging from the least critical (1–2) to the most critical (3–5) Figure 3.5 [6].

3.3.6 Criticality Classes

Depending on the order in which the different temperature levels (described above) follow each other, different types of scenarios arise. These differ by their respective criticality, allowing classification by a criticality index. This index is a

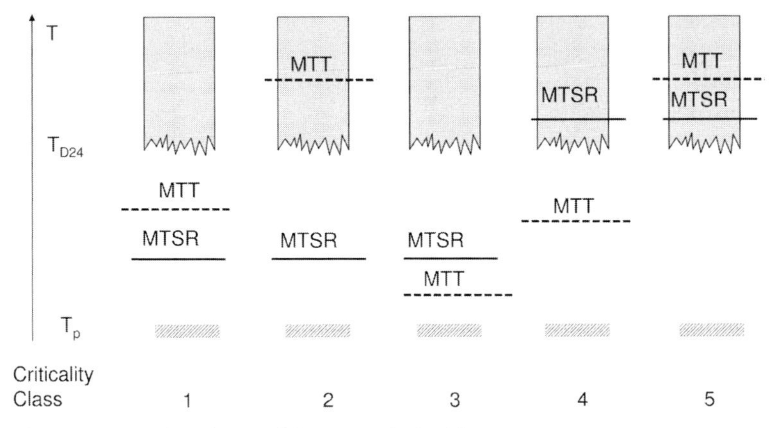

Figure 3.5 Criticality Classes of Scenario, obtained by combining the four temperature levels: *TP*, *MTSR*, T_{D24} and *MTT*.

useful tool, not only for the risk assessment, but also for the choice and the definition of adequate risk reducing measures. This point is presented in more detail in Chapter 10.

3.3.6.1 **Criticality Class 1**

After loss of control of the synthesis reaction, the technical limit ($MTSR < MTT$) cannot be reached and the decomposition reaction cannot be triggered, since the *MTSR* stays below T_{D24}. Only if the reaction mass is maintained for a long time under heat accumulation conditions, can the *MTT* be reached. Then the evaporative cooling may serve as an additional safety barrier. The process presents a low thermal risk.

Therefore, no special measure is required for this class of scenario, but the reaction mass should not be held for a longer time under heat accumulation conditions. The evaporative cooling or the emergency pressure relief could serve as a safety barrier as far as their design is appropriate.

3.3.6.2 **Criticality Class 2**

After loss of control of the synthesis reaction, the technical limit cannot be reached ($MTSR < MTT$) and the decomposition reaction cannot be triggered ($MTSR < T_{D24}$). The situation is similar to class 1; nevertheless, since the level *MTT* is above T_{D24}, if the reaction mass is maintained for a longer time under heat accumulation conditions, the decomposition reaction could be triggered and reach the *MTT*. In this case, reaching the boiling point could be a hazard if the heat release rate at *MTT* is too high. As long as the reaction mass is not kept for a longer time under heat accumulation conditions, the process presents a low thermal risk.

If heat accumulation can be avoided, no special measure is required. If heat accumulation conditions cannot be excluded, evaporative cooling or the emergency pressure relief could eventually serve as a barrier. Therefore, these measures must be designed for that purpose.

3.3.6.3 **Criticality Class 3**

After loss of control of the synthesis reaction, the technical limit ($MTSR > MTT$) will be reached, but the decomposition reaction cannot be triggered ($MTSR < T_{D24}$). In this situation, the safety of the process depends on the heat release rate of the synthesis reaction at the MTT.

The first measure is to use the evaporative cooling or controlled depressurisation to keep the reaction mass under control. The distillation system must be designed for such a purpose and has to function, even in the case of failure of utilities. A backup cooling system, dumping of the reaction mass, or quenching could also be used. Alternatively, a pressure relief system may be used, but this must be designed for two-phase flow that may occur, and a catch pot must be installed in order to avoid any dispersion of the reaction mass outside the equipment. Of course, all these measures must be designed for such a purpose and must be ready to work immediately after the failure occurs. The use of thermal characteristics of the scenario for the choice of technical measures is presented in detail in Chapter 10.

3.3.6.4 **Criticality Class 4**

After loss of control of the synthesis reaction, the technical limit will be reached ($MTSR > MTT$) and the decomposition reaction could theoretically be triggered ($MTSR > T_{D24}$). In this situation, the safety of the process depends on the heat release rate of both the synthesis reaction and decomposition reaction at the MTT. The evaporative cooling or the emergency pressure relief may serve as a safety barrier. This scenario is similar to class 3, with one important difference; if the technical measures fail, the secondary reaction will be triggered.

Thus, a reliable technical measure is required. It may be designed in the same way as for class 3, but the additional heat release rate due to the secondary reaction has also to be taken into account. The use of thermal characteristics of the scenario in the design of technical measures is presented in detail in Chapter 10.

3.3.6.5 **Criticality Class 5**

After loss of control of the synthesis reaction, the decomposition reaction will be triggered ($MTSR > T_{D24}$) and the technical limit will be reached during the runaway of the secondary reaction. In such a case, it is unlikely that the evaporative cooling or the emergency pressure relief can serve as a safety barrier. This is because the heat release rate of the secondary reaction at the temperature level MTT may be too high and result in a critical pressure increase. Thus, it is a critical scenario.

Hence, in this class there is no safety barrier between the main and secondary reaction. Therefore, only quenching or dumping can be used. Since, in most cases, the decomposition reactions release very high energies, particular attention has to be paid to the design of safety measures. It is worthwhile considering an alternative design of the process in order to reduce the severity or at least the probability of triggering a runaway. As an alternative design, the following possibilities should be considered: reduce the concentration, change from batch to semi-batch, optimize semi-batch operating conditions in order to minimize the accumulation, change to continuous operation, and so on.

3.3.6.6 Remarks Concerning the Use of MTT as a Safety Barrier

In the scenarios corresponding to classes 3 and 4, the technical limit (*MTT*) plays an important role. In an open system, this limit may be the boiling point. In such a case, the distillation or reflux system must be designed for this purpose, so that its capacity is sufficient for the vapor flow rate produced at this temperature during the runaway. Special care must be taken if there is possibility of flooding of the vapor tube or of the reaction mass swelling, which both may lead to increased head loss. The condenser must also provide sufficient capacity, even at a vapor velocity that may be relatively high. Further, the reflux system must be designed to operate with an independent cooling medium to avoid common mode failure. This point is explained in detail in Section 9.3.5.

In a closed system, the technical limit may be the temperature at which the pressure in the reactor reaches the set pressure of the relief system. In such a case, it may be possible to depressurize the reactor in a controlled way before the set pressure is reached. This allows tempering a reaction at a temperature where it is still controllable.

If the pressure is allowed to increase to the set pressure of the emergency relief system with a reacting system, the pressure increase rate may be fast enough to cause two phase flow and corresponding high discharge flow rates. The design of the system, which may be a safety valve or a bursting disk, must be performed using the techniques developed in the frame of the Design Institute of Emergency Relief Systems (DIERS) [7–9]. These points will be developed in more detail in Chapters 9 and 10.

3.4 Assessment Procedures

3.4.1 General Rules for Thermal Safety Assessment

At first glance, the data and concepts used for the assessment of thermal risks may appear complex and difficult to overlook. In practice, however, two rules simplify the procedure and reduce the amount of work to the required minimum:

1. Simplification: As a first approach the problem should be simplified as far as possible. This allows limiting the volume of data to the essential minimum. It results in an economic problem solution.

2. Worst-case approach: In case approximations are to be made, one should ensure that the approximation is conservative, in the sense that it leads to majoring the risks in the assessment procedure.

When this approach is used, if the results obtained by the simplified approach allow the intended operation, it should be ensured that sufficiently large safety margins are applied. In case it leads to a negative assessment, where the conclusion is that the operation cannot be performed safely under these simplified assumptions, it means that more precise data are required to make the final decision. By doing so, the data set is adjusted to the difficulty of the problem. These rules are systematically used in the worked examples in this book.

3.4.2 Practical Procedure for the Assessment of Thermal Risks

The six key questions described in the cooling failure scenario allow us to identify and assess the thermal risks of a chemical process. The first steps allow building a failure scenario, which is easy to understand and serves as a base for the assessment. The proposed procedure (Figure 3.6) is based on the separation of severity and probability, taking into account the economic aspects of data determination in a safety laboratory. In a second step, based on the scenario, the criticality index can be determined to help in the choice and design of risk-reducing measures.

In order to ensure an economic assessment, by only determining the required data, a practical systematic procedure was elaborated (Figure 3.7). In the first part of the procedure, worst-case conditions are assumed, for example, for the reaction, the accumulation is considered to be 100%. This allows an assessment of the worst case.

The first step of the assessment is screening for the energy potential of a sample of a reaction mass, where a reaction has to be assessed, or of a sample of a substance, where the thermal stability has to be assessed. This may be obtained from a dynamic DSC experiment on samples of the reaction mass taken before, during, and after the reaction. Obviously, when the thermal stability of a sample has to be assessed, this is reduced to a representative sample of the reacting mass. If there is no significant energy potential, such as if the adiabatic temperature rise is less than 50 K and there is no overpressure, the study can be stopped at this stage.

If a significant energy potential is found, one must find out if it stems from the main reaction or from a secondary reaction:

- If the potential stems from the desired reaction, all aspects concerning the heat release rate, cooling capacity, and the accumulation, i.e. MTSR, must be studied.
- If the potential stems from secondary reactions, their kinetics must be studied in order to determine the TMR_{ad} at MTSR.

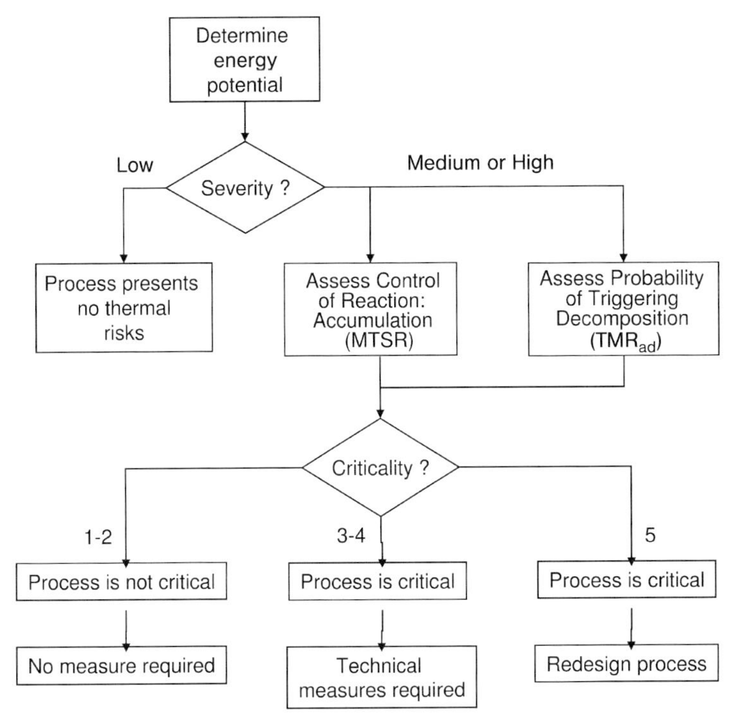

Figure 3.6 Overview of an assessment procedure showing the different steps and the data required for building the failure scenario and for the assessment using the criticality index.

Thus, the MTSR plays a key role as it is the result of the loss of control of the main reaction, and is the temperature for which the thermal stability must be ensured.

This study may be carried out in steps of increasing degree of detail, in order to provide the required data in the required accuracy:

Estimation:

- As an example the main reaction may first be considered as a batch, that is, with an accumulation of 100%. The MTSR is calculated for a batch reaction.
- The temperature, at which TMR_{ad} is 24 hours (T_{D24}), may be estimated, as shown in Chapter 11.

If the consequences of worst-case assumptions are not acceptable, the data must be determined with more accuracy:

- The accumulation of reactants during the main reaction is determined, for example, by reaction calorimetry. This allows the determination of the true accumulation and consequently of the true MTSR. The questions arising from the control of the reaction, that is, the maximum heat release rate must be compared to the cooling capacity of the reactor. Moreover, the gas release rate, if applicable, must be compared to the gas treatment capacity of the reactor.

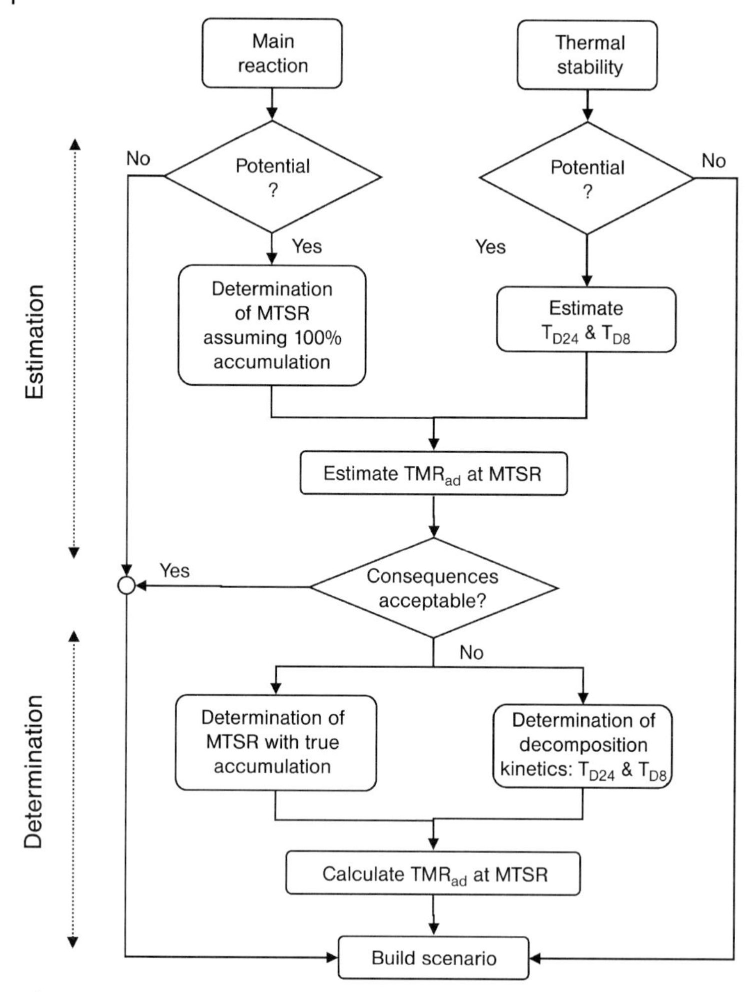

Figure 3.7 Assessment using the principles of increasing accuracy.

- Following the same principle of increasing complexity, the TMR_{ad} must be determined as a function of temperature from the kinetics of the secondary reaction. This allows determination of the temperature at which the TMR_{ad} is 24 hours (T_{D24}).

These data are summarized in graphical form (Figure 3.8), a fast check of the nature of thermal risks linked to a given process. This procedure is based on the principle of increasing accuracy, where only the data required for the assessment are determined. If the assessment is based on simple tests, the safety margins remain large. Nevertheless, with more accurate data, these safety margins may be reduced without impinging on the thermal safety of the process under study.

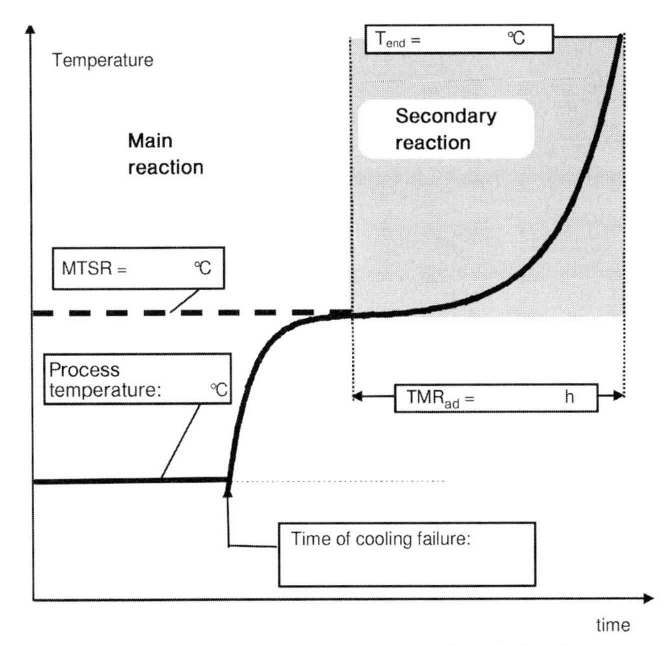

Figure 3.8 Graphical summary of the thermal risks linked to the performance of an industrial process.

Worked Example 3.1: Amination (RX19)

A chloro-aromatic compound is to be converted to the corresponding aniline compound in a $1\,m^3$ autoclave with a maximum allowed working pressure of 100 bar g. A large stoichiometric excess (4:1) of ammonia (30% in water) is used. This allows the neutralization of the resulting hydrochloric acid and maintaining an alkaline pH in order to avoid corrosion. The reaction reaches a conversion of 90% within 8 hours at 180 °C:

$$Ar-Cl+2\,NH_3 \xrightarrow{180^\circ C} Ar-NH_2+NH_4Cl$$

The charge is 315 kg of chloro-aromatic compound, 2 kmol and 415 kg Ammonia 30%, 8 kmol. Both reactants are initially charged at room temperature and the reactor is heated to the reaction temperature of 180 °C, which is maintained during 12 hours.

Data:
Reaction enthalpy: $-\Delta H_r = 175\,kJ\,mol^{-1}$ (including neutralization)
Specific heat capacity of the reaction mass: $c_P' = 3200\,J\,kg^{-1}\,K^{-1}$
Decomposition energy of the final reaction mass: $Q_D' = 840\,kJ\,kg^{-1}$
Temperature at which the TMR_{ad} of the decomposition is 24 hours:
$T_{D24} = 280\,°C$

Vapor pressure of an ammonia solution 30% (w/w):

$$\ln(P(\text{bar})) = 11.47 - \frac{3385}{T(\text{K})}$$

Vapor pressure of an ammonia solution 19% (w/w):

$$\ln(P(\text{bar})) = 11.62 - \frac{3735}{T(\text{K})}$$

Question:
Assess the thermal risk linked to the performance of this process, and determine the criticality class.

Solution:
The process is performed as a batch reaction at 180 °C. The adiabatic temperature rise is

$$\Delta T_{ad} = \frac{Q'_r}{c'_P} = \frac{175\,\text{kJ}\cdot\text{mol}^{-1} \times 2000\,\text{mol}}{768\,\text{kg} \times 3.2\,\text{kJ}\cdot\text{kg}^{-1}\cdot\text{K}^{-1}} = 143\,\text{K}$$

The severity of a runaway of the amination reaction is "medium."
The temperature that will be reached if the amination gets out of control is

$$MTSR = T_P + \Delta T_{ad} = 180 + 143 = 323°\text{C} = 596\,\text{K}$$

The pressure would then be 211 bar. In this calculation, the conversion that takes place during heating the reactor is ignored: this is conservative. This temperature is above T_{D24}, meaning that the decomposition reaction will be triggered, leading to a further temperature rise of

$$\Delta T_{ad} = \frac{Q'_r}{c'_P} = \frac{840\,\text{kJ}\cdot\text{kg}^{-1}}{3.2\,\text{kJ}\cdot\text{kg}^{-1}\cdot\text{K}^{-1}} = 263\,\text{K}$$

The severity of the decomposition reaction is "high."
The maximum allowed pressure of 100 bar will be reached at approximately 240 °C: this temperature will be taken as the maximum temperature for technical reasons (MTT).

Hence, the succession of characteristic temperatures is $T_P < MTT < T_{D24} < MTSR$, which corresponds to a criticality class 4, which requires technical measures. The high latent heat of evaporation of water and ammonia would allow stopping the runaway at 240 °C by depressurizing the reactor in order to use evaporation cooling. This possibility will be analysed in Chapter 10.

Worked Example 3.2: Hydrogenation (AR8)

A ketone is to be hydrogenated to the corresponding alcohol at 30 °C in an aqueous solution at a concentration of $0.1\,\text{mol}\,\text{l}^{-1}$ and at a pressure of 2 bar g in a reactor protected against overpressure by a safety valve with a set pressure of 3.2 bar g. The molecule presents no other reactive functional groups.

Data:
No thermal data are available, but similar reactions have an enthalpy of $200\,\text{kJ}\,\text{mol}^{-1}$.
Typical enthalpies of decomposition reactions are shown in Table 11.1.
The specific heat capacity of the reaction mass is $c_P' = 3.6\,\text{kJ}\,\text{kg}^{-1}\cdot\text{K}^{-1}$.

Questions:
1. Assess the thermal risks linked to this hydrogenation reaction.
2. Assess the thermal risks linked to this decomposition reaction.
3. What other risks should be considered for this hydrogenation?

Solution:
This example shows that with only sparse thermal data it is sometimes possible to assess thermal risks. This is possible due to the low concentration used in this hydrogenation.

(1) The reaction is performed in a diluted aqueous solution. Thus, its density can be assumed to be $1000\,\text{kg}\,\text{m}^{-3}$. Then, the specific heat of reaction is

$$Q_r' = \frac{C(-\Delta H_r)}{\rho} = \frac{0.1\,\text{mol}\cdot\text{l}^{-1} \times 200\,\text{kJ}\cdot\text{mol}^{-1}}{1\,\text{kg}\cdot\text{l}^{-1}} = 20\,\text{kJ}\cdot\text{kg}^{-1}$$

and the corresponding adiabatic temperature rise is

$$\Delta T_{ad} = \frac{Q_r'}{c_P'} = \frac{20\,\text{kJ}\cdot\text{mol}^{-1}}{3.6\,\text{kJ}\cdot\text{kg}^{-1}\cdot\text{K}^{-1}} \simeq 6\,\text{K}$$

Such a weak adiabatic temperature rise cannot lead to a thermal explosion. The severity is low. In case of malfunction of the reactor cooling system, the reaction, providing it is not stopped, will lead to an immediate temperature rise by 6 K reaching the MTSR of 36 °C. The thermal risk linked to this hydrogenation reaction is low.

(2) In order to cause a problem if a decomposition reaction would be triggered, its energy must allow a pressure increase higher than 3.2 bar g. Since the compounds are either a cetone or an alcohol, no gas release is expected. Thus, the pressure in the system is only due to vapor pressure. With the low concentration, we assume the vapor pressure to be that of water. In order to reach, say 3.2 bar g, starting from 2 bar g hydrogen, if we neglect the hydrogen uptake

by the reaction, the temperature must reach 105 °C. Thus, the energy of decomposition should lead to a temperature increase from MTSR to MTT, that is, from 36 to 105 °C or 69 K. The required energy is

$$Q'_D = c'_P \cdot \Delta T_{ad} = 3.6\,\text{kJ} \cdot \text{kg}^{-1} \cdot \text{K}^{-1} \times 69\,\text{K} \simeq 250\,\text{kJ} \cdot \text{kg}^{-1}$$

Taking the concentration into account, the decomposition enthalpy should be $2500\,\text{kJ} \cdot \text{mol}^{-1}$. Such a high decomposition enthalpy cannot be found in the table. Thus, the severity of a eventual decomposition reaction is assessed to be low.

(3) Other risks linked to the performance of this hydrogenation are essentially due to the explosion properties of hydrogen. Leaks must be avoided and the plant must be well ventilated. Further, hydrogenation catalysts are often pyrophoric and the toxicity of the compounds involved on the process must be considered.

3.5 Exercises

► Exercise 3.1

A diazotization is by adding sodium nitrite to an aqueous solution of the amine ($2.5\,\text{mol}\,\text{kg}^{-1}$). The industrial scale charge is 4000 kg of final reaction mass in a stirred tank reactor with a nominal volume of $4\,\text{m}^3$. The reaction temperature is 5 °C and the reaction is very fast. For the safety study, an accumulation of 10% is considered realistic.

Thermal data:

Reaction: $-\Delta H_r = 65\,\text{kJ mol}^{-1}$ $\quad c'_P = 3.5\,\text{kJ}\,\text{kg}^{-1}\,\text{K}^{-1}$

Decomposition: $-\Delta H_{dc} = 150\,\text{kJ}\,\text{mol}^{-1}$ $\quad T_{D24} = 30\,°\text{C}$

Questions:

1. Evaluate the thermal risk linked with the performance of this reaction at industrial scale.
2. Determine the criticality class of the reaction.
3. Are measures required to cope with the thermal risk?

(This problem is continued in Chapter 10)

► Exercise 3.2

A condensation reaction is to be in a stirred tank reactor in the semi-batch mode. The solvent is acetone, the industrial charge (final reaction mass) is 2500 kg, and the reaction temperature is 40 °C. The second reactant is added in a stoichiometric amount at a constant rate over 2 hours. Under these conditions, the maximum accumulation is 30%.

Data:

Reaction:	$Q'_r = 230\,\text{kg}\,\text{kg}^{-1}$	$c'_P = 1.7\,\text{kJ}\,\text{kg}^{-1}\,\text{K}^{-1}$
Decomposition:	$Q'_{dc} = 150\,\text{kJ}\,\text{kg}^{-1}$	$T_{D24} = 130\,°\text{C}$
Physical data:	Acetone	$T_b = 56\,°\text{C}$

Questions:

1. Evaluate the thermal risk linked with the performance of this reaction at industrial scale.
2. Determine the criticality class of the reaction.
3. Are measures required to cope with the thermal risk?

(This problem is continued in Chapter 10)

▶ Exercise 3.3

A sulfonation reaction is performed as a semi-batch reaction in 96% sulfuric acid as a solvent. The total charge is 6000 kg with a final concentration of $3\,\text{mol}\,\text{l}^{-1}$. The reaction temperature is 110 °C and Oleum 20% is added in a stoichiometric excess of 30% at constant rate over 4 hours. Under these conditions, the maximum accumulation of 50% is reached after approximately 3 hours addition.

Data:

Reaction:	$Q'_r = 150\,\text{kJ}\,\text{kg}^{-1}$	$c'_P = 1.5\,\text{kJ}\,\text{kg}^{-1}\,\text{K}^{-1}$
Decomposition:	$Q'_{dc} = 350\,\text{kJ}\,\text{kg}^{-1}$	$T_{D24} = 140\,°\text{C}$

Questions:

1. Evaluate the thermal risk linked with the performance of this reaction at industrial scale.
2. Determine the criticality class of the reaction.
3. Are measures required to cope with the thermal risk?

(This problem is continued in Chapter 10)

References

1 Gygax, R. (1993) *Thermal Process Safety, Data Assessment, Criteria, Measures,* Vol 8 (eds ESCIS), ESCIS, Lucerne.

2 Gygax, R. (1988) Chemical reaction engineering for safety. *Chemical Engineering Science,* **43** (8), 1759–71.

3 Townsend, D.I. (1977) Hazard evaluation of self-accelerating reactions. *Chemical Engineering Progress,* **73**, 80–1.

4 ESCIS (1996) *Loss of Containment,* Vol 12, ESCIS, Lucerne.

5 Zurich (1987) *"Zurich" Hazard Analysis, A brief introduction to the "Zurich" method of Hazard Analysis,* Zurich Insurance, Zurich.

6 Stoessel, F. (1993) What is your thermal risk?. *Chemical Engineering Progress,* October, 68–75.

7 CCPS (1998) *Guidelines for Pressure Relief and Effluent Handling Systems,* CCPS, AICHE.

8 Fisher, H.G., Forrest, H.S., Grossel, S.S., Huff, J.E., Muller, A.R., Noronha, J.A., Shaw, D.A., Tilley, B.J. (1992) *Emergency Relief System Design Using DIERS Technology, The Design Institute for Emergency Relief Systems (DIERS) Project Manual,* AICHE, New York.

9 Etchells, J. and Wilday, J. (1998) *Workbook for Chemical Reactor Relief System Sizing,* HSE, Norwich.

4
Experimental Techniques

Case History "Diazotization"

A blue diazo-dyestuff was produced from 2-chloro-4,6-dinitro-aniline, which was diazotized by nitrosyl-sulphuric acid in sulfuric acid as a solvent. This strong reagent is required for the diazotization of this weakly reactive aniline. Further, a relatively high reaction temperature of 45 °C for a diazotization was necessary. Due to an increasing demand for blue dyestuff, the productivity of the process had to be increased. The chemist in charge of the process decided to increase the concentration by using more reactants and less solvent. Nevertheless, he was conscious of the fact that by doing so, the heat released by the reaction would increase. Therefore, he decided to perform a laboratory experiment to assess this problem. He took a 3-necked flask, placed in a water bath at 45 °C. He monitored the temperature of the bath and of the reaction medium with two thermometers. The diazotization was carried out by progressively adding the nitrosyl-sulfuric acid to the pre-charged aniline in sulfuric acid. During the reaction no temperature difference was observed between bath and reaction mixture. Thus, it was concluded that the exotherm was not significant and could be mastered.

At plant scale, the diazotization led to a dramatic explosion resulting in 5 fatalities and over 30 injured, as well as a huge damage to the production plant.

Lessons drawn

In fact, the simple detection device used in the laboratory was unable to detect the exothermal reaction: At laboratory scale, the heat exchange area is larger by about two orders of magnitude (see Section 2.4.1.2), compared to plant scale. Hence the heat of reaction could be removed without detectable temperature difference, whereas at plant scale the same exotherm could not be mastered. This incident enhanced the necessity of a reaction calorimeter and promoted the development of the instrument, which was under development at this time by Regenass [1]. Later, it became a commercial device (RC1).

Thermal Safety of Chemical Processes: Risk Assessment and Process Design. Francis Stoessel

ISBN: 978-3-527-31712-7

A further positive reaction to this dramatic incident took place in the central research department of the company. A physico-chemist had the idea of using his differential scanning calorimeter (DSC) to look at the energy involved in this reaction. He performed an experiment with the initial concentration and a second with a higher concentration. The thermograms he obtained were different and he realized that he could have predicted the incident (see Exercise 11.1). As a consequence, it was decided to create a laboratory dedicated to this type of experiment. This was the beginning of the scientific approach of safety assessments using thermo-analytic and calorimetric methods. From this time on, many different methods were developed in different chemical companies and became commercially distributed, often by scientific instrument companies.

4.1 Introduction

In this chapter, some of these instruments are reviewed. A first section is a general introduction to calorimetric principles. In a second part, some methods commonly used in safety laboratories are reviewed. This is not an exhaustive review of such instruments, but based only on the experience of the author.

4.2 Calorimetric Measurement Principles

Heat cannot be directly measured. In most cases heat measurement is made indirectly by using temperature measurement. Nevertheless, there are some calorimeters able to measure directly the heat release rate or thermal power. Calorimetry is a very old technique, which was first established by Lavoisier in the 18th century. In the mean time, a huge choice of different calorimeters, using a broad variety of designs and measurement principles, were developed.

4.2.1 Classification of Calorimeters

There are different ways to classify calorimetric and thermo-analytic methods:

- The scale of the sample: micro-calorimetry (mg), macro-calorimetry (g), preparative or bench scale (hg-kg). This classification is essentially useful when the amount of available reactants is limited, or when dangerous reactions have to be studied. In such a case, using only small amounts allows one to run the experiments safely. Of course, larger scales perform more realistic experiments in the sense that they mimic the manufacturing process.

- The device can be designed as a single or a twin calorimeter, also called differential calorimeters. The twin calorimeter technique eliminates the perturbations due to heat loss to the surroundings, for example. They measure the difference of heat release in two symmetrically constructed alorimetric cells.
- Calorimeters may also be classified with respect to their thermal sensitivity, which may range from some $mW\,kg^{-1}$ to $10–20\,W\,kg^{-1}$. The sensitivity is important when the heat release rate has to be measured for the determination of the TMR_{ad}. With very sensitive instruments, the measurement can be made directly at the desired temperature. For assessing the thermal risks at storage, a high sensitivity is required. With less sensitive instruments, the measurement must be made at a higher temperature – accelerating the reaction – in order to obtain a measurable heat release rate. The results must then be extrapolated to the (lower) temperature of interest. The less sensitive instruments are essentially used for screening purposes.
- Calorimeters may also be classified with respect to the way they use the heat balance. In fact, every calorimeter is based on a heat balance (as reactors are). Here we may distinguish ideal accumulation calorimeters or adiabatic calorimeters, from ideal heat flow or isothermal calorimeters and isoperibolic[1] calorimeters.

4.2.2 Operating Modes of Calorimeters

Most of the calorimeters may be used with different temperature control modes. The following temperature control modes are commonly used:

- Isothermal: The temperature of the sample is maintained constant by adjusting the temperature of the surroundings in an appropriate way. The advantage is that the temperature effect, the exponential variation of the reaction rate, is eliminated during the measurement, which gives direct access to the conversion term of the reaction rate. The drawback is that there is no information on the temperature effect from one experiment alone. A series of experiments at different temperatures is required for this purpose (see Sections 11.4.2.1 and 11.4.3.1).
- Dynamic: The temperature of the sample is varied linearly (scanned) over a given temperature range. This type of experiment gives information on the thermal activity over a broad temperature range and allows short measuring times. This method is best suited in determining the energies of reactions. For a kinetic study, both temperature and conversion effects are superposed. Therefore, more sophisticated evaluations techniques are required (see Section 11.4.3.2).

1) Isoperibolic means that the surroundings temperature is maintained constant.

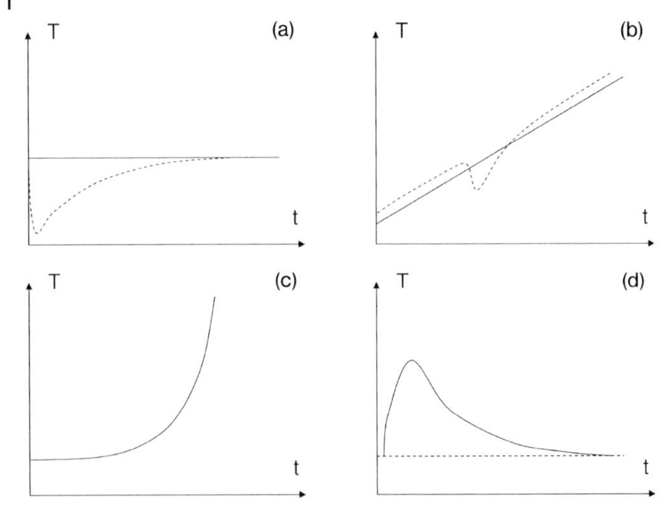

Figure 4.1 Temperature course of the sample (solid line) and of the surroundings (dashed line) as a function of time for the different operating modes: (a) isothermal, (b) dynamic, (c) adiabatic, (d) isoperibolic.

- Adiabatic: The temperature of the sample results from its thermal activity. This technique gives direct access to the thermal runaway curve. The results must be corrected by the adiabacity coefficient, since a part of the heat released in the sample is used to increase the temperature of the calorimetric cell. This rends the kinetic evaluation complex.
- Isoperibolic: The temperature of the surroundings is maintained constant, whereas the sample temperature varies.

In Figure 4.1, the evolution of the reaction mass and the temperature of the surroundings are compared for the different operating modes described above. These different operating modes are best understood by a closer examination of the heat balance used in calorimeters.

4.2.3 Heat Balance in Calorimeters

In order to determine the heat released by a reaction, the calorimeter can work using a simplified heat balance, as presented in Section 2.4.2. Many calorimeters are designed in such a way as to eliminate one of the three terms of the heat balance, in order to determine the heat release rate by measuring the other term.

$$\begin{Bmatrix} \text{Heat} \\ \text{accumulation} \end{Bmatrix} = \begin{Bmatrix} \text{Heat released} \\ \text{by the reaction} \end{Bmatrix} - \begin{Bmatrix} \text{Heat exchanged with} \\ \text{the surroundings} \end{Bmatrix} \tag{4.1}$$

4.2.3.1 Ideal Accumulation

The totality of the heat released by the reaction under study must be converted into heat accumulation, that is, into a temperature increase, which can be measured. This is obtained by eliminating the heat exchange with the surroundings, achieving adiabatic conditions:

$$q_{ac} = q_{rx} + q_{ex} \tag{4.2}$$

Adiabatic conditions may be achieved either by a thermal insulation or by an active compensation of heat losses. Examples are the Dewar calorimeter, achieving a thermal insulation [2–4] or the Accelerating Rate Calorimeter (ARC) [5] or the Phitec [6], using a compensation heater avoiding the heat flow from the sample to the surroundings. These calorimeters are especially useful for the characterization of runaway reactions.

4.2.3.2 Ideal Heat Flow

In this case, the totality of heat released by the reaction under study is transferred to the surroundings:

$$q_{ac} = q_{rx} + q_{ex} = 0 \tag{4.3}$$

In this category of calorimeters, we find the isothermal calorimeters and the dynamic calorimeters where the temperature is scanned using a constant temperature scan rate. The instrument must be designed in such a way that any departure from the set temperature is avoided and the heat of reaction must flow to the heat exchange system where it can be measured. The instrument acts as a heat sink. In this family we find the reaction calorimeters, the Calvet calorimeters [7], and the Differential scanning calorimeter (DSC) [8].

4.2.3.3 Isoperibolic Methods

In these calorimeters, the temperature of the surroundings is controlled and maintained constant or scanned. So the temperature of the sample is allowed to vary as well as the heat flow to the surroundings. Hence the results are more difficult to evaluate than with the techniques described above. Therefore these instruments are often semi-quantitative.

Remark: Since most instruments only approach ideal conditions, ideal heat flow or ideal accumulation, they should be considered as isoperibolic.

4.3 Choice of Instruments Used in Safety Laboratories

There are numerous calorimeters available on the market. Nevertheless, only a relatively restricted choice may be used for the determination of the data required for safety assessment. These are essentially selected for their robustness with

Table 4.1 Comparison of different common calorimetric methods used in safety laboratories.

Method	Measuring principles	Application range	Sample size	Temperature range	Sensitivity W kg^{-1} [a]
DSC differential scanning calorimetry	Differential, ideal flux, or isoperibolic	Screening, secondary reactions	1–50 mg	−50 + 500 °C	(2)[b]–10
Calvet	Differential, ideal flux	Main and secondary reactions	0.5–3 g	30 à 300 °C	0.1
ARC accelerating rate calorimeter	Ideal accumulation	Secondary reactions	0.5–3 g	30 à 400 °C	0.5
SEDEX sensitive detector of exothermal processes	Isoperibolic, adiabatic	Secondary reactions, storage stability	2–100 g	0–400 °C	0.5[c]
RADEX	Isoperibolic	Screening, secondary reactions	1.5–3 g	20–400 °C	1
SIKAREX	Ideal accumulatipon, isoperibolic	Secondary reactions	5–50 g	20–400 °C	0.25
RC reaction calorimeter	Ideal flux	Main reactions	300–2000 g	–40 à 250 °C	1.0
TAM thermal activity monitor	Differential, ideal flux	Secondary reactions, storage stability	0.5–3 g	30 à 150 °C	0.01
Dewar	Ideal accumulation	Main reactions and thermal stability	100–1000 g	30 à 250 °C	[d]

a) Typical values.
b) Most recent instruments under optimal conditions.
c) Depends on cell used.
d) Depends on volume and Dewar quality.

respect to the specific conditions prevailing for safety studies. They must either be able to reproduce normal operating conditions for the study of desired reactions, or be able to face extreme conditions, as during the study of runaway reactions or of thermal stability. Several instruments were explicitly designed for these purposes. Some of them are presented in Table 4.1, with the criteria described in Section 4.2.

In the following subsections, typical calorimeters, classified according to their operation mode and heat balance, are briefly presented.

4.3.1 Adiabatic Calorimeters

4.3.1.1 On the Evaluation of Adiabatic Experiments

In adiabatic calorimeter, it is unavoidable that a part of the heat released by the sample serves to heat the crucible or the calorimetric vessel. Thus, a correction for the so-called thermal inertia of the vessel is required. The heat balance for an adiabatic calorimeter is expressed as

$$(M_r \cdot c'_{P,r} + c_W)\frac{dT_r}{dt} = q_{rx} \tag{4.4}$$

In this equation, c_W stands for the heat capacity of the calorimetric vessel. Traditionally, this correction was performed via the "Water Equivalent" of the calorimeter. This means that the heat capacity of the calorimetric cell is described by a thermally equivalent mass of water.

$$c_W = c'_{P,H_2O} \cdot M_{H_2O} \tag{4.5}$$

When calculating the heat of reaction from an adiabatic experiment, the result must take the water equivalent into account:

$$\Delta H_r = (M_r \cdot c'_{P,r} + c_W) \cdot \Delta T_{meas} \tag{4.6}$$

Another way of correcting the results is to use the adiabacity coefficient (Φ) or thermal inertia of the cell:

$$\Phi = \frac{M_r \cdot c'_{p,r} + M_{cell} \cdot c'_{p,cell}}{M_r \cdot c'_{p,r}} = 1 + \frac{M_{cell} \cdot c'_{p,cell}}{M_r \cdot c'_{p,r}} \tag{4.7}$$

Ideal adiabatic conditions correspond to a thermal inertia $\Phi = 1$. Under normal operating conditions, this factor lies between 1.05 and 8. The thermal inertia depends on the degree of filling of the cell. The temperature can be corrected by

$$T_f = T_0 + \Phi \cdot \Delta T_{ad} \tag{4.8}$$

An important consequence of the thermal inertia is that for a Φ factor of 4, for example, if the measured adiabatic temperature rise is 100 °C, in reality under true adiabatic conditions it would be 400 °C. Since, over a range of 400 °C different reactions may be observed, or even different reactions may be triggered, leading to a higher energy potential than measured, the calorimeter must be forced to explore the entire temperature range. The great advantage of this calorimeter is to deliver an adiabatic temperature curve together with the pressure rise curve. The drawback is that these curves must be corrected for true adiabatic conditions and that this correction requires an assumption concerning the reaction kinetics that is to be characterized. Moreover, this correction may be important, depending on the thermal inertia of the cell.

Thus, the correction of the temperature (on the temperature axis) is easy and straightforward, but the correction of the reaction dynamics is more complex: the reaction rate is a function of temperature; therefore the temperature achieved at a given conversion in the experiment is not the same as would be achieved under ideal adiabatic conditions. Thus, the reaction rate at this conversion must

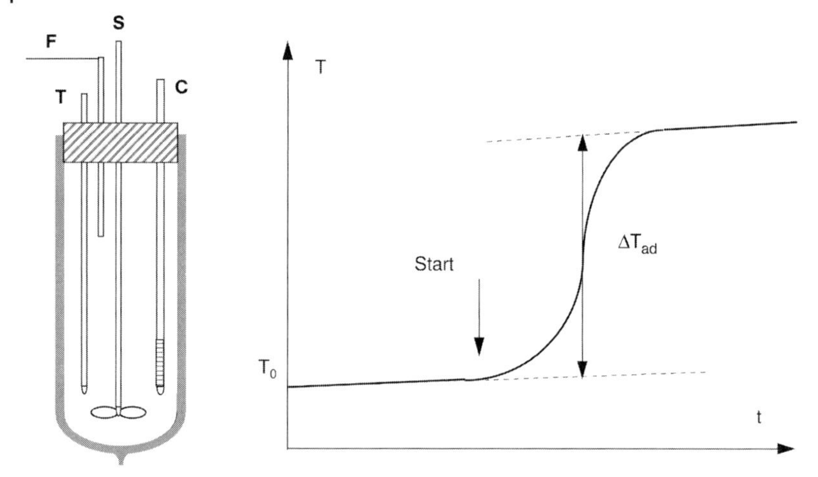

Figure 4.2 Left: Dewar calorimeter equipped with stirrer and calibration heater. T: thermometer, C: calibration heater, S: stirrer, F: feed line. Right: Typical temperature curve.

be corrected. This is only possible by assuming a rate equation, that is, a kinetic model. Hence the dynamic correction becomes an iterative process. This means the kinetic evaluation of adiabatic experiments requires a specific procedure [9–12].

4.3.1.2 **Dewar Calorimeters**

Dewar flasks are often considered as adiabatic vessels. This is not absolutely true since if the heat losses are limited, they are not zero. Nevertheless, over a limited time range, and providing the temperature difference with the surroundings is not too important, the heat losses may be neglected and the Dewar considered being adiabatic.

The reaction is initiated by the addition of a reactant, which must be exactly at the same temperature as the Dewar contents, in order to avoid the sensitive heat effects. Then the temperature is recorded as a function of time. The obtained curve must be corrected for the heat capacity of the Dewar flask and its inserts, respective of their wetted parts, which are also heated by the heat of reaction to be measured. The temperature increase results from the heat of reaction (to be measured), the heat input by the stirrer and the heat losses. These terms are determined by calibration, which may be a chemical calibration using a known reaction or an electrical calibration using a resistor heated by a known current under a known voltage (Figure 4.2). The Dewar flask is often placed into thermostated surroundings as a liquid bath or an oven. In certain laboratories, the temperature of the surroundings is varied in order to track the contents temperature and to avoid heat loss. This requires an effective temperature control system.

Nevertheless, the basic method is very simple and does not require any special equipment. But to become really quantitative, several precautions must be taken.

The larger the Dewar flask, the more sensitive it will be, since heat losses are essentially proportional to the ratio of surface to volume (A/V). A Dewar flask of 1 liter shows heat losses approximately equivalent to a non-stirred $10\,m^3$ industrial reactor, that is, $0.018\,W\,kg^{-1}\,K^{-1}$ [3] (see Section 2.4.1.7).

4.3.1.3 Accelerating Rate Calorimeter (ARC)

The accelerating rate calorimeter (Figure 4.3) is an adiabatic calorimeter, where adiabaticity is not realized by thermal insulation but by an active control of the heat losses by adjusting the oven temperature to the temperature of a thermocouple placed at the external surface of the cell containing the sample. Thus, there is no temperature gradient between the cell and its surroundings, that is, no heat flow. The sample is placed in a spherical cell of $10\,cm^3$ volume made of titanium (S), which may hold between 1 and 10 g. This cell is mounted in the center of an oven (Th), whose temperature is accurately adjusted by a sophisticated temperature control system (H). The cell is also linked to a pressure sensor (P) allowing for pressure measurement [13]. This instrument can work following two different modi:

1. Heat, wait, and search: the temperature at which the exothermal reaction is detectable is searched using a defined series of temperature steps. At each step, the system is stabilized for a defined time, then the controller is switched to the adiabatic mode. If the temperature increase rate surpasses a level (typically $0.02\,K\,min^{-1}$), the oven temperature follows the sample temperature in the adiabatic mode. If the temperature increase rate remains below the level, the next temperature step is achieved (Figure 4.4).

2. Isothermal age mode: The sample is directly heated to the desired initial temperature. At this temperature, the instrument seeks for an exothermal effect as above.

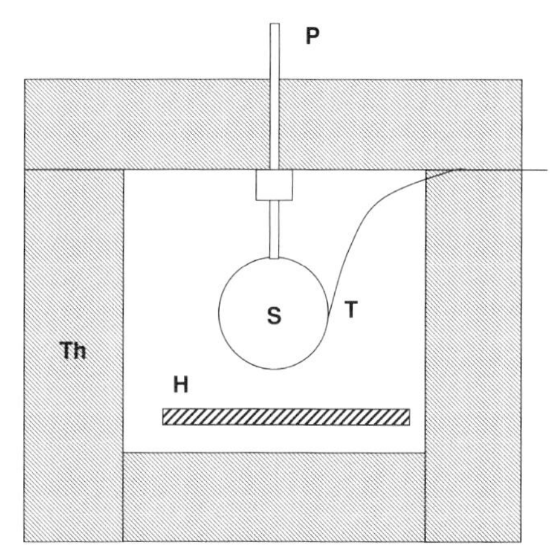

Figure 4.3 Principles of the accelerating rate calorimeter showing the oven with the sample holding cells in its center. T: thermocouple, H: heater, Th: thermostat, P: pressure transducer.

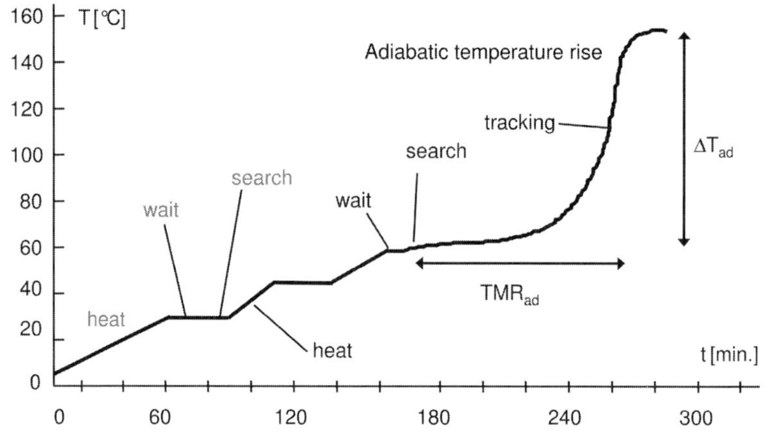

Figure 4.4 Typical temperature curve obtained in an accelerating rate calorimeter using the heat, wait, and search mode.

By this technique, the temperature directly "tracks" the exothermal process under pseudo-adiabatic conditions. Pseudo, because a part of the heat released in the sample serves to heat the cell itself. Nevertheless, essentially in the USA, it became a very popular method as a screening technique. Concerning its sensitivity, for a well-tuned instrument, able to detect a self heating rate of $0.01\,\mathrm{K\,min^{-1}}$, with a sample mass of 2 g, the sensitivity is as low as $0.5\,\mathrm{W\,kg^{-1}}$.

There are other types of adiabatic calorimeters available on the market [14, 15], such as the VSP (Vent Sizing Package) [16], PHITEC [6], and RSST (Reactive System Screening Tool). These instruments are essentially designed for vent sizing requirements [17–20] and present a lower thermal inertia than the ARC.

4.3.2
Micro Calorimeters

4.3.2.1 Differential Scanning Calorimetry (DSC)

Differential Scanning Calorimetry (DSC) was used for a long time in the field of process safety [21–23]. This is essentially due to its versatility for screening purposes. The small amount of sample required (micro-calorimetric technique) and the fact that quantitative data are obtained, confer on this technique a number of advantages. The sample is contained in a crucible placed into a temperature controlled oven. Since it is a differential method, a second crucible is used as a reference. This may be empty or contain an inert substance.

The "true" DSC uses a heating resistor placed under each crucible, controlling the crucibles' temperature and maintaining them as equal [8]. The difference in heating power between these heating resistors directly delivers the thermal power of the sample. Thus, it is a method following the principles of ideal heat flux (see Section 4.2.3.2).

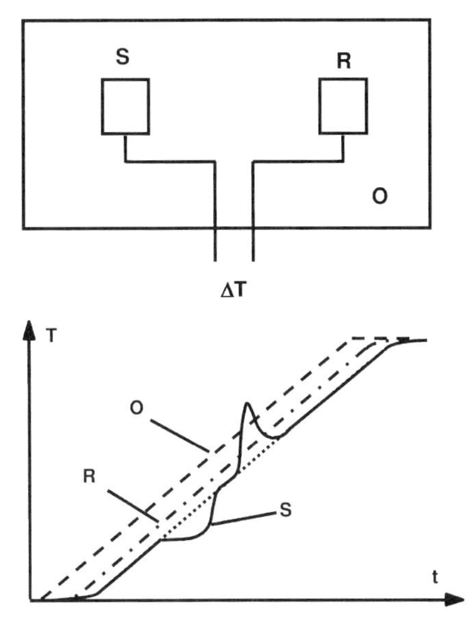

Figure 4.5 Principles of a DSC, after Boersma [8]. S: sample, R: reference, O: oven.

Another measurement principle is the DSC, after Boersma [8]. In this case, no compensation heating is used and a temperature difference is allowed between sample crucible and reference crucible (Figure 4.5). This temperature difference is recorded and plotted as a function of time or temperature. The instrument must be calibrated in order to identify the relation between heat release rate and temperature difference. Usually this calibration is by using the melting enthalpy of reference substances. This allows both a temperature calibration and a calorimetric calibration. In fact, the DSC after Boersma works following the isoperibolic operating mode (see Section 4.2.2). Nevertheless, the sample size is so small (3 to 20 mg) that it is close to ideal flux.

The oven temperature may be controlled in two ways: the dynamic mode also called scanning mode, where the oven temperature is varied linearly with time, and the isothermal mode, where the oven temperature is maintained as constant. Since the DSC uses only small sample sizes in the order of some milligrams, very exothermal phenomena may be studied, even under extreme conditions, without any risk either to the laboratory personal, or to the instrument. Moreover, a scanning experiment from ambient temperature to 500 °C with a scan rate of $4\,\mathrm{K\,min^{-1}}$ takes only about 2 hours. Thus, the DSC became a very popular instrument for screening purposes [21, 24, 25].

The sensitivity of the instrument is governed by the following parameters:

- Construction of the measurement head: where the materials as well as the number of thermocouples may differ.

- The type of crucible used: for safety purposes, relatively heavy pressure resistant crucibles are used, which impinge the sensitivity.
- The experimental conditions: essentially the scanning rate.

Therefore the sensitivity usually ranges between 2 and 20 W kg^{-1}. This heat release rate corresponds to a temperature increase rate of about 4 to 40 °C hour^{-1} under adiabatic conditions. This also means that an exothermal reaction is detected at a temperature where the time to explosion (TMR_{ad}) is in the order of magnitude of one hour only.

Since samples may contain some volatile compounds, during heating in scanned experiments, these compounds may evaporate, which has two consequences:

1. Evaporation is endothermal and adds a negative contribution to the heat balance, that is, to the measured signal that may mask an exothermal reaction.
2. A part of the sample is lost during the experiment, giving a false interpretation of the results.

Thus, for the determination of the energy potential of a sample, it is essential to use closed pressure resistant crucibles for these experiments. This is true for DSC, but holds also for other instruments. Experience has shown that gold plated crucibles with a volume of 50 µl and resistance to pressures up to 200 bar are best suited for safety studies. These crucibles are commercially available.

DSC is best suited for the determination of heat of decomposition (see Chapter 11). Overall heats of reaction may also be determined if the reactants are mixed at too low a temperature in order to slow down the reaction and starting the temperature scan from low temperature. By doing so, one must be aware that in DSC the sample cannot be stirred and there is no way of adding a reactant during reaction. Nevertheless, the dimensions of DSC crucibles allow for a short diffusion time, thus mixing is achieved by diffusion even without stirring.

The aim of such scanning experiments is to simulate worst-case conditions: the sample is heated to 400 or 500 °C, a temperature range where most of the organic compounds are forced to decompose. Moreover, this kind of experiment is carried out in a closed vessel, from which no decomposition product may escape, that is, under total confinement conditions. Such a thermogram shows the energetic potential of a sample as an "energy finger print" of the sample (see Section 11.3). Since the measurement is quantitative, the corresponding adiabatic temperature rise, allowing the assessment of the severity of a runaway reaction (see Section 3.3.2), may be obtained in an easy way. This kind of screening experiment is useful for the identification of potentially hazardous mixtures.

4.3.2.2 **Calvet Calorimeters**

The Calvet calorimeters have their roots in the work of Tian [26] and the later modifications by Calvet [7]. Presently this calorimeter type is commercially available from Setaram and the models C80 and BT215 are particularly well adapted for safety studies. It is a differential calorimeter that may be operated isothermally or in the scanning mode as a DSC in the temperature range from room temperature to 300 °C for the C80 and −196 to 275 °C for the BT215. They show a high

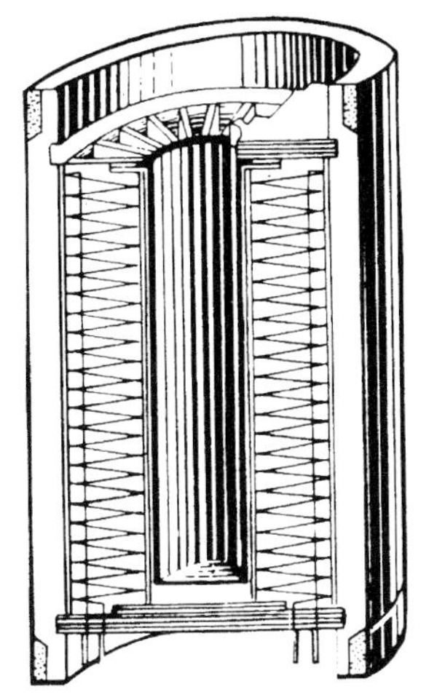

Figure 4.6 Vertical section of the measuring zone showing the radial arrangement of thermocouples. (Courtesy of Setaram).

sensitivity of 0.1 W kg^{-1} or even better, which is essentially due to the measurement principles, based on a pile of thermocouples totally surrounding the cell containing the sample (Figure 4.6). The calorimeter may be used with a closed cell, also allowing for pressure measurement, or mixing cells well adapted for the study of reactions and for the assessment of several safety problems. They are useful if there is an accidental intrusion of a cooling medium into a reaction mass and for assessing the efficiency of safety measures as quenching of reactions or dumping reaction masses (see Sections 10.4.3 and 10.4.4). Obviously, the heat of reactions and the thermal stability of reaction masses can also be studied in these instruments. A particularly efficient experimental combination is illustrated in Figure 4.7. First the reaction was performed at 30 °C and after the signal returned to the base line, a temperature scan determined the heat of decomposition as well as the

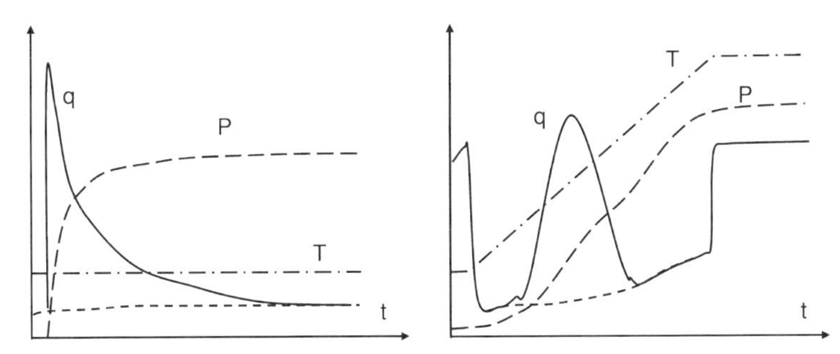

Figure 4.7 Example of a typical study of a reaction in the isothermal mode (upper thermogram), followed by the study of the thermal stability of the final reaction mass in dynamic mode (lower thermogram).

corresponding pressure effect. Thus, the main energetic characteristics of a reaction can be studied in one combined experiment. The sample size is in the order of magnitude of 0.5 to 1.0 g.

4.3.2.3 Thermal Activity Monitor

The thermal activity monitor was initially developed in Sweden by Suurkuusk and Wadsö [27, 28] for the study of biological systems. It is a differential calorimeter with a high sensitivity, able of measuring in the order of magnitude of μW. With a sample size of 1 g, this corresponds to $1\,mW \cdot kg^{-1}$. This sensitivity is achieved by a battery of thermocouples surrounding the sample and by a high precision thermostat controlling the temperature with an accuracy of 0.1 mK. This instrument is well suited for the study of long-term stability at storage. For example, decomposition with energy of $500\,kJ\,kg^{-1}$, and with a heat release rate of $3\,mW\,kg^{-1}$, would reach a conversion of 2.5% in one month. This is a typical heat release rate that can be measured by this instrument. Thus, it finds its applications in the field of process safety for the study of thermal confinement problems (see Chapter 13), or when an extrapolation of the heat release rate is measured by isothermal experiments in DSC (see Section 11.4.2.1). An example of such an application is illustrated in Figure 4.8. The heat release rate measured by DSC between 170 and 200 °C (90 to $500\,W\,kg^{-1}$) is extrapolated to lower temperatures by using the heat release rate measured at 65 and 75 °C in the TAM (20 and $50\,mW\,kg^{-1}$). This is a useful confirmation of the validity of the extrapolation.

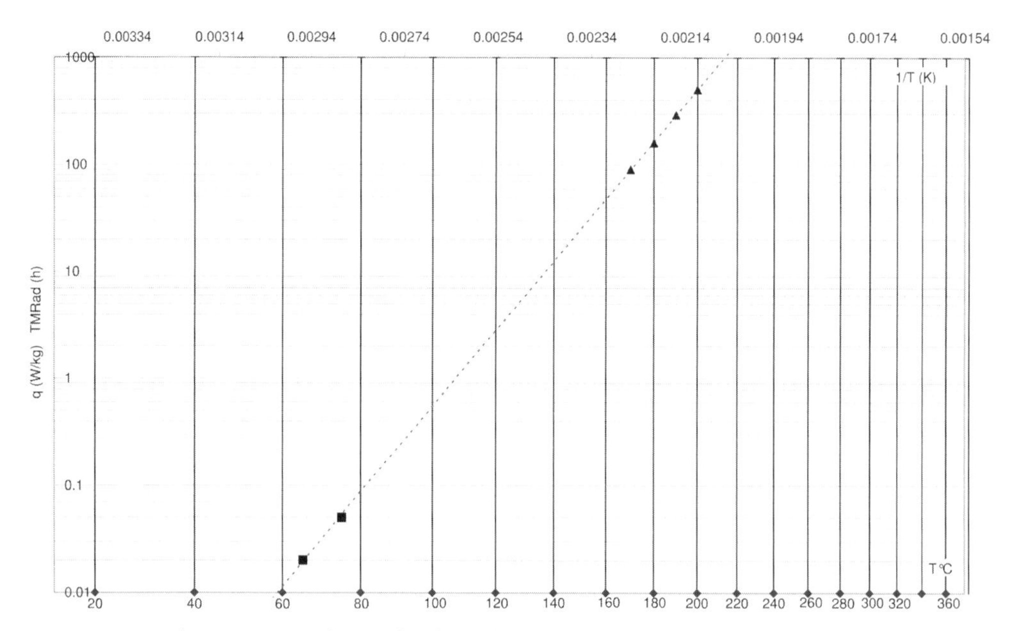

Figure 4.8 Extrapolation of isothermal DSC data and confirmation by TAM-measurement at low temperature.

4.3.3
Reaction Calorimeters

Basically, reaction calorimeters are designed in such a way that they perform a reaction under conditions that are as close as possible to plant operation conditions. This means that the temperature of a reaction calorimeter may be controlled in the isothermal mode or in a temperature programmed mode. Moreover, a reaction calorimeter must allow addition of reactants in a controlled way, distillation, or refluxing [29–31] and reactions with gas release [32]. In summary, they must perform the same operations as an industrial stirred tank reactor, with a major difference: they allow tracking the thermal phenomena occurring during these operations. A review of the working principles of reaction calorimeters is given in [33, 34]. The reaction calorimeters were initially developed for safety analysis [1, 35], but very soon their benefits for process development and scale-up were recognized [6, 36–46]. The precise temperature control and the measurement of the heat release rate also allowed the identification of reaction kinetics [34, 47–50]. This becomes even more effective in combination with other analytical methods [51]. Thus, reaction calorimetry was applied to a great diversity of different reactions, including polymerization [52–54], Grignard reactions [55, 56], nitrations [57–61], hydrogenations [62–65] epoxydations [66], and more [49, 67, 68]. Under pressure from the pharmaceutical industry, small-scale calorimeters [69] and an isoperibolic differential calorimeter [70] were developed. As an example, the

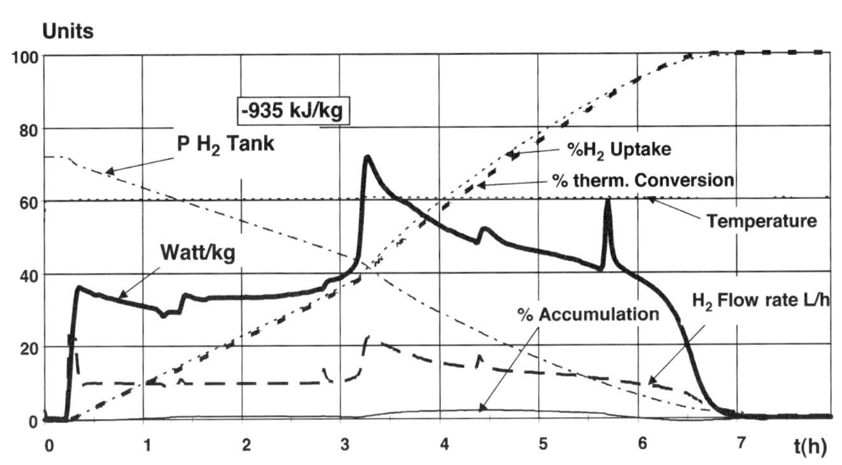

Figure 4.9 Thermogram obtained in a reaction calorimeter (Mettler–Toledo RC1) during the study of a catalytic hydrogenation. The heat release rate suddenly increases after 3 hours, which indicates a complex reaction scheme involving several steps in series. The second sharp peak after 5.8 hours is due to crystallization of the final product. The hydrogen uptake measured as pressure decrease in a calibrated reservoir and as flow rate follows the heat release rate. The difference between hydrogen uptake (chemical conversion and thermal conversion) indicates a small accumulation.

kinds of data that can be obtained from a reaction calorimeter are illustrated by a catalytic hydrogenation (Figure 4.9). In this thermogram, the heat release rate of the reaction is obtained under the conditions as they are intended for plant operation, that is, at a pressure of 20 bar hydrogen and a temperature of 60 °C. The heat release rate profile is particularly interesting in this case: after a period of 3.5 hours, where it remains stable at $35\,W\,kg^{-1}$, it suddenly increases to $70\,W\,kg^{-1}$ before decreasing. This is due to a multiple step reaction. The hydrogen uptake, proportional to the chemical conversion, is obtained from the pressure decrease in a calibrated hydrogen reservoir. The chemical conversion can be compared to the thermal conversion obtained by integration of the heat release rate curve: the difference gives the accumulation [63]. The use of reaction calorimeters in the study of process safety will be illustrated by numerous examples in Chapters 6 and 7.

4.4 Exercises

► Exercise 4.1

A catalytic reaction must be performed in aqueous solution at industrial scale. The reaction is initiated by addition of catalyst at 40 °C. In order to evaluate the thermal risks, the reaction was performed at laboratory scale in a Dewar flask. The charge is 150 ml solution in a Dewar of 200 ml working volume. The volume and mass of catalyst can be ignored. For calibration of the Dewar by Joule effect, a heating resistor with a power of 40 W was used in 150 ml water. The resistor was switched on for 15 minutes and the temperature increase was 40 K. During the reaction, the temperature increased from 40 to 90 °C within approximately 1.5 hours. The specific heat capacity of water is $4.2\,kJ\,kg^{-1}\,K^{-1}$.

Questions:

1. Determine the specific heat of reaction ($kJ\,kg^{-1}$).
2. What recommendations would you make to the plant?

► Exercise 4.2

A sample of a final reaction mass is analysed by two different calorimetric methods, DSC and Dewar-Calorimetry. The reaction is to be performed in the batch mode and the *MTSR* is 120 °C. The DSC thermogram recorded with a scan rate of $4\,K\,min^{-1}$ presents two exothermal peaks. The first, with an energy of $200\,kJ\,kg^{-1}$ is detected in the temperature range from 140 to 180 °C. The second, with an energy of $800\,kJ\,kg^{-1}$ is detected in the temperature range from 200 to 270 °C. The Dewar test is operated with an adiabacity coefficient of $\Phi = 2$. The temperature rises from 120 to 170 °C within 80 minutes and then remains stable. The specific heat capacity of the reaction mass is $2\,kJ\,kg^{-1}\,K^{-1}$.

Questions:

1. Comment on these thermograms.
2. What precautions should be taken when using adiabatic calorimetry?
3. What do you think about the thermal risks linked with an industrial performance of this process?

► Exercise 4.3

The thermal risks of a synthesis step with an exothermal bimolecular reaction (A + B → P) must be assessed. For this, the required thermal data have to be determined in a safety laboratory equipped with a reaction calorimeter and a DSC.

The process can be described as follows:

1. Charge the total amount of A solution (10 kmol) into the reactor. The solvent is mesitylene with a boiling point of 165 °C.
2. Heat this solution to the process temperature of 140 °C.
3. At this temperature, add 12 kmol of B at a constant rate for 4 hours.
4. Heat to 150 °C.
5. Maintain this temperature for 4 hours.
6. Cool to 80 °C. The product P precipitates.
7. Transfer to filter.

Questions:

1. Describe the required experiments, allowing for the determination of the criticality class.
2. What data will you obtain from the different experiments?
3. Which results do you anticipate for the criticality?

References

1 Regenass, W. (1997) The development od Stirred-Tank Heat Flow Calorimetry as a Tool for Process Optimization and Proces safety. *Chimia*, **51**, 189–200.

2 Steinbach, J. (1995) *Dewar and Other Methods for the Characterisation of Exothermic Runaway Reactions*, Chemische Sicherheitstechnik, VCH, Weinheim.

3 Rogers, R.L. (1989) The advantages and limitations of adiabatic dewar calorimetry in chemical hazards testing. *Plant Operation Progress*, **8** (2), 109–12.

4 Wright, T.K. and Rogers, R.L. (1986) Adiabatic dewar calorimeter, in *Hazards IX*, Institution of Chemical Engineers, Manchester.

5 Townsend, D.I. and Tou, J.C. (1980) Hazard evaluation by an accelerating rate calorimeter. *Thermochimica Acta*, **37**, 1–30.

6 Singh, J. (1993) Reliable scale-up of thermal hazards data using the Phi-Tec II calorimeter. *Thermochimica Acta*, **226** (1–2), 211–20.

7 Calvet, E. and Prat, H. (1956) *Microcalométrie: applications physico-chimiques et biologiques*, Masson, Paris.

8 Höhne, G., Hemminger, W. and Flammersheim, H.-J. (1995) *Differential Scanning Calorimetry an Introduction for Practitioners*, Springer, Berlin.

9 Leonhardt, J. and Hugo, P. (1997) Comparison of thermokinetic data

obtained by isothermal, isoperibolic, adiabatic and temperature programmed experiments. *Journal of Thermal Analysis*, **49**, 1535–51.

10 Zhan, S., Lin, J., Qin, Z. and Deng, Y. (1996) Studies of thermokinetics in an adiabatic calorimeter. II Calorimetric curve analysis methods for ireversible and reversible reactions. *Journal of Thermal Analysis*, **46**, 1391–401.

11 Wilcock, E. and Rogers, R.L. (1997) A review of the Phi factor during runaway conditions. *Journal of Loss Prevention in the Process Industries*, **10** (5–6), 289–302.

12 Snee, T.J., Bassani, C. and Ligthart, J.A.M. (1993) Determination of the thermokinetic parameters of an exothermic reaction using isothermal adiabatic and temperature programed calorimetry in conjunction with spectrophotometry. *Journal of Loss Prevention in the Process Industries*, **6** (2), 87.

13 Townsend, D.I., ed. (1981) *Accelerating Rate Calorimetry*. Industrial chemical Engineering Series, Vol. 68. IchemE.

14 Grewer, T. (1994) *Thermal Hazards of Chemical Reactions.*. Industrial safety series, Vol. 4, Elsevier, Amsterdam.

15 Barton, A. and Rogers, R. (1997) *Chemical Reaction Hazards*. Institution of Chemical Engineers, Rugby.

16 Gustin, J.L. (1991) Calorimetry for emergency relief systems design. In Benuzzi, A. and Zaldivar, J.N. (eds.) *Safety of Chemical Batch Reactors and Storage Tanks*, ECSC, EEC, EAEC, Brussels, pp. 311–54.

17 CCPS (1998) *Guidelines for Pressure Relief and Effluent Handling Systems*. CCPS, AICHE.

18 Fisher, H.G., Forrest, H.S., Grossel, S.S., Huff, J.E., Muller, A.R., Noronha, J.A., Shaw, D.A. and Tilley, B.J. (1992) *Emergency Relief System Design Using DIERS Technology, The Design Institute for Emergency Relief Systems (DIERS) Project Manual*, AICHE, New York.

19 Schmidt, J. and Westphal, F. (1997) Praxisbezogenes Vorgehen bei der Auslegung von Sicherheitsventilen und deren Abgaseleitungen für die Durchströmung mit Gas/Dampf-Flüssigkeitsgemischen, Teil 2. *Chemie Ingenieur Technik*, **69** (8), 1074–91.

20 Etchells, J. and Wilday, J. (1998) *Workbook for Chemical Reactor Relief System Sizing*, HSE, Norwich.

21 Brogli, F., Gygax, R. and Meyer, M.W. (1980) DSC a powerful screening method for the estimation of the hazards inherent in industrial chemical reaction, in *International Conference on Thermal Analysis*, Birkhäuser Verlag, Basel.

22 Gygax, R. (1980) Differential scanning calorimetry – Scope and limitations of its use as tool for estimating the reaction dynamics of potential hazardous chemical reactions, in *6th International Conference on Thermal Analysis and Calorimetry*, Birkhäuser Verlag, Basel.

23 Eigenmann, K. (1976) Sicherheitsuntersuchungen mit thermoanalytischen Mikromethoden, in *Int. Symp. on the Prevention of Occupational Risks in the Chemical Industry*. IVSS, Frankfurt a.M.

24 Eigenmann, K. (1976) Thermische Methoden zur Beurteilung der chemischen Prozess-Sicherheit. *Chimia*, **30** (12), 545–6.

25 Frurip, D.J. and Elwell, T. (2005) Effective use of differential scanning calorimetry in reactive chemicals hazard assessment, in *NATAS*. Conference proceedings.

26 Calvet, E. and Prat, H. (1963) *Recent Progress in microcalorimetry*, Pergamon Press, Oxford.

27 Suurkuusk, J. and Wadsö, I. (1982). *Chemica Scripta*, **20**, 155–63.

28 Bäckman, P., Bastos, M., Briggner, L.E. Hägg, S., Hallén, D., Lönnbro, P., Nilsson, S.O., Olofsson, G., Schön, A., Suurkuusk, J., Teixeira, C., and Wadsö, I. (1994) A system of microcalorimeters. *Pure and Applied Chemistry*, **66** (3), 375–82.

29 Wiss, J., Stoessel, F. and Killé, G. (1990) Determination of heats of reaction under refluxing conditions. *Chimia*, **44** (12), 401–5.

30 Wiss, J. (1993) A systematic procedure for the assessment of the thermal safety and for the design of chemical processes at the boiling point. *Chimia*, **47** (11), 417–23.

31 Nomen, R., Sempere, J. and Lerena, P. (1993) Heat flow reaction calorimetry under reflux conditions. *Thermochimica Acta*, **225** (2), 263–76.

32 Lambert, P., Amery, G. and Watts, D.J. (1992) Combine reaction calorimetry with gas evolution measurment. *Chemical Engineering Progress*, October, 53–59.

33 Regenass, W. (1997) The development of heat flow calorimetry as a tool for process optimization and process safety. *Journal of Thermal Analysis*, **49**, 1661–75.

34 Zogg, A., Stoessel, F., Fischer, U. and Hungerbuhler, K. (2004) Isothermal reaction calorimetry as a tool for kinetic analysis. *Thermochimica Acta*, **419**, 1–17.

35 Hoppe, T. (1992) Use reaction calorimetry for safer process designs. *Chemical Engineering Progress*, September, 70–4.

36 Schildknecht, J. *Development and application of a "Mini Pilot Reaction Calorimeter"*. 2nd international Synposium on Loss prevention & Safety Promotion in the process industries.

37 Giger, G., Aichert, A. and Regenass, W. (1982) Ein Wärmeflusskalorimeter für datenorientierte Prozess-Entwicklung, *Swiss Chemistry*, **4** (3a), 33–6.

38 Riesen, R. and Grob, B. (1985) Reaktionskalorimetrie in der chemischen Prozess-Entwicklung. *Swiss Chemistry*, **7** (5a), 39–42.

39 Grob, B., Riesen, R. and Vogel, K. (1987) Reaction calorimetry for the development of chemical reactions. *Thermochimica Acta*, **114**, 83–90.

40 Rellstab, W. (1990) Reaktionskalorimetrie, ein Bindeglied zwischen Verfahrensentwicklung und Prozesstechnik. *Chemie-Technik*, **19** (5), 21–5.

41 Landau, R.N. and Williams, L.R. (1991) Benefits of reaction calorimetry. *Chemical Engineering Progress*, December, 65–9.

42 Bollyn, M., Bergh, A.V.d. and Wright, A. (1996) Schneller Scale-up, Reaktionskalorimetrie und Reaktorsimulation in Kombination. *Chemieanlagen und Verfahren*, **4**, 95–100

43 Regenass, W. (1996) Calorimetric instrumentation for process optimization and process safety. History, present status, potential for future development, in *8th RC1 User Forum*. Mettler-Toledo, Hilton Head SC, USA.

44 Stoessel, F. (1997) *Applications of reaction calorimetry in chemical engineering*. Journal of Thermal. *Analysis*, **49**, 1677–88.

45 Bou-Diab, L., Lerena, P. and Stoessel, F. (2000) A tool for process development. Safety investigations with high-pressure reaction calorimetry. *Chemical Plants and Processing*, **2**, 90–4.

46 Singh, J., Waldram, S.P. and Appleton, N.S. (1993) *Simultaneous Determination of Thermo-chemical and Heat Transfer Changes During an Exothermic Batch Reaction*, Hazard Evaluation Laboratory Ltd., Barnet Herts.

47 Girgis, M.J., Kiss, K., Ziltener, C.A., Prashad, K., Har, D., Yoskowitz, R., Basso, B., Repic, O., Blacklock, T.J. and Landau, R.N. (2001) Kinetic and calorimetric considerations in the scale-up of the catalytic reduction of a substituted nitrobenzene. *Organic Process Research & Development*, **1** (5), 339–49.

48 Yih-Shing, D., Chang-Chia, H., Chen-Shan, K. and Shuh, W.Y. (1996) Applications of reaction calorimetry in reaction kinetics and thermal hazard evaluation. *Thermochimica Acta*, **285** (1), 67–79.

49 Silva, C.F.P.M.e. and Silva, J.F.C.d. (2002) Evaluation of kinetic parameters from the synthesis of triaryl phosphates using reaction calorimetry. *Organic Process Research & Development*, **6**, 829–32.

50 Zaldivar, J.M., Hernandez, H. and Barcons, C. (1996) Development of a mathematical model and a simulator for the analysis and optimisation of batch reactors: Experimental model characterisation using a reaction calorimeter. *Thermochimica Acta*, **289**, 267–302.

51 Zogg, A., Fischer, U. and Hungerbuehler, K. (2004) A new approach for a combined evaluation of calorimetric and online infrared data to identify kinetic and thermodynamic parameters of a chemical reaction. *Chemometrics and Intelligent Laboratory Systems*, **71**, 165–76.

52 Moritz, H.U. (1989) Polymerisation calorimetry- a powerful tool for reactor control, in *Third Berlin International Workshop on Polymer Reaction Engineering*, VCH, Weinheim, Berlin.
53 Poersch-Parke, H.G., Avela, A., Reichert, K.H. (1989) Ein Reaktionskalorimeter zur Untersuchung von Polymerisationen. *Chemie Ingenieur Technik*, **61** (10), 808–10.
54 Gugliotta, L.M., Leiza, J.R., Arotçarena, M., Armitage, P.D. and Asua, J.M. (1995) Copolymer composition control in unseeded emulsion polymerization using calorimetric data. *Industrial Engineering Chemistry Research*, **34**, 3899–906.
55 Ferguson, H.D. and Puga, Y.M. (1997) Development of an efficient and safe process for a Grignard reaction via reaction calorimetry. *Journal of Thermal Analysis*, **49**, 1625–33.
56 Tilstam, U. and Weinmann, H. (2002) Activation of Mg metal for safe formation of Grignard reagents at plant scale. *Organic Process Research & Development*, **6**, 906–10.
57 Hoffmann, W. (1989) Reaction calorimetry in safety: the nitration of a 2,6-disubstituted benzonitrile. *Chimia*, **43**, 62–7.
58 Luo, K.M. and Chang, J.G. (1998) The stability of toluene mononitration in reaction calorimeter reactor. *Journal of Loss Prevention in the Process Industries*, **11**, 81–7.
59 Grob, B. (1987) Data-oriented hazard assessment: the nitration of benzaldehyde, in *Mettler Documentation*, Mettler, Toledo.
60 Ubrich, O. and Lerena, P. (1996) Methodology for the assessment of the thermal risks applied to a nitration reaction, in *Nitratzioni sicure in laboratorio e in impianto industriale*, Stazione sperimentalo combustibili, Milano.
61 Wiss, J., Fleury, C. and Fuchs, V. (1995) Modelling and optimisation of semi-batch and continuous nitration of chlobenzene from safety and technical viewpoints. *Journal of Loss Prevention in the Process Industries*, **8** (4), 205–13.
62 Roth, W.R. and Lennartz, H.W. (1980) Bestimmung von Hydrierwärmen mit einem isothermen Reaktionskalorimeter. *Chemische Berichte*, **113**, 1806–17.
63 Stoessel, F. (1993) Experimental study of thermal hazards during the hydrogenation of aromatic nitro compounds. *Journal of Loss Prevention in the Process Industries*, **6** (2), 79–85.
64 Cardillo, P., Quattrini, A. and Pava, E.V.d. (1989) Sicurezza nella produzione della 3,4-dichloranilina. *La chimia e l'industria*, **71** (7–8), 38–41.
65 Fuchs, R. and Peacock, L.A. (1979) Gaseous heat of hydrogenation of some cyclic and open chain alkenes. *Journal of Physical Chemistry*, **83** (15), 1975–8.
66 Murayama, K., Ariba, K., Fujita, A. and Iizuka, Y. (1995) The thermal behaviour analysis of an epoxidation reaction by RC1, in *Loss Prevention and Safety Promotion in the Process Industries*, Elsevier Scirnec, Barcelona.
67 Liang, Y., Qu, S.-S., Wang, C.X., Zou, G.L., Wu, Y.X. and Li, D.H. (2000) An on-line calorimetric study of the dismutation of superoxid anion catalyzed by SOD in batch reactors. *Chemical Engineering Science*, **55**, 6071–8.
68 Clark, J.D., Shah, A.S. and Peterson, J.C. (2002) Understanding the large-scale chemistry of ethyl diazoacetate via reaction calorimetry. *Thermochimica Acta*, **392–393**, 177–86.
69 Visentin, F., Zogg, A., Kut, O. and Hungerbühler, K. (2004) A pressure resistant small scale reaction calorimeter that combines the principles of power compensation and heat balance (CRC.v4). *Organic Process Research & Development*, **8** (5), 725–37.
70 Andre, R., Bou-Diab, L., Lerena, P., Stoessel, F., Giordano, M. and Mathonat, C. (2002) A new reaction calorimeter for screening purpose during process development. *Organic Process Research & Development*, **6**, 915–21.

Part II
Mastering Exothermal Reactions

5
General Aspects of Reactor Safety

Case History "Process Deviation"

A pharmaceutical intermediate was initially produced at a scale of 500 kg (product) per batch in a 2.5 m^3 reactor. The reaction was the condensation of an amino-aromatic compound with an aromatic chloride to form a di-phenyl amine by elimination of hydrochloric acid. This acid was neutralized in situ by sodium carbonate, forming water, sodium chloride, and carbon dioxide. The manufacturing procedure was very simple: The reactants were mixed at 80 °C, a temperature above the melting point of the reaction mass. Then the reactor was heated with steam in the jacket to a temperature of 150 °C. At this temperature, the steam valve had to be closed and the reaction left to proceed for a further 16 hours. During this time, the temperature increased to a maximum of 165 °C. Several years later, the batch size was increased to 1000 kg per batch in a 4 m^3 reactor. Two years after this a further increase to 1100 kg was decided.

Six months after this final increase, following a break for the Christmas holidays, the first batch of a new campaign had to be started. As one of the reactants was discharging from the storage tank by pumping it into the reactor, the transfer line plugged due to the cold weather. Due to an urgent demand on this product it was decided to charge the reactor by transferring the reactant using drums. The reaction was started by heating as usual, but instead of shutting the steam valve at 150 °C, this was done later as the reactor temperature reached 155 °C. On checking the reactor, the operator saw that the reaction mass was boiling: some refluxing liquid was visible in the riser. Since the condensate was not returned to the reactor, the solvent was distilled off, leading to an increase in the concentration and the boiling point of the reaction mass. This evaporation proceeded so rapidly that the pressure in the reactor increased and led to a discharge through the pressure relief system. A major part of the reaction mass was released to the outside, but the pressure continued to increase. Finally the reaction mass was spread over the entire plant through the gaskets of the riser.

Thermal Safety of Chemical Processes: Risk Assessment and Process Design. Francis Stoessel

ISBN: 978-3-527-31712-7

The consequences were an interruption of the whole plant for two months. For the particular process involved in the incident the interruption was over six months. The material damage was several millions of US dollars.

The inquiry showed that the process was operated in the parametric sensitive range. As the batch size was increased to 1100 kg, the maximum temperature during the holding phase increased to 170 °C. Moreover, the thermometer had a range of 200 °C from −30 °C to +170 °C, because the reactor was multi-purpose equipment also equipped with a brine cooling system. Thus, the technical equipment was not adapted to the process conditions.

Lessons drawn

Neither the process conditions nor the technical equipment of the reactor were adapted to the nature of the reaction. Moreover the effect of the increase in batch sized was overlooked. The process had to be changed to semi-batch operation in order to ensure a safe control of the reaction.

5.1 Introduction

In the present chapter, some important aspects of reactor stability and the corresponding assessment criteria for normal operating conditions will be presented. In a second section, the assessment criteria for deviating conditions, such as cooling failure, are introduced.

A chemical reactor is considered to be safe when the reaction course can be easily controlled. Therefore, with regard to thermal process safety, the control of the reactor temperature will play a key role. For this reason the heat balance of reactors must be thoroughly understood in order for their safe design. The different terms entering into the heat balance were presented in Section 2.4. A well understood heat balance will allow designing safe reactors under normal operating conditions. In some instances, the heat balance may present a so-called parametric sensitivity, that is, the behavior of the reactor may change dramatically for only small changes in the governing operating parameters. The reactor stability can be characterized by some stability criteria, which will be described in the first section.

Moreover, reactor safety also requires fulfilling a more ambitious objective, that is, to design a reactor that will remain stable in case of mal-operation. The result will be a robust process towards deviations from normal operating conditions. This goal can be reached if the accumulation of non-converted reactants is controlled and maintained at a safe level during the course of reaction. The concept of maximum temperature of synthesis reaction (*MTSR*) was introduced for this purpose. This point will be described in the second section. In the last section, examples of reactions that will be used in the following chapters are described.

5.2 Dynamic Stability of Reactors

5.2.1 Parametric Sensitivity

The differential equations governing a non-isothermal batch reaction describe the material balance coupled to the heat balance (see also Section 2.4.2):

$$\begin{cases} \dfrac{dX}{dt} = k(1-X) \\ M_r c_p' \dfrac{dT}{dt} = V(-\Delta H_r)\, C_0 \dfrac{dX}{dt} + UA(T_c - T) \end{cases} \tag{5.1}$$

By rearranging and expressing the kinetic constant as a function of temperature, one obtains:

$$\begin{cases} \dfrac{dX}{dt} = k_\infty e^{-E/RT} C_0(1-X) \\ \dfrac{dT}{dt} = \Delta T_{ad} \dfrac{dX}{dt} + \dfrac{UA(T_c - T)}{M_r c_p'} \end{cases} \tag{5.2}$$

Thus the temperature course in a reactor depends on the following terms:

- the adiabatic temperature rise, which describes the energy contents of a reaction mass;
- the cooling rate defined by the overall heat exchange coefficient, by the heat exchange area, and by the temperature difference between reaction mass and cooling medium;
- the heat production rate by the reaction and its temperature dependence.

Whereas the cooling capacity depends linearly on temperature, the heat production rate depends exponentially following the Arrhenius law. This may result in extremely high temperature maxima, if the control is not appropriate. Thus, it is important to characterize the effect of temperature on the heat balance.

This problem was studied by many authors [1–9]. A comprehensive review has been presented by [10] and [11].

5.2.2 Sensitivity Towards Temperature: Reaction Number B

Since the sensitivity of the reaction rate towards temperature, and therefore of the heat release rate of a reaction, dominates the heat balance, it is important to define a criterion as an indicator of this effect. By differentiation of the reaction rate with temperature one obtains

$$r = k_{\infty} e^{-E/RT} C_0 (1 - X) \tag{5.3}$$

$$\frac{dr}{dT} = k_{\infty} e^{-E/RT} C_0 (1 - X) \cdot \frac{E}{RT^2} = r \cdot \frac{E}{RT^2} \tag{5.4}$$

Thus, the relative variation of the reaction rate with temperature is

$$\frac{d(r)/r}{dT} = \frac{d\ln(r)}{dT} = \frac{E}{RT^2} \tag{5.5}$$

The term $\frac{E}{RT^2}$ is called the temperature sensitivity. By multiplying this with the adiabatic temperature rise, one obtains a dimensionless criterion, also named reaction number or dimensionless adiabatic temperature rise:

$$B = \frac{\Delta T_{ad} \cdot E}{RT^2} \tag{5.6}$$

Thus, the higher the activation energy, the more sensitive to temperature the reaction rate will be. High *B*-values indicate a reaction that is difficult to control (high adiabatic temperature rise and high sensitivity to temperature). Some values, calculated for 100 °C, are represented in Figure 5.1. As an example, P. Hugo [6] showed that reactions with *B*-values above 5 are difficult to control in batch reactors. But this criterion alone gives no information on the heat removal by the

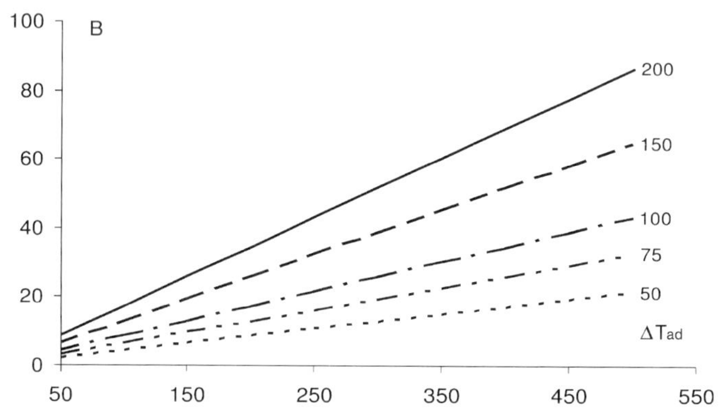

Figure 5.1 *B*-numbers calculated for a temperature of 373 K as a function of the adiabatic temperature rise. The parameter is the activation energy varied from 50 to 200 kJ mol^{-1}.

reactor heat exchange system. Thus, more comprehensive criteria involving a heat balance have to be used.

5.2.3 Heat Balance

In this section we consider different criteria developed on the basis of a heat balance. The criteria described here represent a selection of easy to use, but also more advanced, criteria allowing discrimination between dynamically stable situations and situations where runaway is likely to occur. Their use will be described in Chapters 6 to 8.

5.2.3.1 The Semenov Criterion

In Section 2.5.2, the Semenov diagram was used to show the critical cooling medium temperature. In the same way it allows discrimination of a stable operation conditions from a runaway situation. A stable operation is achieved for a limit value of the Semenov criterion ψ (also called Semenov number):

$$\psi = \frac{q_0 E}{UART_0^2} < \frac{1}{e} \simeq 0.368 \tag{5.7}$$

This condition was established for zero-order reactions, and is valid for highly exothermal reactions, resulting in a high temperature increase even for low degrees of conversion. In this criterion, beside the heat removal properties of the reactor, only the heat release rate at process temperature (q_0) and the activation energy (E) of the reaction are required.

5.2.3.2 Stability Diagrams

A more comprehensive approach consists of studying the variation of the Semenov criterion as a function of the reaction energy. Such an approach is presented in [12], where the reciprocal Semenov criterion is studied as a function of the dimensionless adiabatic temperature rise. This leads to a stability diagram similar to those presented in Figure 5.2 [11, 13]. The lines separating the area of parametric sensitivity, where runaway may occur, from the area of stability is not a sharp border line: it depends on the models used by the different authors. For safe behavior, the ratio of cooling rate over heat release rate must be higher than the potential of the reaction, evaluated as the dimensionless adiabatic temperature rise.

5.2.3.3 Heat Release Rate and Cooling Rate

By considering the heat balance and its variation with temperature, it is obvious that as long as the cooling capacity increases faster with temperature than the heat release rate does, the situation is stable, as was considered in Section 2.5.4 for calculating the critical temperature:

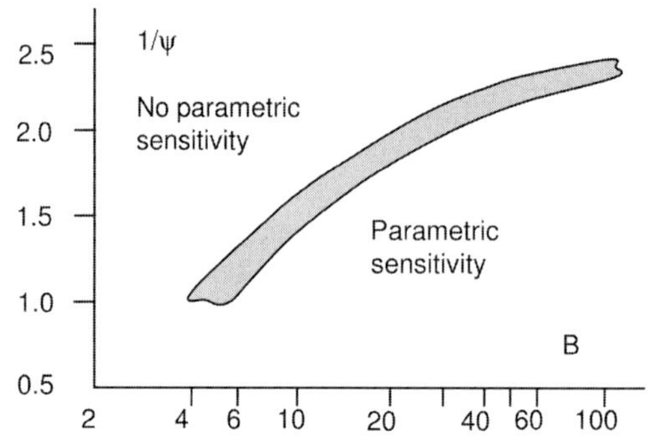

Figure 5.2 Stability diagram presenting the variation of the reciprocal Semenov criterion as a function of the dimensionless adiabatic temperature rise.

$$\frac{\partial q_{ex}}{\partial T} >> \frac{\partial q_{rx}}{\partial T} \tag{5.8}$$

with

$$\frac{\partial q_{ex}}{\partial T} = \frac{\partial[UA(T - T_c)]}{\partial T} = UA \tag{5.9}$$

and

$$\frac{\partial q_{rx}}{\partial T} = \frac{\partial\left[k_0 e^{-E/RT} \cdot f(X) \cdot C_{A0} \cdot (-\Delta H_r)\right]}{\partial T} = \frac{E}{RT^2} \cdot r \cdot (-\Delta H_r) \tag{5.10}$$

which results in

$$\frac{\tau_r}{\tau_c} \frac{RT^2}{\Delta T_{ad} E} >> 1 \tag{5.11}$$

where τ_r is the characteristic reaction time defined in Equation 5.15 and τ_c the thermal time constant of the reactor defined in Equation 5.16. This criterion was confirmed and refined by simulation analysis [13], which considers the initial reaction rate:

$$\frac{\tau_{r0}}{\tau_c} = \frac{\tau_{rc0}}{\tau_c} >> \left(\frac{\Delta T_{ad} E}{RT_c^2}\right)^{1.2} = B^{1.2} \tag{5.12}$$

This criterion is a so-called "sliding" criterion, since it is established for time 0, but may be applied for any time (sliding). It also represents a summary of the stability diagrams presented in Section 5.2.3.2. The exponent 1.2 of B introduces a safety margin. This criterion uses a comprehensive knowledge of the reaction kinetics.

5.2.3.4 Using Dimensionless Criteria

For reactions following an nth-order kinetic scheme, the reaction rate is often characterized by a dimensionless number: the Damköhler criterion [14]:

$$Da = \frac{r_A \tau_r}{C_0} \tag{5.13}$$

or by expressing the reaction rate explicitly:

$$Da = kC_0^{n-1}\tau_r \tag{5.14}$$

This Damköhler criterion is the Damköhler number of type I (Da_I). Other Damköhler numbers were defined [12]: type II used to characterize the material transport at the surface of a solid catalyst, type III used to characterize the convective heat transport at the catalyst surface, and type IV used to characterize the temperature profile in a solid catalyst. For batch reactions, the reaction time τ_r is defined at a reference temperature, the cooling medium temperature, as

$$\tau_r = \frac{1}{k_{(T_c)} \cdot C_0{}^{n-1}} \tag{5.15}$$

The other side of the heat balance, the cooling rate, can be characterized by the thermal time constant of the reactor:

$$\tau_c = \frac{\rho V c_p'}{UA} \tag{5.16}$$

By dividing the reaction time by the thermal time constant, one obtains a dimensionless number, the modified Stanton criterion:

$$St = \frac{UA \cdot \tau_r}{M_R \cdot c_p'} = \frac{UA \cdot \tau_r}{\rho \cdot V \cdot c_p'} = \frac{\tau_r}{\tau_c} \tag{5.17}$$

The modified Stanton criterion compares the characteristic reaction time with the thermal time constant of the reactor. The time can be eliminated from the equations by building the ratio:

$$\frac{Da}{St} = \frac{k \cdot C_0^{n-1} \rho \cdot V \cdot c_p'}{UA} << 1 \tag{5.18}$$

This criterion, as in (5.12), uses a comprehensive kinetic description of the reaction. In fact, this ratio allows comparison of the characteristic time of the reaction rate with the cooling rate. It is strongly affected by a change in the reactor size, as explained in Section 2.4.1.2. Moreover, it varies non-linearly with reactor size. Hence it is especially important to consider its effect during scale-up.

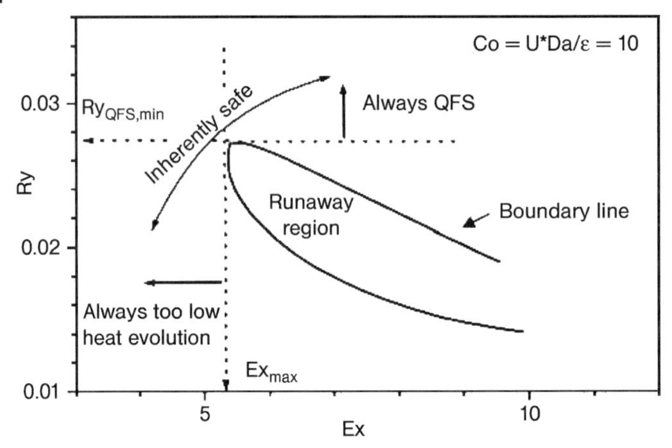

Figure 5.3 Inherently safe operating conditions for a slow reaction in the continuous phase showing the boundary line for runaway conditions.

In a more recent study [15], Westerterp and Molga introduced a set of dimensionless numbers (cooling, reactivity, and exothermicity numbers) characterizing the stability of heterogeneous slow liquid–liquid reactions in the semi-batch reactor. They demonstrated that the key parameter is the cooling number *Co*:

$$Co = \frac{(UA)_0 \cdot t_{fd}}{(V\rho c'_P)_0 \varepsilon}$$

$$Ry = \frac{(\nu_A/\nu_B)M \cdot E/R}{(R_H + U \cdot Da/\varepsilon)\varepsilon} \tag{5.19}$$

$$Ex = \frac{\Delta T_{ad,0} \cdot E/R}{T_c^2(R_H + U \cdot Da/\varepsilon)\varepsilon}$$

In these equations, ε represents the relative volume increase due to the feed and R_H the ratio of the heat capacities of both liquid phases. By representing the reactivity number as a function of the exothermicity number (Figure 5.3), different regions are obtained. The region where runaway occurs is clearly delimited by a boundary line. Above this region, for a high reactivity, the reaction is operated in the QFS conditions (Quick onset, Fair conversion and Smooth temperature profile) and leads to a fast reaction with low accumulation and easy temperature control (see Section 7.6).

5.2.3.5 **Chaos Theory and Lyapunov Exponents**

This criterion is based on a sophisticated mathematical approach, and therefore is not as easy to use as the preceding criteria. Nevertheless, this approach represents the most advanced technique in this field. It was developed by Strozzi and Zaldivar and co-workers [7–9, 16–19].

The Lyapunov exponents describe the behavior of two neighboring points of a system in the phase space as a function of time. If the Lyapunov exponent is positive, these points diverge from each other; if it is negative they converge. This is an indicator of the system's sensitivity. For a batch reactor, the Lyapunov exponents are defined as a function of time:

$$\lambda_j(t) = \frac{1}{t}\log_2\frac{L_j(t)}{L_j(0)}, \quad \text{for } j = 1, 2, \ldots m \tag{5.20}$$

The state variables define an ellipsoid in the state space, where $L_j(0)$ and $L_j(t)$ are the lengths of the j-axis of the ellipsoid at $t = 0$ and $t = t$. This gives the evolution of the m-sphere in the state space. The volume is given by

$$V(t) = 2^{[\lambda_1(t)+\lambda_2(t)+\ldots+\lambda_m(t)]\cdot t} \tag{5.21}$$

The sensitivity towards an input parameter ϕ is given by

$$s(V, \phi) = \frac{\Delta\{\max_t 2^{[\lambda_1(t)+\lambda_2(t)+\ldots+\lambda_m(t)]\cdot t}\}}{\Delta\phi} \tag{5.22}$$

This criterion allows distinguishing two states in a batch reactor, no-runaway and runaway. For the semi-batch reactor, there are four different states, no-ignition, runaway, marginal ignition, and QFS.

The advantage of this criterion is that it can be computed on-line, based only on temperature measurements, without the necessity of a model of the process [9]. The method uses a reconstruction of the phase space using only one state variable. This allows building a warning system for detecting a runaway situation: this aspect is presented in Chapter 10.

5.2.4 Reactor Safety After a Cooling Failure

5.2.4.1 Potential of the Reaction, the Adiabatic Temperature Rise

The first criterion is the adiabatic temperature rise. It is static and gives an indication of the excursion potential of a reaction. The higher the adiabatic temperature rise, the higher the final temperature will be if the cooling system fails. Its value can be obtained from Equation 2.5:

$$T_f = T_p + \Delta T_{ad} < T_{max} \tag{5.23}$$

Here T_{max} represents a maximum allowed temperature. This can be defined to avoid triggering secondary reactions or to avoid high pressures. In fact, it corresponds to the worst case approach presented in Section 3.4.

The heat balance in Equation 2.28 also shows that highly exothermal reactions will be more difficult to control than low exothermal reactions, as even for small increases in conversion, the increase in temperature becomes important (see Section 2.4.3). Further, the severity in the event of an incident may be higher.

5.2.4.2 Temperature in Cases of Cooling Failure: The Concept of MTSR

If a cooling failure occurs while an exothermal reaction is being performed, the non-converted reactants will react away without cooling, causing a temperature rise above the intended reaction temperature. Therefore a temperature range may be reached where secondary reactions could become dominant or where the vapor pressure of the system could surpass the maximum allowed working pressure of the reactor. In order to predict the consequences of the loss of control of a desired reaction, it is necessary to know the Maximum Temperature which can be reached by the Synthesis Reaction (MTSR) under adiabatic conditions [12]. Here, only the heat of the desired (synthesis) reaction is considered. The temperature level (T_{cf}), which can be reached in a cooling failure is a function of the process temperature (T_p) of the degree of accumulation (X_{ac}) and of the total adiabatic temperature rise (ΔT_{ad}):

$$T_{cf} = T_p + X_{ac} \cdot \Delta T_{ad} \tag{5.24}$$

The degree of accumulation is the fraction of the total heat of reaction that has not yet been released at a time t:

$$X_{ac}(t) = 1 - X = \frac{\int_t^\infty q_{rx} d\tau}{\int_0^\infty q_{rx} d\tau} = 1 - \frac{\int_0^t q_{rx} d\tau}{\int_0^\infty q_{rx} d\tau} \tag{5.25}$$

Since the process temperature, as well as the degree of accumulation, may vary during the reaction, the temperature after cooling failure (T_{cf}) depends strongly on the strategy of control of the reaction. The temperature T_{cf} is a function of time. Thus, for the prediction of the behavior of a reactor when there is a cooling failure, the knowledge of the instant at which it is maximum, is an important datum. The assessment of the process safety and the design of safety measures will be based on the MTSR corresponding to the maximum of T_{cf}:

$$MTSR = [T_{cf}]_{\max} \tag{5.26}$$

Note: When working at low temperatures (below ambient), the MTSR may be taken as the ambient temperature in cooling failure, even if the adiabatic temperature rise would not allow it to reach this point. This is because a reactor left at a sub-ambient temperature will equilibrate with its surroundings.

5.3 Example

5.3.1 Example Reaction System

In the following chapters, an example reaction system will be used for illustrating purposes. In order to focus on thermal aspects of reactor safety, no explicit chemistry will be used, but a general reaction scheme is used instead:

$$A + B \xrightarrow{k_1} P \xrightarrow{k_2} S \tag{5.27}$$

The first reaction is the synthesis reaction, a single bimolecular second-order reaction with the rate equation:

$$-r_A = k_1 \cdot C_A \cdot C_B \tag{5.28}$$

The second reaction is a first-order decomposition reaction of the product P, with a rate equation:

$$-r_P = k_2 \cdot C_P \tag{5.29}$$

This reaction scheme is used in two variants, a fast reaction called the addition reaction and a slow synthesis reaction called the substitution reaction. The thermal and kinetic data are summarized in Table 5.1. The decomposition reaction presents a heat release rate of $10\,\mathrm{W\,kg^{-1}}$ at 150 °C. Together with the activation energy, this heat release rate allows calculating the time to explosion (TMR_{ad}) as a function of temperature. The amounts of reactants to be used in discontinuous operations are summarized in Table 5.2. The solvent used has a boiling point of 140 °C at atmospheric pressure.

The reactor to be used is a $4\,\mathrm{m^3}$ stainless steel stirred tank following DIN-Standards [20]. It is equipped with a indirect heating cooling system using a monofluid (water-diethylene glycole mixture) circulating in a heat exchanger

Table 5.1 Thermal and kinetic data for the example reactions used in Chapters 6 to 8.

Data	Substitution reaction	Addition reaction	Decomposition
Enthalpy	$-150\,\mathrm{kJ\,mol^{-1}}$	$-150\,\mathrm{kJ\,mol^{-1}}$	$-575\,\mathrm{kJ\,mol^{-1}}$
Specific heat capacity	$1.7\,\mathrm{kJ\,kg^{-1}\,K^{-1}}$	$1.7\,\mathrm{kJ\,kg^{-1}\,K^{-1}}$	$1.7\,\mathrm{kJ\,kg^{-1}\,K^{-1}}$
Activation energy	$60\,\mathrm{kJ\,mol^{-1}}$	$60\,\mathrm{kJ\,mol^{-1}}$	$100\,\mathrm{kJ\,mol^{-1}}$
Pre-exponential factor	$10^9\,\mathrm{kg\,mol^{-1}\,h^{-1}}$	$10^{11}\,\mathrm{kg\,mol^{-1}\,h^{-1}}$	$7 \cdot 10^{10}\,\mathrm{h^{-1}}$
Concentration C_{A0}	$3\,\mathrm{mol\,kg^{-1}}$	$3\,\mathrm{mol\,kg^{-1}}$	–
Final concentration	$2\,\mathrm{mol\,kg^{-1}}$	$2\,\mathrm{mol\,kg^{-1}}$	–
Molar ratio B/A	1.25	1.25	–

Table 5.2 Charge for discontinuous operation.

Compound	Mass (kg)	Moles	Mole ratio
A	2000	6000	1.0
B	1000	7500	1.25
Total	3000	–	–

Density: $1000\,\mathrm{kg \cdot m^{-3}}$.

Table 5.3 Reactor characteristics.

Characteristics	Values
Nominal volume	$4\,m^3$
Material	Stainless steel
Maximum working volume	$5.1\,m^3$
Maximum heat exchange area	$7.4\,m^2$ at $3.4\,m^3$
Minimum heat exchange area	$3.0\,m^2$ at $1.05\,m^3$
Jacket type	Half welded coils
Overall heat transfer coefficient	$200\,W\,m^{-2}\,K^{-1}$
Maximum heating temperature	150 °C
Minimum cooling temperature	−15 °C with brine
Minimum cooling temperature	+5 °C with water
Heating time constant (jacket)	0.20 h
Cooling time constant (jacket)	0.23 h

system providing 3 temperature levels: steam 5 bar 150 °C, water 5 °C, and brine −15 °C. The temperature control may be jacket mode (Isoperibolic) or a cascade controller, controlling the internal temperature either in the isothermal mode or following a defined temperature gradient. Its characteristics are summarized in Table 5.3.

Worked Example 5.1: Safety Criteria Applied on the Example Reaction

In order to assess the dynamic stability of the example reaction in the reactor described above, we can apply different criteria as described in Section 5.2. The following criteria are used: Semenov Equation 5.7, Villermaux Equation 5.12, and the ratio Da/St Equation 5.18. The reaction number B is also represented. Since they are a function of the cooling system temperature, they were calculated as a function of this temperature. The results are represented in Figure 5.4.

The last criterion Da/St says that this ratio must be significantly smaller than 1. This may be interpreted as smaller than 0.1. This limit is reached for a cooling medium temperature of approximately 30 °C at maximum. The limits corresponding to the other criteria may be directly read from the intercept of the representative curves. A similar interpretation of the Villermaux criterion shows a maximum temperature of approximately 20 °C and finally the Semenov criterion a temperature of 10 °C.

This scatter of the results merits an explanation:

- The Semenov criterion means that for a cooling medium temperature above 10 °C, the initial heat release rate of the reaction cannot be removed by the cooling system. This delivers a broad enough margin for performing the reaction with a cooling system temperature below this level. This is a static criterion.

- The Villermaux criterion and the *Da*/*St* criterion are dynamic stability criteria, meaning that with a cooling medium temperature above the limit level, 20 resp. 30 °C, the reactor will be operated in the instable region and present the phenomenon of parametric sensitivity. If instead of $B^{1.2}$, B is used, both criteria lead to the same result. This should not be surprising since they derive from the same heat balance considerations, that is, the heat release rate of the reaction increases faster with temperature than the heat removal does.

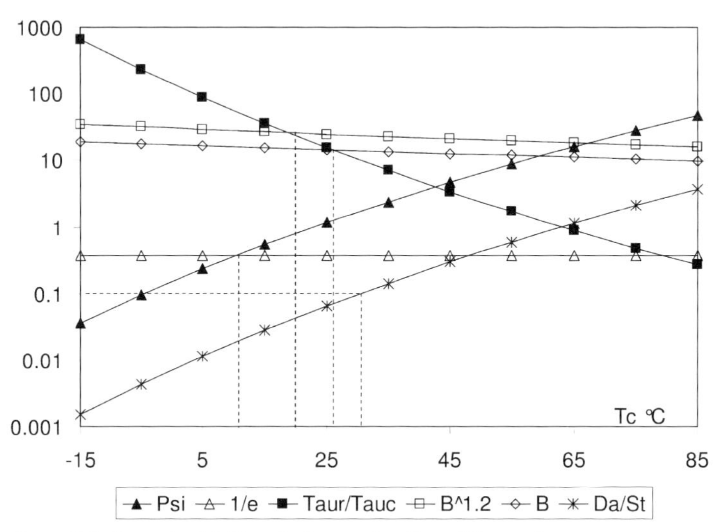

Figure 5.4 Dynamic stability criteria as a function of the cooling system temperature. Squares represent the Villermaux criterion, triangles the Semenov criterion, and stars the ratio *Da* : *St*.

Table 5.4 Thermal data for the safety assessment.

Safety relevant data	Value and assessment
Specific heat of main reaction	300 kJ kg^{-1} [a]
Specific heat capacity of reaction mass	1.7 kJ kg^{-1} K^{-1}
Adiabatic temperature rise of main reaction	176 K
Severity in case of runaway	Medium
Specific heat of secondary reaction	1150 kJ kg^{-1} [a]
Adiabatic temperature rise of secondary reaction	676 K
Severity in case of runaway of secondary reaction	High
High probability for triggering secondary reaction	above T_{D8} = 122 °C
Low probability for triggering secondary reaction	below T_{D24} = 113 °C

a) Reference is final reaction mass.

The thermal data of the reaction are summarized in Table 5.4. It should be noted that these criteria do not use any explicit kinetic data, but only the results of calorimetric experiments. For the decomposition reaction, by taking the activation energy into account, the safety limits of T_{D24} = 113 °C and T_{D8} = 122 °C may be established, according to the assessment criteria presented in Section 3.3.3. The activation energy may be determined, for example, from DSC experiments, as described in Chapter 11. Without knowledge of the process conditions of temperature and feed rates, the assessment remains global, as shown in Table 5.4. More detailed assessment will be provided in the next chapters for different reactor types and process conditions.

References

1 Amundson, N.R. and Bilous, O. (1995) Chemical reactor stability and sensitivity. *American Institution of Chemical Engineers Journal*, **1** (4), 513–21.

2 Aris, R. (1965) *Introduction to the Analysis of Chemical Reactors*. Prentice-Hall, Englewood Cliffs, New Jersey.

3 Eigenberger, G. and Schuler, H. (1986) Reaktorstabilität und sichere Reaktionsführung. *Chemie Ingenieur Technik*, **58** (8), 655–65.

4 Heiszwolf, J.J. and Fortuin, J.M.H. (1996) Runaway behaviour and parametric sensitivity of a batch reactor an experimental study. *Chemical Engineering Science*, **51** (11), 3095–100.

5 Luo, K.M., Lu, K.T. and Hu, K.H. (1997) The critical condition and stability of exothermic chemical reaction in a non-isothermal reactor. *Journal of Loss Prevention in the Process Industries*, **10** (3), 141–50.

6 Hugo, P. (1980) Anfahr- und Betriebsverhalten von exothermen Batch-Prozessen. *Chemie Ingenieur Technik*, **52** (9), 712–23.

7 Alos, M.A., Strozzi, F. and Zaldivar, J.M. (1996) A new method for assessing the thermal stability of semi-batch processes based on Lyapunov exponents. *Chemical Engineering Science*, **51** (11), 3089–96.

8 Alos, M.A., Zaldivar, J.M., Strozzi, F., Nomen, R. and Sempere, J. (1996) Application of parametric sensitivity to batch process. *Chemical Engineering Technology*, **19**, 222–32.

9 Zaldivar, J.M., Cano, J., Alos, M.A., Sempere, J., Nomen, R., Lister, D., Maschio, G., Obertopp, T., Gilles, E.D., Bosch, J. and Strozzi, F. (2003) A general criterion to define runaway limits in chemical reactors. *Journal of Loss Prevention in the Process Industries*, **16** (3), 187–200.

10 Villermaux, J. (1991) *Reviews in Chemical Engineering*, **7** (1), 51–108.

11 Varma, A., Morbidelli, M. and Wu, H. (1999) *Parametric Sensitivity in Chemical Systems*, Cambridge University Press, Cambridge.

12 Baerns, M., Hofmann, H. and Renken, A. (1987) *Chemische Reaktionstechnik*. Georg Thieme, Stuttgart.

13 Villermaux, J. (1993) *Génie de la Réaction Chimique, Conception et Fonctionnement des Réacteurs*. Lavoisier Tec Doc.

14 Hugo, P. (1992) Grundlagen der thermisch-sicheren Auslegung von chemischen Reaktoren, in *Dechema Kurs Sicherheit chemischer Reaktoren*, Dechema, Berlin.

15 Westerterp, K.R. and Molga, E.J. (2004) No more runaways in fine chemical reactors. *Industrial Engineering Chemical Research*, **43**, 4585–94.

16 Strozzi, F., Zaldivar, J.M., Kronberg, A. and Westerterp, K.R. (1997) Runaway prevention in chemical reactors using chaos theory techniques. *American Institution of Chemical Engineers Journal*, **45**, 2394–408.

17 Strozzi, F. and Zaldivar, J.M. (1994) A general method for assessing the thermal stability of chemical batch reactors by sensitivity calculation based on Lyapunov exponents. *Chemical Engineering Science*, **49**, 2681–8.

18 Strozzi, F., Alos, M.A. and Zaldivar, J.M. (1994) A method for assessing thermal stability of batch reactors by sensitivity calculation based on Lyapunov exponents: experimental verification. *Chemical Engineering Science*, **49**, 5549–61.

19 Alos, M.A., Nomen, R., Sempere, J.M., Strozzi, F. and Zaldivar, J.M. (1998) Generalized criteria for boundary safe conditions in semi-batch processes: simulated analysis and experimental results. *Chemical Engineering and Processing*, **37**, 405–21.

20 DIN 28131 #394 (1979) *Rührer für Rührbehälter, Formen und Hauptabmessungen*, Beuth, Berlin, Germany.

6
Batch Reactors

Case History "Nitroaniline"

After 40 years of accident-free production of nitroaniline, an explosion occurred with severe consequences for the building and its surroundings. A part of the autoclave weighing 6 tonnes was catapulted 70 meters [1]. The reaction scheme is depicted in Figure 6.1.

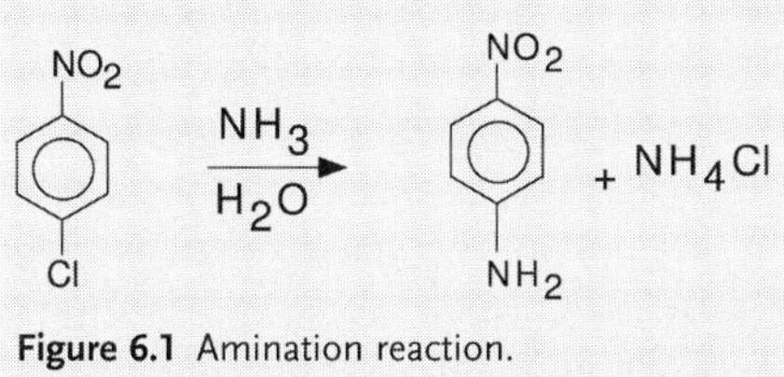

$$\text{Cl-C}_6\text{H}_4\text{-NO}_2 \xrightarrow[H_2O]{NH_3} \text{H}_2\text{N-C}_6\text{H}_4\text{-NO}_2 + NH_4Cl$$

Figure 6.1 Amination reaction.

Subsequent enquiries revealed the following:

- The batch causing the accident had a massive overcharge of chloronitrobenzene and thus an undercharge of ammonia. This raised the reaction energy of the starting material mass and the reaction speed, and lowered the pressure below the specified.
- Kinetic studies of the defective batch showed that the jacket cooling was capable of dissipating the amination heat up to approximately 190 °C.
- Due to an impact at one end of the scale, the temperature registration (0–200 °C) inaccurately indicated 194 °C.
- The autoclave was equipped with a separate pressure relief system consisting of a rupture disk and a safety valve connected to it in series (set pressure 50 bar in each case). It was clear from the debris that these mechanisms had been activated.
- Thermal balance calculations show that the reactor could have been relieved via gas flow, that is, one phase flow, through the rupture disk and safety valve up to 250 °C and 65 bar. When the accident happened, the mass

Thermal Safety of Chemical Processes: Risk Assessment and Process Design. Francis Stoessel

ISBN: 978-3-527-31712-7

was not capable of being relieved even at lower temperatures, because a two phase flow had occurred, since the gas entrained liquid.

- It must be assumed that pressure had built up between the rupture disk and the safety valve due to faulty seals (i.e. in the worst case the actual set pressure of the relief system could have amounted to $2 \times 50 = 100$ bar).
- It may be concluded from thermal studies that the heat release due to decomposition attributed to the nitro group made a substantial contribution to the destructive power of the thermal explosion from 350–400 °C.
- A reconstructed temperature/time profile is represented in Figure 6.2.

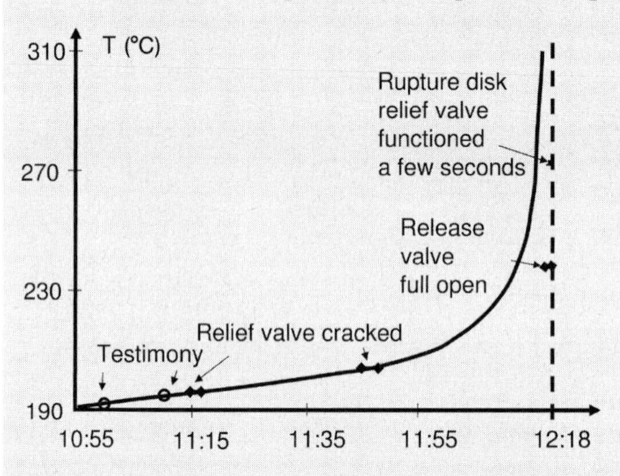

Figure 6.2 Evolution of temperature during the incident batch.

Lessons drawn

- Batch reactors are sensitive to charging errors: In a batch reactor, the reaction course is only governed by the temperature, and charging raw material is equivalent to "charge an energy potential" into the system.
- Protective systems must be designed properly: Here deviations from normal operating conditions must be carefully identified by risk analysis, and protection systems must be designed accordingly.

6.1 Introduction

These points are explained in detail in this chapter. In a first section, the general aspects of reaction engineering for batch reactors are briefly presented. The mass and heat balances are analysed and it is shown that a reliable temperature control is central to the safety of batch reactors. The different strategies of temperature control and their consequences on reactor safety are explained in the following sections. For each strategy, the design criteria and the safety assessment procedure are introduced. The chapter is closed by recommendations for the design of thermally safe batch reactions.

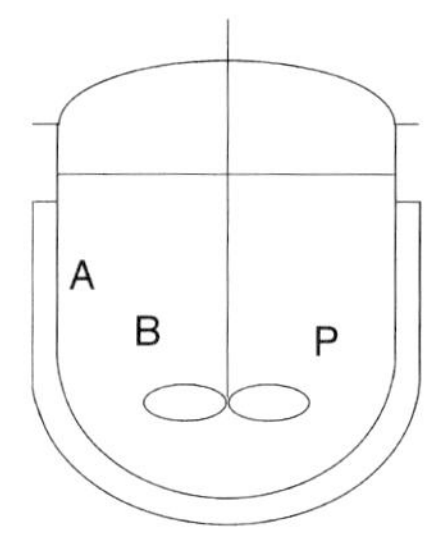

Figure 6.3 Batch reactor: For a single reaction of the type A + B → P, both reactants A and B are charged initially into the vessel. Therefore, temperature control is practically the only way to influence the reaction course.

6.2
Principles of Batch Reaction

6.2.1
Introduction

In chemical reaction engineering, an ideal batch reactor is defined as a closed reactor, meaning there is no addition and no removal of any components during the reaction time. The prototype of this reactor is the autoclave, where all reactants are charged into the reactor at the beginning of the operation (Figure 6.3). The reactor is then closed and heated to reaction temperature, the temperature at which the reaction is allowed to complete or at which a catalyst is added. After the reaction is completed, the reactor is cooled and discharged. It is now ready for a new cycle.

For our purposes, we extend the definition of the batch reaction to reactions where a product is allowed to leave the reactor during the reaction, for example, as a gaseous product or as vapor by distillation. This is because, in terms of thermal process safety, the focus is the mean for reaction rate control. The reaction rate is a function of temperature and concentration, but the concentration cannot be influenced by external means such as progressive addition of a reactant. Thus, in the batch reactor, the only way to control the reaction rate is with temperature. Therefore, the heat exchange system becomes very important and its failure may have serious consequences. Nevertheless, in certain cases of heterogeneous reactions, with mass transfer control, the stirrer may also provide a means of controlling the reaction rate by influencing mixing. Thus, the way in which the temperature is controlled plays a central role. This point will be discussed in detail in this chapter.

6.2.2
Mass Balance

In general, the overall mass balance written for the molar flow of a reactant comprises four terms:

$$\begin{bmatrix}\text{Flow into}\\\text{the reactor}\end{bmatrix}=\begin{bmatrix}\text{Flow out of}\\\text{the reactor}\end{bmatrix}+\begin{bmatrix}\text{Rate of}\\\text{disappearance}\end{bmatrix}+\begin{bmatrix}\text{Rate of}\\\text{accumulation}\end{bmatrix} \quad (6.1)$$

Since, by definition, there is no reactant flow into or out of the reactor, the first two terms are equal to zero. The remaining balance is

$$\begin{bmatrix}\text{Rate of}\\\text{disappearance}\end{bmatrix}=-\begin{bmatrix}\text{Rate of}\\\text{accumulation}\end{bmatrix} \quad (6.2)$$

The rate of disappearance of a reactant A due to the reaction is proportional to the reaction rate and to the volume of the fluid ($-r_A V$). The rate of accumulation is equal to the variation of the number of moles of A present in the reactor per unit of time [2]:

$$\frac{dN_A}{dt}=\frac{d[N_{A0}(1-X_A)]}{dt}=-N_{A0}\frac{dX_A}{dt} \quad (6.3)$$

Thus the material balance becomes

$$-r_A' V = N_{A0}\frac{dX_A}{dt} \Leftrightarrow \frac{dX_A}{dt}=\frac{-r_A}{(C_{A0})} \quad (6.4)$$

and after integration, we obtain

$$t = N_{A0}\int_0^{X_A}\frac{dX_A}{(-r_A)V} \quad (6.5)$$

This expression is also called the performance equation of the reactor and calculates the time required to achieve a certain conversion in a given reactor, or the reactor volume required to achieve a defined conversion in a given time.

6.2.3 Heat Balance

Only a simplified heat balance is considered here:

$$q_{ac}=q_{rx}-q_{ex} \Leftrightarrow \rho V c_P' \frac{dT_R}{dt}=(-r_A)V(-\Delta H_r)-UA(T_r-T_c) \quad (6.6)$$

or rearranged so as to enhance the variation of temperature with conversion [3]:

$$\frac{dT_r}{dt}=\Delta T_{ad}\frac{-r_A}{C_{A0}}-\frac{UA}{\rho V c_P'}(T_r-T_c) \quad (6.7)$$

Dividing Equation 6.7 by 6.4, the equation of the trajectory $T_R = f(X_A)$ can be obtained:

$$\frac{dT_r}{dX_A} = \Delta T_{ad} - \frac{UAC_{A0}}{\rho V c'_P(-r_A)}(T_r - T_c) \tag{6.8}$$

The trajectory is a useful tool in the study of strategies of temperature control. For an adiabatic reaction the trajectory is linear and any cooling results in a deviation from this linear trajectory. This tool is demonstrated in the next section.

6.3 Strategies of Temperature Control

If batch reactions are occasionally at constant temperature (isothermal), most reactions are started at a lower initial temperature and the temperature is increased to its desired value, sometimes by using the heat of reaction: the reaction is performed under non-isothermal conditions. Different strategies of temperature control are technically practiced:

- adiabatic reaction: no heat exchange at all,
- polytropic reaction: with different periods, for example, adiabatic, full cooling, controlled cooling,
- isoperibolic: with constant cooling medium temperature,
- temperature controlled: the temperature of the reaction mass is directly controlled by the heat exchange system.

These strategies are analysed below.

6.4 Isothermal Reactions

6.4.1 Principles

In practice, isothermal reactions are often initiated by a catalyst or by fast addition of one of the reactants. The reaction mixture is brought to the initial reaction temperature, T_0 without catalyst or without one of the reactants. This type of process is often encountered in polymerization reactions.

6.4.2 Design of Safe Isothermal Reactors

To realize isothermal conditions, the heat release rate of the reaction must be exactly compensated by the heat exchange rate by the cooling system:

$$q_{rx} = q_{ex} \Leftrightarrow -r_A V(-\Delta H_r) = UA(T_r - T_c) \tag{6.9}$$

This requires a cooling capacity at least equal to the maximum heat release rate of the reaction. For single *n*th-order reactions, the maximum heat release rate takes place at the beginning of the reaction (Figure 6.4) and can be calculated by the rate equation, using the initial concentrations:

$$q_{rx} = kC_{A0}^n(-\Delta H_r)V \tag{6.10}$$

For single *n*th-order reactions, the required temperature of heat carrier can be calculated by [4]:

$$T_c = T_r - \frac{kC_{A0}^n(-\Delta H_r)V}{UA} \tag{6.11}$$

The required temperature of the cooling system can also be calculated from the maximum heat release rate of the reaction, for instance, measured in a calorimetric experiment, which does not require explicit knowledge of the kinetic parameters. Nevertheless, the maximum cooling capacity of the reactor is required to control the temperature at the beginning of the reaction. In the example shown in Figure 6.4, at the beginning the heat release rate of the reaction is $50\,\mathrm{W\,kg^{-1}}$, whereas 1 hour later it is only the half of the initial value. This results in a "waste" of cooling capacity, causing purely isothermal reactions relatively seldom in indus-

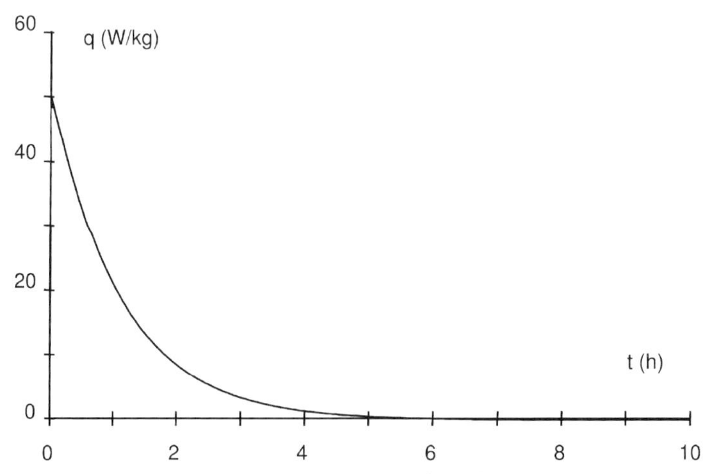

Figure 6.4 First-order reaction under isothermal conditions. Strictly isothermal conditions can only be realized at small scale and with a powerful temperature control system.

trial practice. Moreover, maintaining isothermal conditions during a reaction at industrial scale would require an extremely powerful cooling system, with a very fast response. In order to illustrate this, the temperature course of the substitution reaction example (see Table 5.1) is presented in Worked Example 6.1.

A common way of increasing the cooling capacity of a reactor, while maintaining isothermal conditions, is to perform the reaction at boiling point and use evaporative cooling. This very efficient way of performing reactions presents several advantages: the reaction temperature and hence the reaction rate is at its maximum (at atmospheric pressure). Additionally, the cooling capacity can be increased independently of the geometry of the reactor because the condenser can be designed separately and also because the heat transfer occurs by condensation leading to higher heat transfer coefficients (see Section 9.3.5). In the case of the example reaction, the boiling point is 140 °C, thus the reaction could be isothermal under reflux even at lower temperatures by applying a vacuum. In such a case, loss of vacuum must be considered during the safety analysis: the boiling point shifts towards the normal boiling point and the safety barrier by evaporation cooling may be lost.

Worked Example 6.1: Substitution Reaction in the Isothermal Batch Reactor

The reaction temperature must be chosen in such a way as to obtain an economically reasonable reaction time, that is, shorter than 10 hours, but low enough to limit the heat release rate. Thus, it is a compromise: we choose 40 °C. The reactant B, preheated at 40 °C, is added within 6 minutes. According to Equation 6.11, the substitution reaction requires a cooling medium temperature of –5 °C. In fact, the initial heat release rate at 40 °C is 20 W kg^{-1}, which represents 60 kW for the industrial reactor. Numerical simulations (Figure 6.5) under isothermal conditions show that even with a precooled jacket, the temperature of the reaction mass could only be maintained roughly isothermal. The main problem is not to realize the required cooling power, but to "track" the reaction dynamics with the heating cooling system of the industrial reactor. Thus, this type of temperature control can only be used for weak exothermal reactions. A method anticipating the dynamic behavior of the industrial reactor [5, 6] is presented in Section 9.5.

Another problem may arise in the case of loss of control: starting from 40 °C, the *MTSR* is 216 °C. This temperature is much higher than the two limits of T_{D24} = 113 °C and T_{D8} = 122 °C. This means that the secondary reaction is immediately triggered. Thus, a lack of control of the reactor temperature results in a thermal explosion. Moreover, the boiling point of 140 °C is reached during runaway, which would result in a pressure increase. Eventually the reactor will burst and there will be a flammable vapor release that may lead to a secondary room explosion. The data are summarized in the scenario presented in Figure 6.6.

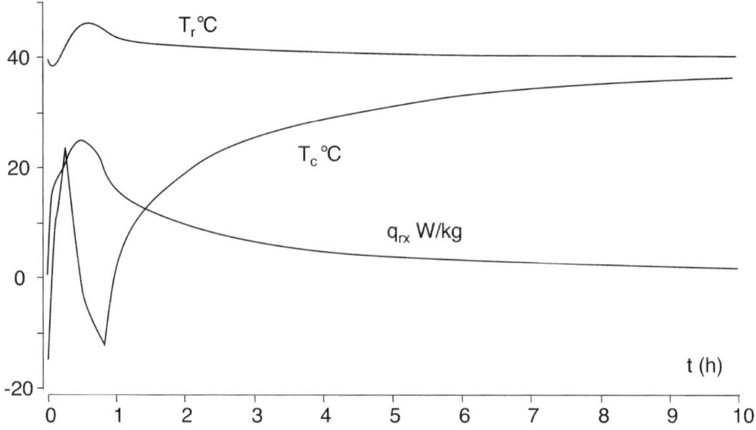

Figure 6.5 Substitution reaction under quasi isothermal conditions: The jacket temperature (T_c) must follow drastic dynamic changes in order to maintain the reaction medium temperature (T_r) only roughly constant.

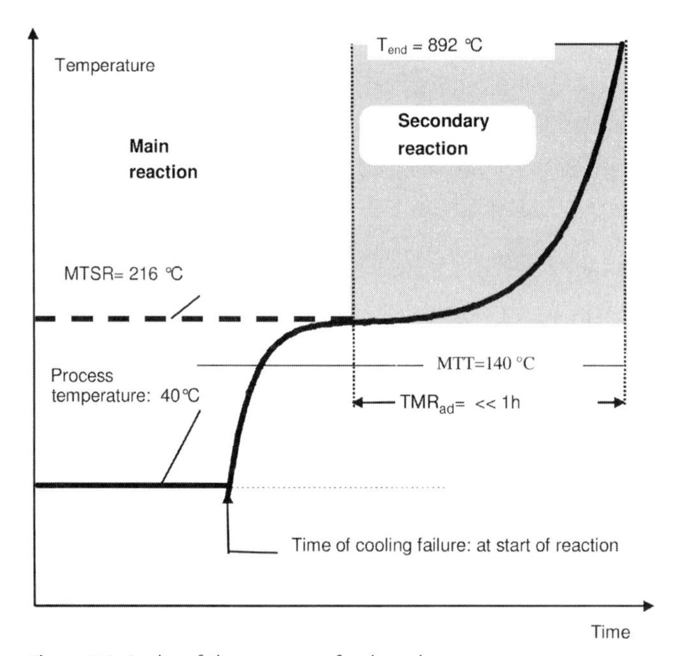

Figure 6.6 Cooling failure scenario for the substitution reaction performed in an isothermal batch reactor.

6.4.3
Safety Assessment

The thermal potential is at its maximum at the beginning of the reaction, when conversion has not yet occurred and it decreases as the reactants convert. Thus, the MTSR is given by

$$MTSR = T_0 + \Delta T_{ad} \tag{6.12}$$

Hence the knowledge of the adiabatic temperature rise is sufficient to calculate the MTSR. The data required for the safety assessment are the maximum heat release rate of the reaction at the desired temperature (q_{rx}) and the reaction energy (Q_{rx}). The first datum is needed to calculate the required cooling capacity of the industrial reactor. The second calculates the adiabatic temperature rise necessary to assess the behavior of the reactor in case of cooling failure. The calorimetric techniques used for batch reactors are presented in Section 6.9.1.

6.5
Adiabatic Reaction

6.5.1
Principles

A reaction is performed under adiabatic conditions, if there is no heat exchange with the surroundings, that is, no cooling. This means that the heat of reaction is converted into a temperature variation: for exothermal reactions into a temperature increase:

$$q_{rx} = q_{ac} \Leftrightarrow M_r \cdot q'_{rx} = M_r \cdot c'_P \cdot \frac{dT_r}{dt} \tag{6.13}$$

The final temperature can be calculated from the initial temperature T_0, from the specific enthalpy of reaction, and from the specific heat capacity or from the adiabatic temperature rise:

$$T_f = T_0 + \frac{Q'_{rx}}{c'_P} X_A = T_0 + \Delta T_{ad} \cdot X_A \tag{6.14}$$

Besides these purely static aspects, the dynamic behavior of an adiabatic batch reactor must also be considered. The adiabatic temperature course is a function of the thermal properties of the reaction mixture. The adiabatic temperature increase influences the final temperature as well as the rate of the temperature increase. For highly exothermal reactions, even for small increase in conversion, the increase in temperature is important (see Section 2.4.3).

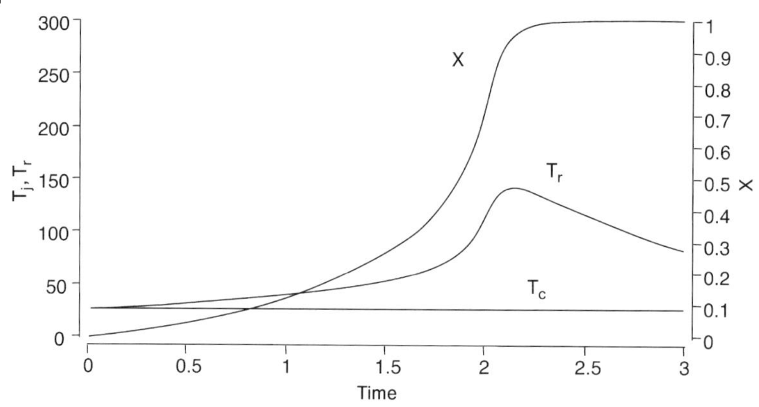

Figure 6.10 Example substitution reaction in the Isoperibolic batch reactor starting from 25 °C with a constant cooling system temperature (T_c) at 25 °C. Reactor temperature (T_r °C) and conversion as a function of time (h).

the cooling system temperature is maintained as constant, while the reaction mass temperature is allowed to change according to the heat balance.

6.7.2
Design of Isoperibolic Operation, Temperature Control

This strategy is simple to use at industrial scale: The reactants are heated to reaction temperature by the heating/cooling system, the temperature of which is maintained as constant. The reaction starts "smoothly" from this temperature, runs over a maximum, and then decreases until it again reaches the temperature of the cooling system (Figure 6.10). During the whole course of the reaction, the heat balance can be described by Equation 6.9. The great advantages of this strategy are the easy control of the initiation of the reaction and the simplicity of the required temperature control system. A drawback is that the choice of the cooling system temperature is critical, because the reactor may be sensitive towards this parameter. This is because the reaction rate is an exponential function of temperature and the cooling capacity only a linear function, thus maximum temperatures attained by the reaction mass may vary strongly, depending on the choice of the cooling medium temperature (Figure 6.11).

6.7.3
Safety Assessment

The safety assessment for isoperibolic reactions is essentially the same as for isothermal reactions. Since the initial temperature of the reaction mass is often equal to the temperature of the cooling system, the MTSR may be calculated in the same way by using Equation 6.12. The thermal stability of the reaction mass must be ensured at this temperature (MTSR).

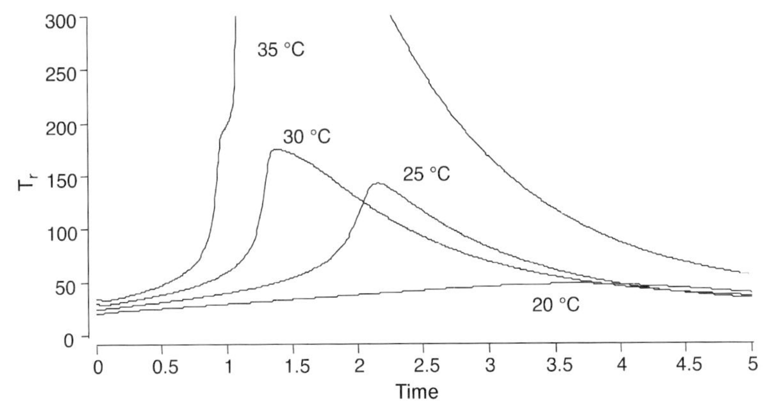

Figure 6.11 Reactor temperature (°C) as a function of time (h) for the substitution reaction example in the isoperibolic batch reactor for different cooling medium temperatures indicated as parameter.

6.8 Temperature Controlled Reaction

6.8.1 Principles

With a cascade temperature controller (see Section 9.2.4.3), it is possible to control the reaction medium temperature by adjusting the jacket temperature. This increases the temperature of the reactor linearly with time until the desired level is reached and maintains a constant temperature during the reaction time. In such a case, it is possible to start at a low temperature where the reaction rate is very low and to initiate the reaction by the increasing temperature (thermal initiation). When the reaction rate increases, the cooling intensity increases too. Depending on the requirement of the temperature increase rate and on the heat released by the reaction, the jacket temperature is adapted and may glide from cooling to heating, and reversely, during the reaction course. This is shown by the example substitution reaction. The mixture is first maintained at 25 °C for 1 hour and then heated to 100 °C at 10 °C h^{-1} (Figure 6.12). This procedure gives the heat release rate, and consequently the temperature, a smooth profile.

6.8.2 Design of Temperature Controlled Reaction

The progressively increasing temperature brings the reaction to a certain conversion before it accelerates due to the increasing temperature. Thus, a subtle balance between decreasing reaction rate due to conversion and increasing reaction rate due to temperature may be realized. It becomes obvious that with this strategy,

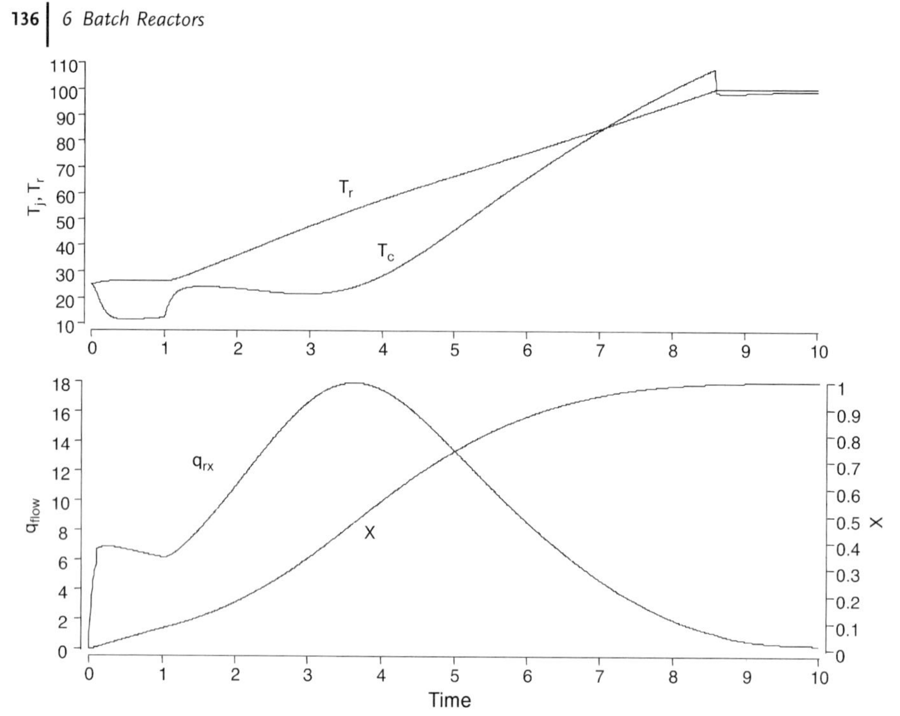

Figure 6.12 Temperature controlled reaction, with the example substitution reaction starting from 25 °C, then heating at 10 °C h^{-1} to 100 °C. Upper plot temperatures (T_r, T_c), lower plot heat release rate in W kg^{-1} and conversion versus time (h).

the choices of the initial temperature and of the heating rate are very important. The "scale down" approach is useful for the development of such processes [5, 6]. It allows predicting the behavior of large-scale reactors from small-scale experiments. In order to illustrate the effect of the heating rate, the substitution reaction example was simulated in a temperature controlled reactor heated at different heating rates (Figure 6.13).

This type of process is much less sensitive to process parameter than the isoperibolic or polytropic reactors. By increasing the heating rate from 10 to 20 °C h^{-1}, the temperature departs from its set point by some degrees. At 30 °C h^{-1} the set temperature is significantly surpassed and at 40 °C h^{-1} there is a significant overshoot of the maximum temperature of 100 °C. The disadvantage of this policy is that the initiation of the reaction is difficult to detect. Nevertheless, it may be detected by observing the temperature difference between jacket and reaction medium.

6.8.3 Safety Assessment

The determination of the temperature that may be reached in the case of a cooling failure (T_{cf}) is an important safety parameter. The MTSR is the maximum of T_{cf}.

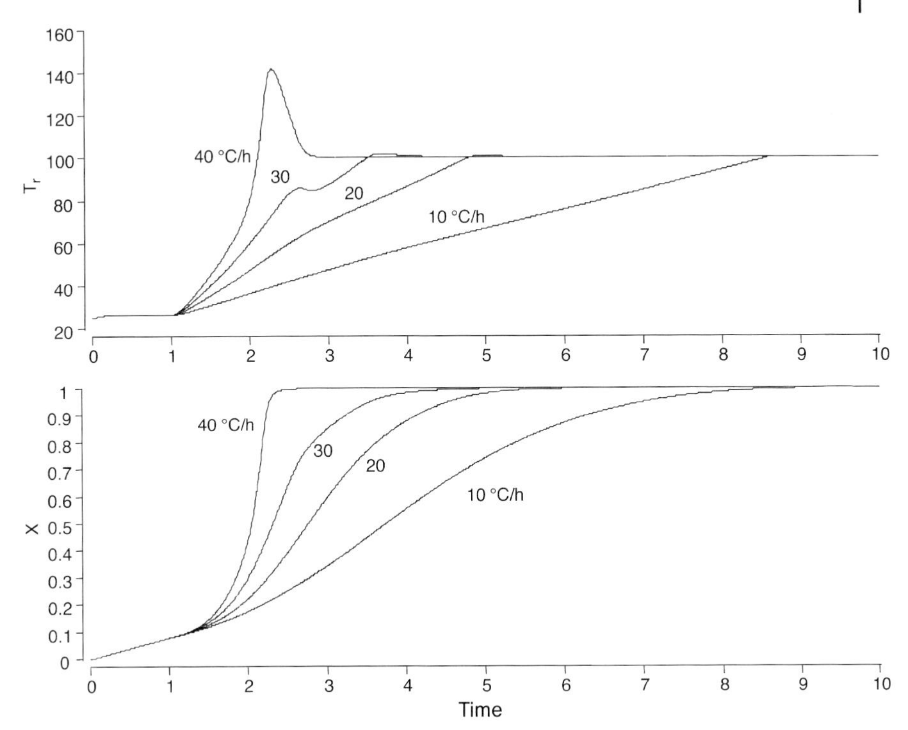

Figure 6.13 Temperature controlled reaction with the substitution reaction example at different heating rates between 10 and 40 °C h^{-1}. Upper graph, temperature of reactor (°C), lower graph conversion vs. time (h).

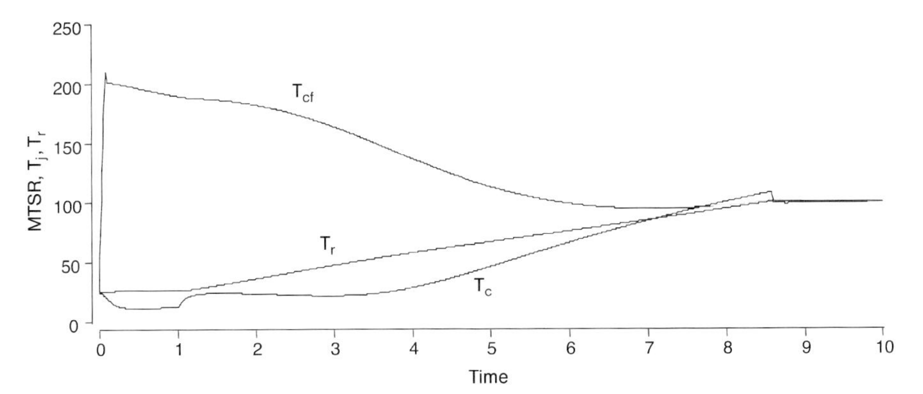

Figure 6.14 Reactor and jacket temperatures (°C) in the example substitution reaction together with the temperature reached in case of cooling failure (T_{cf}) as a function of time (h).

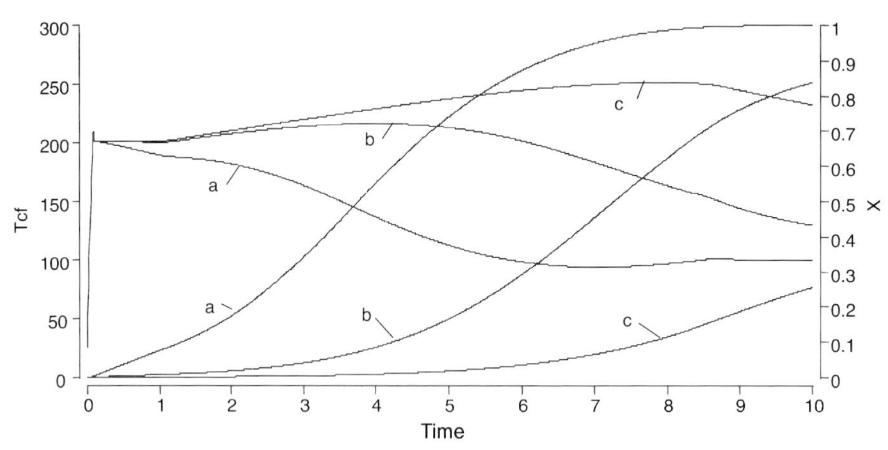

Figure 6.17 T_{cf} (left axis in °C) and conversion (right axis) as a function of time (in hours) for three reactions. The first (a) corresponds to the substitution example reaction. The two other curves (b, c) were obtained under the same conditions, but with 10 (b) and 100 (c) times smaller pre-exponential factors.

Worked Example 6.2: Substitution Reaction Example in a Batch Reactor

Since the process may be carried out under the same temperature conditions as those in industry, a complete set of data may be evaluated from the experiment. The overall heat of reaction ($300\,kJ\,kg^{-1}$) is obtained through integration of the heat release rate over time. The maximum heat release rate of $18\,W\,kg^{-1}$ is reached after 3.6 hours at a temperature of 53 °C. The specific heat capacity ($1.7\,kJ\,kg^{-1}\,K^{-1}$) is calculated from the steps in heat released rate at the beginning and end of the temperature program. As an example, we consider a cooling failure at 3 hours. The reactor temperature is 47 °C and the thermal conversion is 0.35. Thus, the temperature after cooling failure can be calculated as

$$T_{cf} = T_P + (1 - X_{th})\Delta T_{ad} = 47 + (1 - 0.35)\frac{300}{1.7} = 162°C \tag{6.22}$$

This calculation can be performed for any time during the process. The resulting curve is represented in the Figure 6.16, showing that in this case, the MTSR of 200 °C is reached at the beginning of the reaction.

6.9.2
Rules for Safe Operation of Batch Reactors

The thermal behavior of a batch reactor strongly depends on the reaction energy. The adiabatic temperature rise depends on the reactants concentration. Therefore, the charge (i.e. the amount of reactants charged) must be strictly respected. Also the quality of the reactants must be strictly controlled, since impurities may catalyse secondary reactions leading to an increase of heat release with possibly dramatic consequences.

The temperature course must be strictly controlled, thus the choice of key temperatures as initial and final temperatures, and jacket temperature and heating rate if applicable, are essential. Any overheating must be avoided during the preheating phase to reach the starting temperature. The temperature increase must not be too fast, in order to avoid mechanical stress of the reactor construction material. The reactor must be designed for the final pressure reached by the reaction mass, especially if the boiling point of volatiles is reached during the reaction course. If gaseous products are formed, either the reactor must be designed to resist the total pressure (closed system) or the venting/scrubbing system must be designed to cope with the maximum gas release rate.

The reaction mass must be thermally stable in the temperature range of the process, that is, no secondary exothermal reaction must take place in the temperature range where the reactor is operated. Moreover, the reaction mass should be stable in the temperature range between T_r and $MTSR$. The MTSR can be calculated from Equations 5.24 and 5.26, therefore depending on the mode of initiation.

For catalytically initiated reactions, the initial temperature that must be taken into account is the reactor temperature at the instant the catalyst is added. The accumulation is 1 at this instant.

For thermally initiated reactions, T_{cf} is a function of time. Its course can be determined experimentally by measuring the thermal conversion as a function of time, while the reaction proceeds under normal operating conditions. These experiments can be carried out with DSC or, preferably with a Reaction Calorimeter. The T_{cf} curves can be obtained in the evaluation of the thermogram by using Equation 5.24. Its maximum (MTSR) can be searched from the T_{cf} curve.

If the thermal stability is not sufficient at the MTSR, emergency measures must be taken to avoid a runaway (see Chapter 10).

Thus, the following *Golden Rules for Safe Batch Reaction* can be formulated:

- Charge: Guarantee the amounts and quality of reactants.
- Temperature control: strictly maintain the defined heating rates and avoid unnecessary high heating system temperatures.
- Provide emergency measures.

Worked Example 6.3: Fast Reaction in a Batch Reactor

A reaction $A \xrightarrow{k} P$ is to be performed in a batch reactor. The reaction follows first-order kinetics and at 50 °C, the conversion reaches 99% in 60 seconds (the rate constant is $k = 0.077\,s^{-1}$. The charge will be $5\,m^3$ in a reactor with a heat exchange area of $15\,m^2$ and an overall heat transfer coefficient of $500\,W\,m^{-2}\,K^{-1}$. The maximum temperature difference with the cooling system is 50 K.

Data:

$C_A\ (t = 0) = C_{A0} = 1000\,mol\,m^{-3}$ $\quad \rho = 900\,kg\,m^{-3}$

$c'_P = 2000\,J\,kg^{-1}\,K^{-1}$ $\quad -\Delta H_r = 200\,kJ\,mol^{-1}$

Questions:

1. Do you think adiabatic reaction is possible? What could the limiting factors be?
2. Is isothermal reaction possible? What would happen in case of cooling failure?
3. Suggest process improvements.
4. What would your answers be for a 100 times slower reaction?

Solution:

1. Adiabatic reaction

Under adiabatic conditions, no heat exchange occurs. Therefore, the heat produced by the reaction is transformed into a temperature rise. For a conversion of 100%, the adiabatic temperature rise is

$$\Delta T_{ad} = \frac{C_{A0} \cdot (-\Delta H_r)}{\rho \cdot c'_P} = \frac{1000\,mol\,m^{-3} \times 200\,kJ\,mol^{-1}}{900\,kg\,m^{-3} \times 2\,kJ\,kg^{-1}\,K^{-1}} = 111\,K$$

With an initial temperature of 50 °C, the temperature will reach 161 °C at the end of the reaction. If no secondary reaction is triggered at this temperature, and if no pressure rise occurs, this reaction is theoretically possible. The problem is due to mechanical stress in the reactor wall due to the fast temperature change from 50 to 161 °C within 1 minute.

2. Isothermal reaction: An overall heat balance gives

$$\text{Heat production: } Q_{rx} = V \cdot C_{A0}(-\Delta H_r) = 5\,m^3 \times 1000\,mol \cdot m^{-3} \times 200\,kJ\,mol^{-1} = 10^6\,kJ$$

$$\text{Heat removal: } q_{ex} = U \cdot A \cdot \Delta T = 500\,W\,m^{-2}\,K^{-1} \times 15\,m^2 \times 50\,K = 375\,kW$$

Since the conversion is 99% in 1 minute, the heat that can be removed, is

$$Q_{ex} = q_{ex} \cdot t_r = 375\,kW \times 60\,s = 22500\,kJ$$

Hence only about 2% of the reaction energy can be removed by the heat exchange system. Thus, the reactor behaves quasi adiabatic. In other words, the reaction is so fast that the heat exchange system is unable to remove any significant heat.

3. Cooling failure
Since the heat exchange is so small compared to the heat production, the cooling failure plays no significant role.

4. Slow reaction
For a 100 times slower reaction ($k = 0.00077\,s^{-1}$), the adiabatic temperature rise will be the same. But taking the acceleration of the reaction with temperature into account, the reaction time may be acceptable.

The initial heat release rate is

$$q_0 = k \cdot C_{A0} \cdot (-\Delta H_r) = 0.00077\,s^{-1} \times \frac{1000\,mol\,m^{-3}}{900\,kg\,m^{-3}} \times 200000\,J\,mol^{-1} = 171\,W\,kg^{-1}$$

The reaction time can be estimated by

$$TMR_{ad} = \frac{c'_P \cdot R \cdot T^2}{q'_0 \cdot E} = \frac{2000 \times 8.314 \times 323^2}{171 \times 100'000} \cong 100\,s$$

For isothermal conditions, the overall balance gives a heat removal of $2.25 \cdot 10^6\,kJ$, which is more than twice the heat production. Nevertheless, this consideration is erroneous since for a first-order reaction, where the maximum heat release rate takes place at the beginning and decreases exponentially with time, the differential form of the heat balance must be used:

$$\begin{aligned} q_0 &= k \cdot C_{A0} \cdot V \cdot (-\Delta H_r) = 0.00077\,s^{-1} \times 1000\,mol\,m^{-3} \times 5\,m^3 \times 200\,kJ\,mol^{-1} \\ &= 700\,kW \end{aligned}$$

$$q_{ex} = 375\,kW$$

In order to ensure smooth control of the reaction, it must be started at a lower temperature, perhaps under adiabatic conditions, followed by a period with maximal cooling when the desired temperature level is reached. Therefore, batch reactions are often performed in the so-called polytropic mode of operation.

In conclusion, a fast and exothermal reaction such as this cannot be performed in a batch reactor. This reaction will also be studied in other reactor types in the following chapters.

6.10 Exercises

▶ Exercise 6.1

For the initiation of a Grignard reaction, magnesium is charged in the solvent Tetrahydrofurane (THF). A small amount, that is, 2% of the total of bromide reactant is charged. One considers performing the initiation under adiabatic conditions in order to observe the temperature increase, which confirms the success of the initiation. Therefore, the cooling system is stopped before the addition of the initiation reactant. This initiation is performed at 30 °C in a reaction calorimeter, in order to measure the thermal data of this operation. The energy of the reaction is found to be $70\,kJ\,kg^{-1}$ with a maximum heat release rate of $260\,W\,kg^{-1}$. The specific heat capacity of the reaction mixture is $1.9\,kJ\,kg^{-1}\,K^{-1}$. The boiling point of THF is 66 °C. Secondary decomposition reactions become significant at a temperature of 150 °C, that is, T_{D24} = 150 °C. Evaluate the thermal risk of this operation.

Answer: MTSR = 67 °C, just around the boiling point of the solvent. Since the reaction mass is relatively small compared to the volume of the reactor, the reactor cannot be considered to be adiabatic. Thus, the boiling point will not be reached at plant scale. The severity is negligible, as triggering the secondary reaction is almost impossible. This operation can be considered thermally safe, as far as the amount of halogenated reactant used for the initiation is strictly limited to the intended 2%.

▶ Exercise 6.2

A substituted phenol is prepared through hydrolysis of the corresponding chloroaromatic compound (Ar–Cl → Ar–OH) with caustic soda at 50% concentration. The reaction is to be performed in a batch reactor. The total charge is 7.5 kmol of choro-aromatic compound and 17.5 kmol of caustic soda, representing a total mass of 5800 kg. In a first stage, the reactor is heated to 80 °C, then the temperature is stabilized at 110–115 °C and the pressure reaches 2 bar (abs).

Questions:

1. The specific heat of reaction is $125\,kJ\,mol^{-1}$ (aromatic compound) and the specific heat capacity of the reaction mixture is $2.8\,kJ\,kg^{-1}\,K^{-1}$. What maximum temperature (MTSR) could the reaction mass reach if the heating cooling system fails to stabilize at 125 °C?
2. What would the pressure then be? Hint: the reaction mass is aqueous, thus Regnault's approximation can be used: $P(\text{bar}) = \left(\frac{T(°C)}{100}\right)^4$ (absolute pressure).
3. In case of cooling failure, could the temperature be stabilized by controlled depressurization? (Latent heat of evaporation of water $\Delta H'_v = 2200\,kJ\,kg^{-1}$).

4. What recommendations would you make to the process manager? Are there other potential problems to be considered?

▶ Exercise 6.3

A second-order dimerization reaction is to be performed in a batch reactor. The initial temperature is 50 °C, and the desired reaction temperature 100 °C, whereas the maximum allowed temperature is 120 °C.

Data:
$\Delta H_r = -100\,\text{kJ mol}^{-1}$
$c'_P = 2\,\text{kJ kg}^{-1}\,\text{K}^{-1}$
$C_0 = 4\,\text{mol kg}^{-1}$; $E = 100\,\text{kJ mol}^{-1}$

Questions:
1. Do you think the temperature control of the reactor will be easy?
2. What other temperature control strategy could you propose?

▶ Exercise 6.4

A process for the synthesis of a secondary amine has to be transferred to a new plant. In the former plant, the reaction was performed in a 25 m³ reactor with a total charge of 25 000 kg. In the new plant, the total charge will be 6000 kg in a 6.3 m³ reactor. The reaction is performed as batch reaction, the reactants being mixed at room temperature and heated to the process temperature of 95 °C. The concentration of the default compound (Ar–Br) of 0.4 mol kg⁻¹ is left unchanged. The solvent is water: 4600 kg in the new plant.

Reaction:

$$\text{Ar}-\text{Br}+\Phi-\text{NH}_2+\text{NaHCO}_3 \rightarrow \text{Ar}-\text{NH}-\Phi+\text{NaBr}+\text{CO}_2+\text{H}_2\text{O}$$

Technical information: The gas relief was a 300 mm diameter pipe in the former plant (25 m³) and will be a 150 mm pipe in the new plant (6.3 m³).

Questions:
There are no thermal data available for this process, nevertheless the risks linked with the performance of the process should be evaluated:

- Build a cooling failure scenario: As worst case the temperature would increase to boiling point and the solvent evaporate. What would the required energy be to evaporate the water from the reaction mass? (Latent heat of evaporation of water $\Delta H'_v = 2200\,\text{kJ kg}^{-1}$).
- Referring to Tables 2.1 and 11.1, do you think the process transfer could be made without additional data measurement?

References

1 Vincent, G.C. (1971) Rupture of a nitroaniline reactor. *Loss Prevention*, **5**, AICHE, New York.

2 Levenspiel, O. (1972) *Chemical Reaction Engineering*, Wiley-VCH, Weinheim.

3 Westerterp, K.R., Swaij, W.P.M.v. and Beenackers, A.A.C.M. (1984) *Chemical Reactor Design and Operation*, Wiley-VCH, Weinheim.

4 Baerns, M., Hofmann, H. and Renken, A. (1987) *Chemische Reaktionstechnik*, Georg Thieme, Stuttgart.

5 Zufferey, B., Stressel, F. and Groth, U. (2007) *Method for simulating a process plant at laboratory scale*, European Patent Office, Pat. Nr. EP 1764662A1.

6 Zufferey, B. and Stoessel, F. (2007) Safe scale up of chemical reactors using the scale down approach, in *12th International Symposium Loss Prevention and Safety Promotion in the Process Industries*, IchemE, Edinburgh.

7 Hugo, P., Konczalla, M. and Mauser, H. (1980) Näherungslösungen für die Auslegung exothermer Batch-Prozesse mit indirekter Kühlung. *Chemie Ingenieur Technik*, **52** (9), 761.

7
Semi-batch Reactors

Case History

The semi-batch reaction that caused a severe incident was performed for many years without any problems. On the incident day, the first reactant (a nitro compound) was charged as usual. The stirrer was then stopped to take a sample for analysis before heating up the reactor to process temperature and starting to feed the second reactant. In the incident batch, the operator forgot to restart the stirrer. After shift change, a second operator started the feed of the second reactant, also omitting to verify that the stirrer was switched on. At the end of the addition, a second sample was taken for quality control and this showed a strange aspect, which led the operator to ask the shift supervisor for advice. Since it night-time, it was decided to cool the reactor and wait for instructions from the chemists in the morning. On returning to the reactor to cool it, the operator noticed that the stirrer was not working so he switched it on to help the cooling process. However, he was not conscious that in doing so he brought both reactants, which were separately layered in the reactor, into a sudden reaction. The reaction course had no chance of being controlled and led to a steep temperature and pressure increase. Even though the pressure relief system was activated, the relief line ruptured and instead the reaction mass was transferred to a catch tank, over ten tons of reaction mass were discharged directly into the atmosphere and caused a huge spillage of over the residential area located nearby. Nobody was injured, but damage was huge and caused a loss of image, also resulting in severe financial consequences.

What the enquiry revealed:

- It is common practice to stop the stirrer for sampling. Nevertheless, it should be restarted immediately afterwards. At least during the heating phase, the operator should have noticed the stirrer was inactive.
- The reaction can easily be mastered by the cooling system of the reactor, provided the stirrer works and the addition of reactants is at the nominal rate.

Thermal Safety of Chemical Processes: Risk Assessment and Process Design. Francis Stoessel

ISBN: 978-3-527-31712-7

- The initially charged reactant had a higher density than the reactant that was fed in later in the process. Consequently, the fed reactant layered above the nitro compound and practically no reaction took place for as long as the stirrer was switched off.
- As the stirrer was switched on, the reaction was suddenly started, but took place as a batch reaction, since all the reactants were charged.
- Under these conditions (batch reaction), the reaction temperature increases rapidly and other exothermal reactions are triggered that increase the thermal potential of the reaction.
- The pressure relief system was not designed for two-phase flow. Thus, the resulting mass flow in the relief line was too high and could not resist such a high mechanical load. As a consequence, the relief line ruptured, causing the spillage.

Lessons drawn

- Semi-batch operation spreads the heat release over time. Moreover, it provides the opportunity to stop the reaction in case of malfunction. This, of course, supposes that the feed is stopped or at least not started while a malfunction exists (see Section 7.8).
- Pressure relief systems for reactors should be designed for two-phase flow that will probably occur if the reactor is filled to a high level (Section 10.5.1).

7.1 Introduction

This chapter first presents the general aspects of reaction engineering of semi-batch reactors. The mass and heat balances are analysed to show that in addition to temperature control, feed rate is a central point in the safety of batch reactors. Thus, a separate section is devoted to the accumulation of reactants. Then the different strategies of temperature control and their consequences on reactor safety are described in the following sections. For each strategy, the design criteria and the safety assessment are described. A separate section is devoted to the different feed strategies that may be implemented. A further section considers the optimization of semi-batch reactors with respect to safety and economy. This chapter is closed with the presentation of an advanced feed strategy, maximizing the productivity under safety constraints.

7.2 Principles of Semi-batch Reaction

7.2.1 Definition of Semi-batch Operation

As with the batch reactor, the semi-batch reactor operates discontinuously. The difference with true batch operation is that for the semi-batch reactor, at least one of the reactants is added as the reaction proceeds (Figure 7.1). Consequently, the material balance as well as the heat balance will be affected by the progressive addition of one of the reactants. Also, as with the batch reactor, there is no steady state. There are essentially two advantages in using a semi-batch reactor instead of a batch reactor:

- For exothermal reactions, the addition controls the heat production rate and therefore adjusts the reaction rate to the cooling capacity of the reactor.
- For multiple reactions, the progressive addition of one of the reactants maintains its concentration at a low level and therefore reduces the rate of a secondary reaction compared to the main reaction.

These two factors mean the semi-batch reactor is a commonly-used reactor type in the fine chemicals and pharmaceutical industries. It retains the advantages of flexibility and versatility of the batch reactor and compensates its weaknesses in the reaction course control by the addition of, at least, one of the reactants.

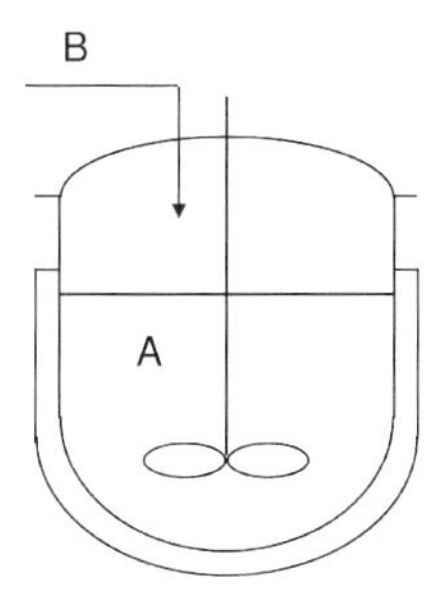

Figure 7.1 Semi-batch reactor: compound A is initially charged and B is fed during the reaction, providing additional control of the reaction course.

7.2.2 Material Balance

For an irreversible bimolecular second-order reaction as the example reaction, the rate equation is

$$A + B \rightarrow P \quad \text{with} \quad -r_A = k \cdot C_A \cdot C_B \tag{7.1}$$

By convention in this chapter, the reactant A will be initially charged into the reactor, whereas B will be added with a constant molar feed rate F_B during the feed time t_{fd}. The variation of the concentration of the different species, which are present in the reactor, results from both the reaction and the variation of the reaction mixture volume due to the feed. At constant feed rate, the volume varies as a linear function of time

$$V = V_0 + \dot{v}_0 t = V_0(1 + \varepsilon t) \tag{7.2}$$

where ε is the volume expansion factor defined as

$$\varepsilon = \frac{V_f - V_0}{V_0} \tag{7.3}$$

The mole balance on A can be written as

$$\frac{-dN_A}{dt} = -r_A V = k\frac{N_A N_B}{V} = k\frac{N_A N_B}{V_0 + \dot{v}_0 t} \tag{7.4}$$

The mole balance on B is

$$\frac{dN_B}{dt} = -r_A V + F_B \tag{7.5}$$

Equations 7.4 and 7.5 form a system of differential equations for which no analytical solution is known. Thus, the description of the behavior of the semi-batch reactor with time requires the use of numerical methods for the integration of the differential equations. Usually, it is convenient to use parameters which are more process-related to describe the material balance. One is the stoichiometric ratio between the two reactants A and B:

$$M = \frac{N_{B \cdot tot}}{N_{A0}} \tag{7.6}$$

The reaction rate can also be expressed as a function of the conversion:

$$-r_A = C_{A0}\frac{dX_A}{dt} = kC_{A0}^2(1 - X_A)(M - X_A) \tag{7.7}$$

The molecular flow rate of B (F_B) can also be expressed as a function of the stoichiometric ratio (M) and the feed time, t_{fd}:

$$F_B = \frac{N_{A0} M}{t_{fd}} \tag{7.8}$$

The initial reaction rate is often characterized by a dimensionless number, the Damköhler number [1, 2]:

$$Da_0 = \nu_A \cdot k \cdot C_{A0} \cdot M \cdot t_{fd} \tag{7.9}$$

7.2.3 Heat Balance of Semi-batch Reactors

The heat balance was explained in a general way in Section 2.4. Here we specify the most important terms for a single bimolecular second-order reaction.

7.2.3.1 Heat Production

The heat production corresponds to the heat release rate by the reaction, as expressed in Equation 2.17. Under isothermal conditions, by combining with Equation 7.7, the heat release rate becomes

$$q_{rx} = k\frac{N_{A0}^2}{V(t)}(1-X_A)(M-X_A)(-\Delta H_r) \tag{7.10}$$

This expression enhances the fact that the heat release rate is a function of the conversion, but also of the varying volume. The dilution of the reaction mass by the feed will contribute to slow the reaction. Usually with a constant feed rate and as long as the volume varies in the cylindrical part of the reactor, $V(t)$ is a linear function of time. In addition to the pure heat of reaction, the mixing effect of the feed with the reaction mass can be accompanied by thermal effects, for example, dilution enthalpy or mixing enthalpy.

7.2.3.2 Thermal Effect of the Feed

As explained in Section 2.4.1.5, if the feed is not at the same temperature as the reaction mixture, it will also produce a thermal effect proportional to the temperature difference between feed (T_{fd}) and reaction mass (T_r), to its specific heat capacity $c'_{P,fd}$ and to the mass flow rate $\dot{m}_{fd}$

$$q_{fd} = \dot{m}_{fd} \cdot c'_{P_{fd}} \cdot (T_{fd} - T_r) \tag{7.11}$$

If the volume of the feed is important compared to the initial charge, that is, great volume expansion factor (ε), the thermal effect of the feed may become comparable, in absolute value, to the heat of reaction.

7.2.3.3 Heat Removal

The heat exchanged with a heat carrier across the reactor wall by forced convection is expressed in the classical way by

$$q_{ex} = UA_{(t)}(T_c - T_r) \tag{7.12}$$

The heat exchange area (A) may vary with time due to the volume increase by the feed. This variation is determined by the geometry of the reactor, especially by its height covered by the heat exchange system (jacket, internal coils, or welded half-coils). In case there is a significant change in the physical chemical properties of the reaction mixture, the overall heat exchange coefficient (U) will also be a function of time.

7.2.3.4 **Heat Accumulation**

The overall heat balance of a semi-batch reactor can be written by using the three terms mentioned above. If the heat exchange does not compensate exactly, the other terms (heat production, effect of the feed, temperature) will vary as

$$\frac{dT_r}{dt} = \frac{q_{rx} + q_{fd} + q_{ex}}{M_r \cdot c'_p} \tag{7.13}$$

The heat balance of an isothermal semi-batch reaction is represented graphically in Figure 7.2. The maximum heat exchange rate ($q_{ex,\max}$) calculated for a constant temperature of the heat carrier is also represented in the diagram. It increases linearly with time until the upper limit of the jacket is reached. In this example, the upper limit of the jacket is not reached during the feed time of four hours.

The sum ($q_{rx} + q_{fd}$) represents the heat that has to be removed from the reaction mass by the heat exchange system to maintain its temperature constant. At the beginning of the addition of reactant B, during a short period of time, the cooling effect of the feed dominates over the reaction ($q_{ex} < 0$). At the end of the addition, as the cooling effect of the feed stops, the heat to be removed by the heat exchange system suddenly increases.

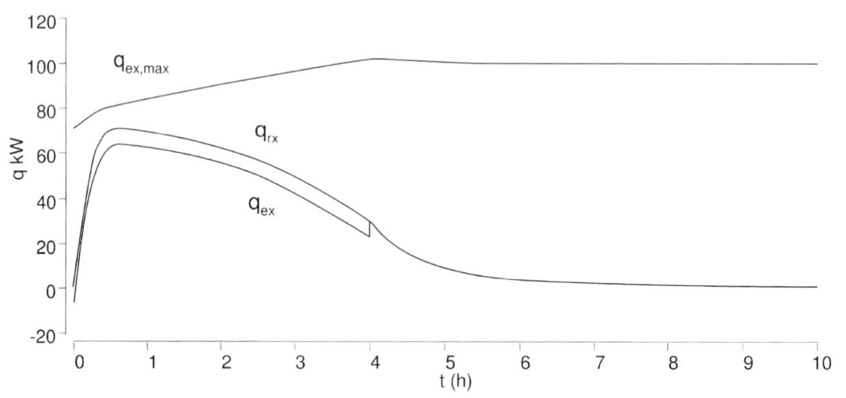

Figure 7.2 The different terms of the heat balance of an isothermal semi-batch reactor (in kW) as a function of time. The maximal cooling capacity of the reactor ($q_{ex,max}$) obtained with cold water at 5 °C is also represented. The difference between both curves q_{rx} and q_{ex} represents the cooling effect by the feed. Its disappearance at the end of the feed at 4 hours is visible.

7.3 Reactant Accumulation in Semi-batch Reactors

The advantages of a semi-batch reaction, that is, a better selectivity in the case of multiple reactions or a better control of the reaction course in the case of exothermal reactions, are obtained if the reaction rate is controlled by the progressive addition of one or more reactants. Indeed, this objective can only be achieved if the added reactant is immediately converted and does not accumulate in the reactor [3]. Often a reaction is said to be feed controlled only because a reactant is fed. This is not always the case, since the feed rate must be adapted to the reaction rate, and the concentration of the added compound (B) is maintained at a low level during the reaction.

This non-converted reactant B is called the reactant accumulation. It results from the mass balance, that is, the feed rate as input and the reaction rate as consumption. In other words, a low accumulation is obtained when the feed rate of B is slower than the reaction rate. Since, as defined in Equation 7.4, the reaction rate depends on both concentrations C_A and C_B, this means that both reactants must be present in the reaction mixture in a sufficiently high concentration. For fast reactions, such as those with a high rate constant, even for low concentrations of the reactant B, the reaction will be fast enough to avoid the accumulation of unconverted B in the reactor. For slow reactions, a significant concentration of B is required to achieve an economic reaction rate. Thus, two cases have to be considered: fast reactions and slow reactions.

7.3.1 Fast Reactions

For fast reactions, since the added reactant is immediately converted to the product, no significant accumulation of reactant B occurs and the rate of reaction is limited by the rate of addition of B:

$$-r_A = kC_A C_B = \frac{F_B}{V} \tag{7.14}$$

This can be shown with the example addition reaction (Figure 7.3). The concentration of B increases only after the stoichiometric point has been reached. In this example, the stoichiometric excess is not required since the reaction is completed before the end of feed.

In fast reactions, in case the reaction takes an abnormal course, the control can immediately be recovered by adjusting the feed rate. In extreme situations, the reaction can be stopped quasi instantaneously, by shutting the feed (Figure 7.4). This provides an excellent safety measure since it gives additional control by technical means. This advantage can be used to maintain control of the temperature in very exothermal reactions or to adapt a gas release to the technical characteristics

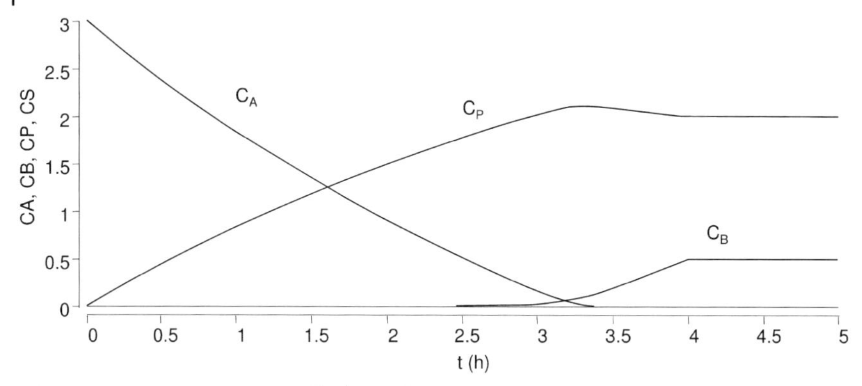

Figure 7.3 Concentrations in mol kg^{-1} as a function of time in a semi-batch reactor with the fast addition reaction. Compound B is fed at constant rate within 4 hours in a stoichiometric excess of 25%. B is fed in stoichiometric amounts.

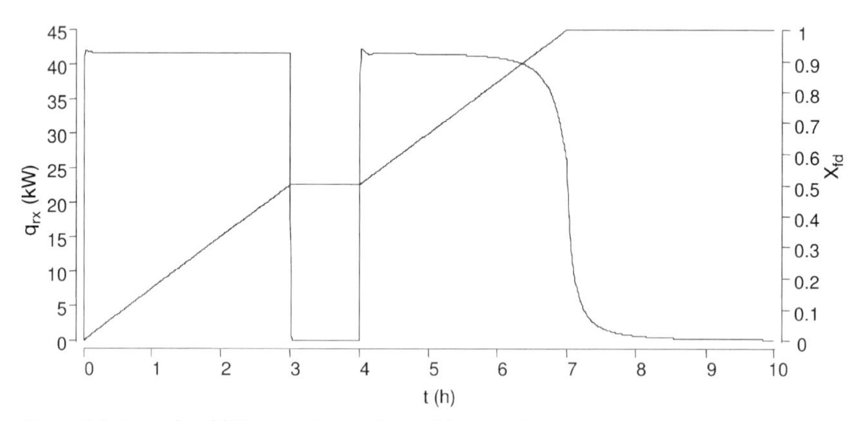

Figure 7.4 Example addition reaction performed in a semi-batch reactor under isothermal conditions at 80 °C with a feed time of 6 hours. The feed was interrupted between 3 and 4 hours. The heat release rate immediately decreases to zero and recovers its initial value after resuming the feed.

of the equipment. If the gas release stems from the same reaction that also causes the heat release, the heat release rate q_r or the gas release rate $\dot{v}_{gas}$ can directly be calculated from the feed time t_{fd}

$$q_r = \frac{Q_r}{t_{fd}} \tag{7.15}$$

$$\dot{v}_g = \frac{V_g}{t_{fd}} \tag{7.16}$$

Obviously, these equations are only valid for a fast reaction rate compared to the feed rate. In fact, the reaction rate is implicitly taken to equal the feed rate.

Worked Example 7.1: Fast Reaction in a SBR

This example is continued from Worked Example 6.3.

A reaction $A \xrightarrow{k} P$ is to be performed in a semi-batch reactor. The reaction follows first-order kinetics, and carried out as a batch reaction at 50 °C, the conversion reaching 99% in 60 seconds. The final volume is $5\,m^3$ in a reactor with a heat exchange area of $15\,m^2$ (assumed to remain constant) and an overall heat transfer coefficient of $500\,W\,m^{-2}\,K^{-1}$. The compound A is fed as a concentrate solution to a reactor containing an inert solvent. The maximum temperature difference with the cooling system is 50 K.

Data:

$C_A\ (t=0) = C_{A0} = 1000\,mol\,m^{-3}$ $\quad\rho = 900\,kg\,m^{-3}$

$c'_P = 2000\,J\,kg^{-1}\,K^{-1}$ $\quad -\Delta H_r = 200\,kJ\,mol^{-1}$

Questions:

1. Isothermal reaction: Which feed time is required to assure isothermal conditions at 50 °C?
2. Slow reaction: What happens with a 100 times slower reaction?
3. Cooling failure: What happens in case of cooling failure for both fast and slow reactions?

Solution:

1. Isothermal reaction:

The heat balance is the same as for the batch reactor.

$$\text{Heat production: } Q_{rx} = V \cdot C_{A0}(-\Delta H_r) = 5\,m^3 \times 1000\,mol \cdot m^{-3} \times 200\,kJ\,mol^{-1} = 10^6\,kJ$$

$$\text{Heat removal: } q_{ex} = U \cdot A \cdot \Delta T = 500\,W\,m^{-2}\,K^{-1} \times 15\,m^2 \times 50\,K = 375\,kW$$

For a fast reaction, the feed rate can be adapted in such a way as to give the heat exchange system enough time to remove the heat of reaction:

$$q_{rx} = \frac{V \cdot C_{A0} \cdot (-\Delta H_r)}{t_{fd}} = \frac{Q_{rx}}{t_{fd}} = q_{ex}$$

Thus:

$$t_{fd} = \frac{Q_{rx}}{q_{ex}} = \frac{10^6\,kJ}{375\,kW} = 2667\,s \cong 45\,min$$

2. Slow reaction:
From the point of view of the heat balance, the feed time could be the same as above. Nevertheless, for a slow reaction, a slower addition is required in order to limit accumulation.

3. Cooling failure:
For the fast reaction, if the feed is immediately stopped after a cooling failure has occurred; the reactor reaches a safe state. Thus, the SBR is a practicable solution for this fast exothermal solution.

For the 100 times slower reaction, the behavior will result from accumulation of non-converted reactants. The temperature increase could trigger secondary reactions. This example will be continued in the following chapter.

7.3.2
Slow Reactions

With slower reactions, the compound B fed to the reactor does not immediately react away, leading to accumulation. In fact, the concentration of reactant B must first increase to a certain level before the rate of reaction becomes appreciable. Thus, during the first few seconds after the feed has been started, the concentration of B increases rapidly, whereas practically no conversion takes place. In Figure 7.5 the concentration of B increases to about 0.2 mol kg^{-1} during this period. Then C_B increases only slowly, the reaction rate is quasi constant, and the situation looks like a quasi steady state until C_A becomes lower, and consequently the reaction slows down. This produces a further increase of C_B to 0.5 mol kg^{-1}. Both lines C_A and C_B cross when a stoichiometric amount of B has been introduced into the reactor.

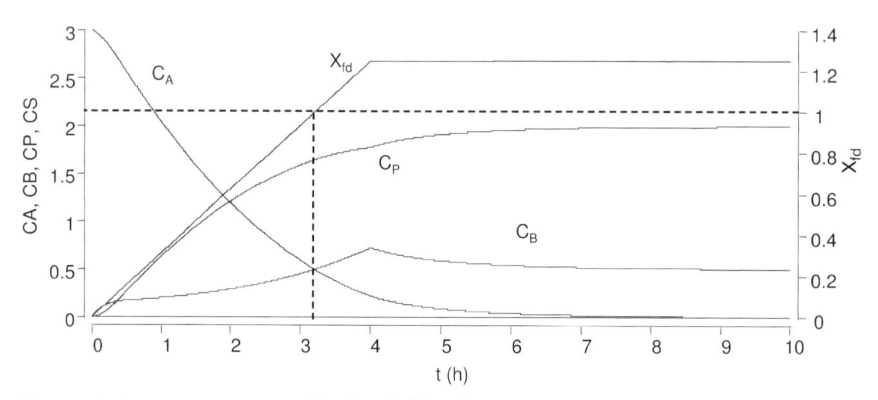

Figure 7.5 Concentrations (in mol kg^{-1} on left-hand scale) and fractional feed (on right-hand scale) as a function of time for the substitution reaction example, in a semi-batch reactor. Compound B is fed at a constant rate within 4 hours. The stoichiometric excess is 25%.

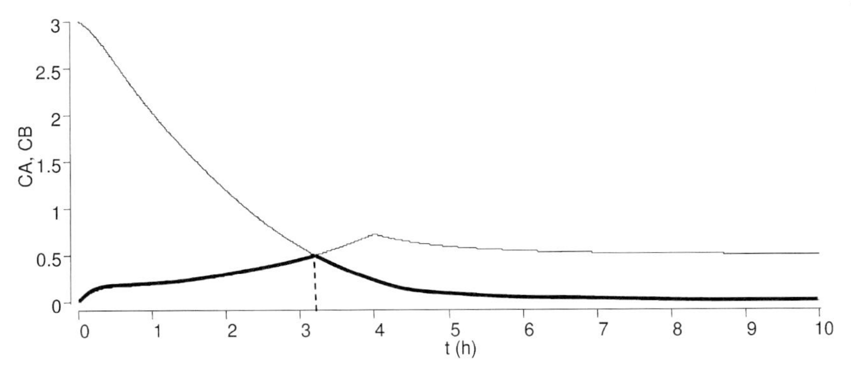

Figure 7.6 Concentration profiles in a semi-batch reactor showing the accumulation (bold line). The total feed time is 4 hours and B is fed in 25% stoichiometric excess. Hence the accumulation is at its maximum at the stoichiometric point reached after 3.2 hours.

Since the accumulation is determined by the compound with the lowest concentration, the conversion is also limited by the amount of this compound. Let us consider a cooling failure after two hours of feed: at this time the compound B presents the lowest concentration. Thus, the behavior of the reaction mass after the cooling failure is governed by the disappearance of compound B. If we now consider a cooling failure after 3.5 hours, the situation is reversed: the behavior will be governed by the disappearance of the default compound A. The switch between these two situations happens at the stoichiometric point. In this context, the stoichiometic point plays an essential role when the number of moles of B still fed to the reactor is equal to the number of moles required to complete the reaction with A, which was initially charged. Hence, before the stoichiometric point is reached, the conversion will be limited by B, and after this point, by A (Figure 7.6). One practical consequence is that the feed rate has no effect on accumulation after the stoichiometric point has been passed, that is, after this time, the feed rate can be set as high as possible with respect to accumulation (but could be limited by the cooling capacity). Then the reaction may run to completion, but the rate of reaction is limited by the lowering concentrations of C_A and C_B. For a constant feed rate, the time t_{st}, at which a stoichiometric amount of B has been fed to the reactor, can be calculated as

$$t_{st} = \frac{t_{fd}}{M} \tag{7.17}$$

In the example shown in Figure 7.5, a 25% molar excess of B has been added ($M = 1.25$). The function of this excess is to increase the reaction rate when C_A becomes small. It is often called the “kinetic excess,” because it increases the reaction rate at high conversion (Equation 7.7).

In this situation, the reaction cannot immediately be stopped by shutting the feed and further, the feed cannot be used to directly control the heat release rate or the gas release rate of a reaction. If, after a deviation from the design conditions, one decides to shut down the feed, the amount of accumulated B will react away despite the feed being stopped. If the reaction is accompanied by a gas release, gas production will continue and if the reaction is exothermal, heat will be released even after the interruption of the feed.

If the deviation was an uncontrolled temperature increase, the temperature increase will continue and accelerate the reaction until the accumulated reactant has been converted. Therefore, it is important to know quantitatively the degree of reactant accumulation during the reaction course, as it predicts the degree of conversion, which may occur after interruption of the feed. This can be done by chemical analysis or by using a heat balance, for example from an experiment in a reaction calorimeter [4]. Since the accumulation is the result of a balance between the amount of reactant B introduced by the feed and the amount converted by the reaction, a simple difference between these two terms calculates the accumulation [5, 6].

Before the stoichiometric point has been reached:

$$X_{ac} = X_{fd} - X = \frac{M \cdot t}{t_{fd}} - X \tag{7.18}$$

After the stoichiometric point has been reached:

$$X_{ac} = 1 - X_A \tag{7.19}$$

For a single reaction, the determination of reactant accumulation can be done directly by using calorimetric methods: the conversion is replaced by the thermal conversion defined by

$$X_{th}(t) = \frac{\int_0^t q_r \cdot d\tau}{\int_0^\infty q_r \cdot d\tau} \tag{7.20}$$

The thermal conversion X_{th}, at the instant t, is the fraction of the total heat of reaction Q_r that has been released before t. Since the added reactant corresponds to an input of energy, the balance between the added and converted reactant delivers the accumulation.

7.4
Design of Safe Semi-batch Reactors

In semi-batch operation, many elements determine the process safety. Among them we mention the temperature control strategy, the feed control strategy, and also the choice of reactant(s) to be initially charged and the reactant(s) to be fed.

With respect to process safety, the most stable reactant should be initially charged and the less stable added progressively in order to limit its accumulation and possible side reactions. This is favorable to process safety, as well as process economy. The decomposition potential may also be a criterion for this choice: feeding the reactant showing the highest decomposition potential limits the energy potential in the reactor during the operation. With complex reactions, the choice may be governed by the selectivity. Unfortunately, there is no general rule for this choice and the decision must be made on a case by case basis.

Concerning the temperature control strategy, semi-batch reactions are often at constant temperatures (isothermal). Another simple temperature control strategy is the isoperibolic mode, where only the jacket temperature is controlled. In rare cases, other temperature control strategies, such as adiabatic or non–isothermal, are used.

The feed may also be controlled in different ways: constant feed rate, by portions, governed by the reactor temperature, and so on.

These different temperature and feed control strategies and their impact on reactor safety, together with general rules for assessing and improving process safety, are presented below. The choice of the reactor temperature and feed rate is also of primary importance for safety and this point will be discussed in the last section of this chapter.

7.5 Isothermal Reaction

7.5.1 Principles of Isothermal Semi-batch Operation

A reliable control of the reaction course can be obtained by isothermal operation. Nevertheless, to maintain a constant reaction medium temperature, the heat exchange system must be able to remove even the maximum heat release rate of the reaction. Strictly isothermal behavior is difficult to achieve due to the thermal inertia of the reactor. However, in actual practice, the reaction temperature (T_r) can be controlled within ±2 °C, by using a cascade temperature controller (see Section 9.2.3). Isothermal conditions may also be achieved by using reflux cooling (see Section 9.2.3.3), provided the boiling point of the reaction mass does not change with composition.

7.5.2 Design of Isothermal Semi-batch Reactors

With single irreversible second-order reactions, the maximum of the heat release rate is reached at the beginning of the feed. At this stage, the heat exchange area may only be partially used, due to the increasing volume. This limits the effective available cooling capacity. Therefore, the knowledge of the maximum heat release rate and the instant at which it is reached, is essential for the design of well controlled isothermal semi-batch reactor. If the maximum heat removal capacity is exceeded by the heat release rate, the latter can be reduced by using slower feed rates. A too high feed rate will cause a temperature increase, in turn accelerating the reaction and may cause a runaway situation. For this reason, the limitation of the maximum feed rate is essential for the thermal control and consequently for safe operation of the reactor. Methods to control the feed rate are presented in Section 7.8.

Besides the heat release rate, the feed rate also affects the maximum reactant accumulation, a further important safety related parameter. The accumulation governs the temperature (T_{cf}) which may be reached in the case of a cooling failure. If the feed is immediately halted at the instant the failure occurs, the attainable temperature is expressed by

$$T_{cf} = T_r + X_{ac}\Delta T_{ad}\frac{M_{r,f}}{M_{r(t)}} \tag{7.21}$$

$M_{r,f}$ represents the mass of the reaction mixture at the end of the feed, $M_{r(t)}$ the instantaneous mass of reactant present in the reactor, and X_{ac} the fraction of accumulated reactant. The ratio of both masses accounts for the correction of the specific energy, since the adiabatic temperature rise is usually calculated using the final reaction mass, that is, the complete batch. In Equation 2.5, the concentration corresponds to the final reaction mass; this is also the case for the specific heat of reaction obtained from calorimetric experiments, which is also expressed for the total sample size. Since in the semi-batch reaction, the reaction mass varies as a function of the feed, the heat capacity of the reaction mass increases as a function of time and the adiabatic temperature rise must be corrected accordingly.

Another question is important for the safety assessment: "At which instant is the accumulation at maximum?" In semi-batch operations the degree of accumulation of reactants is determined by the reactant with the lowest concentration. For single irreversible second-order reactions, it is easy to determine directly the degree of accumulation by a simple material balance of the added reactant. For bimolecular elementary reactions, the maximum of accumulation is reached at the instant when the stoichiometric amount of the reactant has been added. The amount of reactant fed into the reactor (X_{fd}) normalized to stoichiometry minus the converted fraction (X), obtained from the experimental conversion curve delivered by a reaction calorimeter ($X = X_{th}$) or by chemical analysis, gives the degree of accumulation as a function of time (Equation 7.18). Afterwards, it is easy to determine the maximum of accumulation $X_{ac,max}$ and the MTSR can be obtained by Equation 7.21 calculated for the instant where the maximum accumulation occurs [7]:

$$MTSR = T_r + X_{ac,\max}\Delta T_{ad}\frac{M_{rf}}{M_{r,\max}} \tag{7.22}$$

where $M_{r,max}$ represents the mass of the reaction mixture at the instant of maximum accumulation. Figure 7.7 shows the substitution example reaction at 80 °C with an additional 4 hours for the reactant B with 25% of stoichiometric excess. Thus, the maximum of accumulation 23% was reached after 3.2 hours (3 hours and 12 minutes). The T_{cf} curve is obtained from Equation 7.21.

Since the accumulation is determined by a balance between feed rate and reaction rate (reactant depletion), it can be influenced by using different feed rates or different temperatures. This offers the possibility of optimizating the process conditions (discussed in Section 7.9).

If the reaction is complex, that is, if intermediates are formed during the reaction, an indirect method has to be used. Samples of the reaction mass are taken at defined stages of the reaction and analysed either chemically or thermally, for example, by DSC. This approach is also recommended when unstable intermediates are present in the reaction mixture: the stability of the reaction mass may pass through a minimum. Another method is to stop the feed during an experimental run in a reaction calorimeter and to measure the heat evolved after the interruption: it is proportional to the accumulation (Figure 7.4).

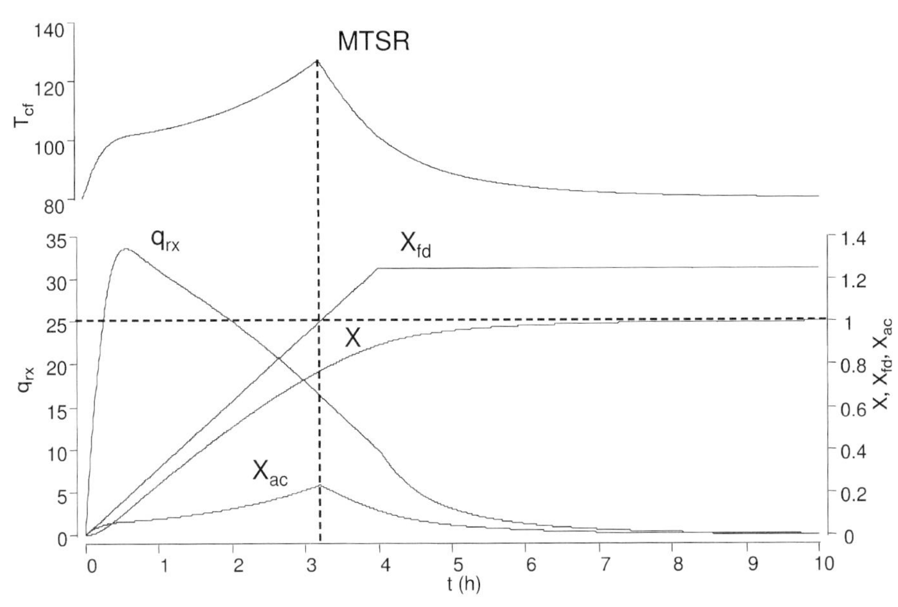

Figure 7.7 The maximum heat release rate and the heat of reaction can be directly determined in the original thermogram obtained by reaction calorimetry. In order to determine the accumulation, the feed X_{fd} must be corrected for the stoichiometry (125% in the present case). The difference between feed and conversion (X) gives the accumulation (X_{ac}). This determines the temperature that may be reached in case of failure (T_{cf}) and its maximum: *MTSR*.

Worked Example 7.2: Accumulation in a SBR

Using the thermogram represented in Figure 7.7, assess the thermal safety of the substitution reaction example A + B → P (see Section 5.3.1) performed as an isothermal semi-batch reaction at 80 °C with a feed time of 4 hours. At industrial scale, the reaction is to be in a 4 m^3 stainless steel reactor with an initial charge of 2000 kg of reactant A (initial concentration 3 mol kg^{-1}). The reactant B (1000 kg) is fed with a stoichiometric excess of 25%.

Solution:

Assessing the thermal risks of the process means answering the six questions in the cooling failure scenario (see Section 3.3.1). The overall energy potential of the reaction is calculated from the molar reaction enthalpy of 200 kJ mol^{-1}. The concentration to be used is that of the final reaction mass (2 mol kg^{-1}), since the reactant B must be added to allow the reaction:

$$Q'_{rx} = C'_{A0} \cdot (-\Delta H_r) = 2\,\text{mol}\,\text{kg}^{-1} \times 150\,\text{kJ}\,\text{mol}^{-1} = 300\,\text{kJ}\,\text{kg}^{-1}$$

Since the specific heat capacity of the reaction mass and feed is 1.7 kJ kg^{-1} K^{-1}, the adiabatic temperature rise is

$$\Delta T_{ad} = \frac{Q'_{rx}}{c'_P} = \frac{300\,\text{kJ}\,\text{kg}^{-1}}{1.7\,\text{kJ}\,\text{kg}^{-1}\,\text{K}^{-1}} = 176\,\text{K}$$

This represents a medium energy potential, severity medium using the criteria presented in Section 3.3.2) Therefore, care must be devoted to the reaction course control, which presents two aspects: heat removal during normal operation (Question 1 in the scenario Section 3.3.1) and the MTSR (Question 2 in the scenario Section 3.3.1). By considering 100% accumulation, the reaction temperature would reach 80 + 176 = 256 °C. Thus, the secondary reaction with a high energy potential (T_{D24} = 113 °C; Table 5.4) would be triggered (Question 3 in the cooling failure scenario).

Temperature control: The maximum heat release rate of the reaction under the conditions specified above, directly read from Figure 7.7 is 31 W kg^{-1} and is reached after 30 minutes of feed. At this instant the reaction mass is

$$M_r = 2000\,\text{kg} + \left(\frac{0.5}{4} \times 1000\,\text{kg}\right) = 2125\,\text{kg} \quad \text{thus the heat release rate is}$$

$$q_{rx} \cong 66\,\text{kW}$$

The cooling capacity of the industrial reactor is given by $q_{ex} = U \cdot A \cdot \Delta T$, which requires the knowledge of the heat exchange area. With 2125 kg of reaction mass, that is, 2.125 m^3 of volume, the heat exchange area is (data from Table 5.3)

$$A = 3.0\,\text{m}^2 + (2.125\,\text{m}^3 - 1.05\,\text{m}^3) \times \frac{7.4\,\text{m}^2 - 3.0\,\text{m}^2}{3.4\,\text{m}^3 - 1.05\,\text{m}^3} \cong 5\,\text{m}^2$$

If we assume an average jacket temperature of 15 °C, with cold water as a coolant (brine is also an option), and with the given heat transfer coefficient of 200 W m^{-2} K^{-1}, the cooling capacity is

$$q_{ex} = 200\,\text{W}\,\text{m}^{-2}\,\text{K}^{-1} \times 5\,\text{m}^2 \times (80 - 15) = 65\,\text{kW}$$

Additionally we could also account for the convective cooling due to the cold feed:

$$q_{fd} = \dot{m} \cdot c'_P (T_r - T_{fd}) = \frac{1000\,\text{kg}}{4\,\text{h} \times 3600\,\text{s}\,\text{h}^{-1}} \times 1.7\,\text{kJ}\,\text{kg}^{-1}\,\text{K}^{-1} \times (80 - 25) = 6.5\,\text{kW}$$

Thus, the temperature can be controlled using cold water as a coolant, but the reaction requires practically the full available cooling capacity of the reactor (Question 1 in the cooling failure scenario Section 3.3.1).

The MTSR (Question 2 in the cooling failure scenario) can be directly determined using Equation 7.22, by reading the data from the thermogram (Figure 7.7). The accumulation of 25% is reached at the stoichiometric point, that is, after 3.2 hours of feed (Question 4 in the cooling failure scenario):

$$MTSR = T_r + T_{ac,max} \cdot \Delta T_{ad} \cdot \frac{M_{rf}}{M_{r,st}} = 80°\text{C} + 0.25 \times 176°\text{C} \times \frac{3000\,\text{kg}}{2800\,\text{kg}} = 127°\text{C}$$

At 127 °C, the decomposition reaction is critical, that is, the time to maximum rate (Question 6 in the cooling failure scenario) is shorter than 8 hours (see Table 5.4).

Hence the intended process belongs to the criticality class 5:

$$T_p = 80°\text{C} < T_{D24} = 113°\text{C} < MTSR = 127°\text{C} < MTT = 140°\text{C}$$

Thus, there are two reasons for modifying the process: the maximum heat release rate of the reaction and the accumulation of reactants are too high. There are different means for solving the problem: first is to increase the feed time (Figure 7.8), the second to increase the reaction temperature (Figure 7.9). This worked example is continued in Section 7.8.2.

7.6 Isoperibolic, Constant Cooling Medium Temperature

This is the simplest way of temperature control for a semi-batch reactor: only the temperature of the cooling medium is controlled. The temperature of the reaction

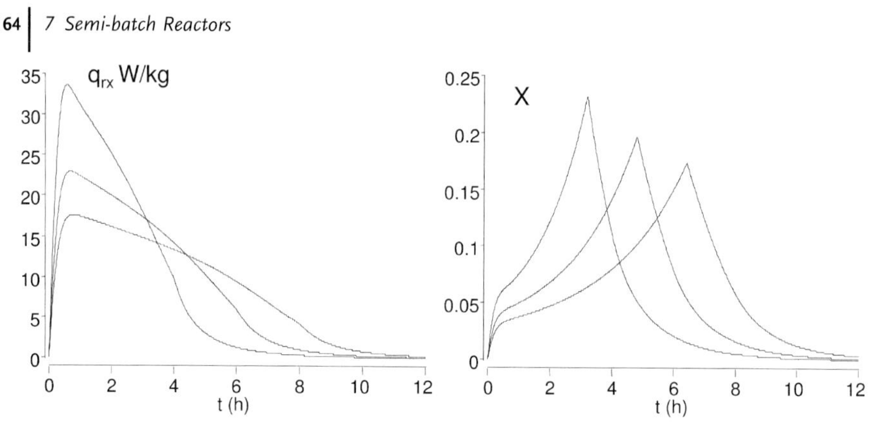

Figure 7.8 Influence of the feed rate 4, 6, and 8 hours on the heat release rate (left) and on the accumulation (right).

mass results from the heat balance of the reactor. Here the most important terms, the heat production by the reaction, the heat removal by the cooling system, and the heat effect due to the feed, must be considered. The greatest disadvantage of this strategy is that no direct control of the reaction temperature is possible. This type of reaction control was intensively studied by Steinbach [1, 2, 8, 9]. The choices of the initial temperature (T_0) and of the temperature of the cooling system (T_c) are critical. If the initial temperature of the reactor is set too low, the reaction is slow at the start of the feed and an important accumulation of the added reactant results. As its concentration increases, the reaction rate and consequently the heat production increase up to a point, which may surpass the cooling capacity of the reactor, resulting in a runaway. In fact, with too low a temperature, the accumulation becomes so high that the reactor behaves like a batch reactor. At the opposite, too high an initial temperature may result in an uncontrollable reaction course. In both cases no steady state, in the sense of a constant reaction temperature, can be reached.

Additionally the semi-batch reactor with constant cooling medium temperature, also in cases where a stationary temperature can be achieved, shows a high sensitivity to its control parameters, that is, initial temperature and coolant temperature. This means that even for small changes in these temperatures, the behavior of the reactor may suddenly change from a stable situation into a runaway course.

The example represented in Figure 7.9 shows the extreme sensitivity of the temperature course towards the initial temperature. A change of only 1 °C from 103 to 104 °C results in a runaway. For too low temperatures, for example, 50 °C, a significant accumulation builds up. This results in a sudden acceleration of the reaction (ignition), but the conversion is only 0.95 after 10 hours. Thus, the different states presented in Section 5.2.3.5, that is, runaway above 100 °C, marginal ignition below 60 °C, and QFS (Quick on-set, Fair conversion and Smooth profile) between 70 and 90 °C are represented [10].

This type of temperature control strategy may be very critical for highly exothermal reactions. In such cases, the choice of the operating conditions, that is, initial

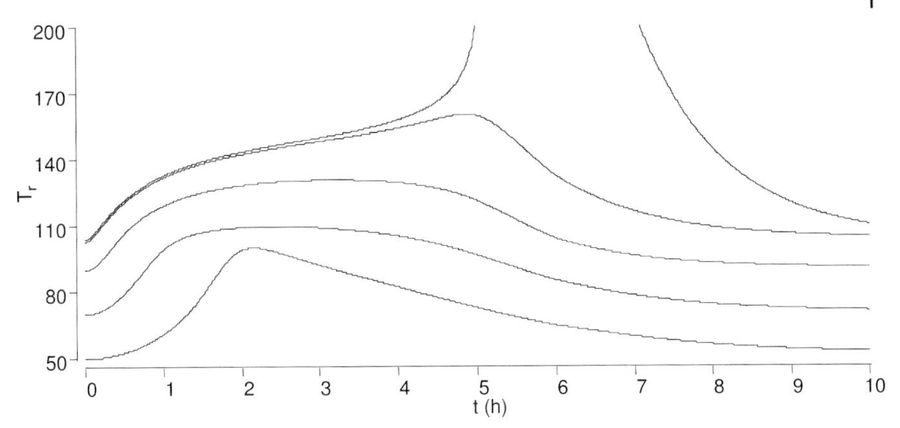

Figure 7.9 Semi-batch reactor with the example slow reaction and constant cooling medium temperature at 50, 70, 90, 103, and 104 °C. The feed time is 6 hours; initial and cooling medium temperatures are equal.

temperature, coolant temperature, and feed time is extremely important. Hugo and Steinbach established the following criterion for safe control of a semi-batch reactor with constant cooling temperature:

$$Da(T_c) > St \Leftrightarrow \frac{k(T_c)N_{A0}M_{rf}c'_P}{V_r UA} > 1 \tag{7.23}$$

This criterion expresses the fact that the reaction rate must be high enough to avoid the accumulation of reactant, even if the reaction is performed at the temperature of the coolant, and the cooling capacity must be sufficient to control the temperature. If this criterion is fulfilled, the reactor will not be parametric sensitive. The set temperature of the reactor can be chosen using the following criterion:

$$T_r - T_c = \frac{\Delta T_{ad}}{St + \dfrac{\varepsilon}{1+\varepsilon}} \tag{7.24}$$

In Equation 7.24, the adiabatic temperature rise is calculated for the final reaction mass. This criterion calculates the active temperature difference for an efficient cooling as a function of the reaction enthalpy.

In practice, however, besides the "normal operating conditions," one also has to consider deviation from these conditions. Important and current deviations from a practical point of view are:

- Initial temperature (T_0): Before starting the feed, the reactor has to be heated up to its initial temperature. If only the heat carrier's temperature can be actively controlled, deviations of the reaction mixture temperature from the set temperature are likely to occur. It is therefore important not to operate such a reactor in the so-called parametric sensitive domain.

- Cooling medium temperature (T_c): This temperature was assumed to be constant. However, for lower flow rates, a temperature difference between inlet and outlet of the heat carrier will be observed. In such a case, the cooling capacity must be calculated using the logarithmic mean to estimate the active temperature difference between reaction mixture and cooling medium. It is recommended to monitor both inlet and outlet temperature of the cooling system during the reaction.
- Feed rate (F_B): The feed time does not explicitly appear in Equation 7.23. However, it directly influences the amount of reactant B accumulated in the reactor, and hence the maximum temperature, which can be reached after a cooling failure has occurred during the course of the reaction. Therefore, the feed rate is a key parameter, which has to be monitored and technically limited to the highest allowable value. This will be discussed in Section 7.8.

7.7 Non-isothermal Reaction

There are different ways of controlling a semi-batch reactor with a non-isothermal reaction course:

- The isoperibolic mode: the temperature of the cooling medium is maintained constant. This type of temperature control was described in (Section 7.6).
- The adiabatic mode: the reaction is performed without any exchange at all. This means the heat of reaction will be converted into a temperature increase. The temperature course can be calculated from the heat balance of the reactor:

$$T_{(t)} = T_0 + \frac{Q_{r(t)} + Q_{fd(t)}}{M_r \cdot c'_P} \tag{7.25}$$

  Since the different terms are a function of time, the temperature profile must be calculated by numerical methods. The final temperature is also a function of the feed rate.
- The polytropic mode: this is a combination of different types of control. As an example, the polytropic mode can be used to reduce the initial heat release rate by starting the feed and the reaction, at a lower temperature. The heat of reaction can then be used to heat up the reactor to the desired temperature. During the heating period, different strategies of temperature control can be applied: adiabatic heating until a certain temperature level is reached, constant cooling medium temperature (isoperibolic control), or ramped to the desired reaction temperature in the reactor temperature controlled mode. Almost after the

heating period, the reaction is ended in this mode. The design of a polytropic semi-batch reactor requires numeric simulation in order to optimize the feed rate, initial temperature, and cooling rate, which govern the temperature and conversion profiles. This mode generally allows a higher productivity compared to the purely isothermal mode, but also requires more design effort.

7.8 Strategies of Feed Control

7.8.1 Addition by Portions

This mode of addition is a traditional way of limiting the accumulation. It is also used for practical reasons, when a reactant is delivered in drums or containers. Nevertheless, the amount of reactant, which can be added in one portion, can also be limited for safety reasons. In this case, the addition must be controlled by the conversion that is, the next portion is only added if the previous portion has been consumed by the reaction. Different criteria can be used to follow the reaction: the temperature, the gas evolution (where applicable), the aspect of the reaction mass, chemical analysis, and so on. For a well designed process, where the reaction kinetics are known to a certain accuracy, the successive additions can also be performed on a time basis.

7.8.2 Constant Feed Rate

This is the most common mode of addition. For safety or selectivity critical reactions, it is important to guarantee the feed rate by a control system. Here instruments such as orifice, volumetric pumps, control valves, and more sophisticated systems based on weight (of the reactor and/or of the feed tank) are commonly used. The feed rate is an essential parameter in the design of a semi-batch reactor. It may affect the chemical selectivity, and certainly affects the temperature control, the safety, and of course the economy of the process. The effect of feed rate on heat release rate and accumulation is shown in the example of an irreversible second-order reaction in Figure 7.8. The measurements made in a reaction calorimeter show the effect of three different feed rates on the heat release rate and on the accumulation of non-converted reactant computed on the basis of the thermal conversion. For such a case, the feed rate may be adapted to both safety constraints: the maximum heat release rate must be lower than the cooling capacity of the industrial reactor and the maximum accumulation should remain below the maximum allowed accumulation with respect to MTSR. Thus, reaction calorimetry is a powerful tool for optimizing the feed rate for scale-up purposes [3, 11].

Worked Example 7.3: Slow Reaction with Different Feed Rates

This is the continuation of Worked Example 7.2.

The substitution reaction was carried out in a reaction calorimeter using different feed rates (Figure 7.8). From these experiments we can read the following data:

- feed time 6 hours: maximum heat release rate $22\,W\,kg^{-1}$; accumulation 0.21
- feed time 8 hours: maximum heat release rate $17\,W\,kg^{-1}$; accumulation 0.18

Questions:

1. Calculate the heat release rate of the reaction performed at industrial scale with the same reactor as described in Worked Example 7.2. Compare it with the available cooling capacity.
2. Calculate the *MTSR* from these data and compare it to the characteristic temperatures of the decomposition reaction (Table 5.4). What are the conclusions?

Solution:

Cooling capacity:
The maximum heat release rate is reached at the same time of 0.5 hours. The heat release rate at industrial scale is calculated using the same method as previously:

- with a feed time of 6 hours: $q_{rx} = q'_{rx} \cdot M_r = 22\,W\,kg^{-1} \times 2125\,kg \cong 47\,kW$
- with a feed time of 8 hours: $q_{rx} = q'_{rx} \cdot M_r = 17\,W\,kg^{-1} \times 2125\,kg \cong 36\,kW$

Since the cooling capacity of the reactor is obviously the same as calculated previously, 65 kW, increasing the feed time easily solves the problem of the cooling capacity, however, at the cost of cycle time.

Accumulation:
Since the maximum accumulation is reached at the stoichiometric point, it occurs always with the same mass in the reactor. Therefore, the calculation is the same as previously.

At 6 hours feed time:

$$MTSR = T_r + X_{acc,max} \cdot \Delta T_{ad} \cdot \frac{M_{rf}}{Mr, st} = 80°C + 0.21 \times 176°C \times \frac{3000\,kg}{2800\,kg} \cong 120°C$$

At 8 hours feed time:

$$MTSR = T_r + X_{acc,max} \cdot \Delta T_{ad} \cdot \frac{M_{rf}}{Mr, st} = 80°C + 0.18 \times 176°C \times \frac{3000\,kg}{2800\,kg} \cong 114°C$$

Thus, the longer feed time reduces the accumulation and the MTSR, but even with 8 hours of feed time, the MTSR of 114 °C remains slightly above the T_{D24} of 113 °C. Nevertheless, this would be an acceptable process as far as the golden rules for the safe SBR are respected (see Section 7.8.3):

- limitation of the feed time,
- interlock with temperature (too low and too high),
- interlock with the stirrer.

This process can be further optimized, since besides the feed time, the process temperature may also be increased (see Sections 7.9 and 7.10).

7.8.3
Interlock of Feed with Temperature

An often-used method for the limitation of the heat release rate is an interlock of the feed with the temperature of the reaction mass. This method consists of halting the feed when the temperature reaches a predefined limit. This feed control strategy keeps the reactor temperature under control even in the case of poor dynamic behavior of the reactor temperature control system, should the heat exchange coefficient be lowered (e.g. fouling crusts) or feed rate too high.

Numeric simulations were performed with the substitution example reaction and several deviations from normal conditions (Table 7.1). For this set of parameters, the feed time should not be shorter than 8 hours in order to maintain the heat release rate below the cooling capacity. For safety reasons, the *MTSR* must be limited to 113 °C so as not to trigger a secondary decomposition reaction. To show the effect of an interlock between reactor temperature and feed, simulations were performed with a flow rate of the feed corresponding to a feed time of one hour (i.e. eight times too high). The effects of wrong controller gain, decreased heat transfer coefficient, and wrong set temperature are shown in Table 7.1.

The temperature controller is a cascade controller, as described in Section 9.2.3. The interlock of the feed with the reactor temperature avoids critical temperature

Table 7.1 Slow example reaction performed in a reactor equipped with feed-temperature interlock. The behavior of the system under different deviations from normal operating conditions is shown.

Temperature switch °C	Heat transfer $W\,m^{-2}\,K^{-1}$	Gain	*MTSR* °C	Actual feed time h	*X* = 99% after h
85	200	10	145	2.7	5.6
85	200	2	128	6.3	8.6
85	100	10	132	4.4	7.1
90	200	10	156	1.8	4.9
82	200	10	122	6.5	9.1

excursions as long as the cooling system is working. But in case of its failure, the temperature reached (MTSR) is too high and would lead to a runaway reaction. This may be avoided by choosing a temperature alarm level closer to the desired operating temperature, for example, 82 °C. If this strategy is followed, it is important to fix the temperature switch level: if the switch level is too high, the feed will be halted at too high a temperature and the MTSR increases. Thus, the choice of the switch level is a critical parameter.

On the other hand, a deviation of the process temperature towards lower temperatures leads to a slower reaction rate and therefore a higher accumulation. At the extreme, the reaction stops and if the feed is not halted, the reactor behaves like a batch reactor. A critical situation may occur if, after having accumulated reactant at a low temperature, the temperature is then corrected by heating, a runaway situation may occur. A similar situation can be caused by stirrer failure. For reaction mixtures with high viscosity, for heterogeneous systems, or if the reactants show a great difference in density, layering may occur. Here too, if the stirrer is switched on, the accumulated reactants may react suddenly, leading to a runaway. For these reasons, a safe design of a semi-batch reactor with constant feed rate must fulfil the following *"Golden Rules"*:

- optimization of temperature and feed rate with regard to heat removal and accumulation,
- limitation of the maximum feed rate by technical means,
- interlock of feed with temperature controlled by high and low temperature switch levels,
- interlock of feed with stirrer.

Following these rules ensures a safe semi-batch reaction, even in cases of technical deviations or deviations from the normal operating conditions.

7.8.4
Why to Reduce the Accumulation

There are essentially two reasons to minimize the accumulation of the fed reactant. First, if the reactant B enters in side reactions, its concentration has to be maintained as low as possible for selectivity reasons. Some examples of such reaction schemes are given below:

$$\begin{cases} A + B \rightarrow P \\ P + B \rightarrow S \end{cases} \quad \begin{cases} A + B \rightarrow P \\ B \rightarrow S \end{cases}$$

It is clear that a low concentration of B will maximize the formation of the desired product P and minimize the formation of the secondary product S. This goal can be achieved with a semi-batch reactor where B is added progressively to the reaction mass.

The second reason, for which the concentration of B should be maintained at a low level, is a safety related reason. A low concentration of B during the reaction course adjusts the reaction rate and accordingly the heat release rate to the cooling power of the reactor, thus avoiding a potentially dangerous temperature increase, should a cooling or stirrer failure occur.

Therefore, the optimization of a semi-batch reactor, for selectivity reasons as well as for safety reasons, will often result in a reduction of the accumulation. Some hints to achieve this goal are given below.

7.9 Choice of Temperature and Feed Rate

The accumulation or the concentration of B results from competition between the rate of addition and the reaction rate that governs the reactant depletion. Thus, the accumulation can be reduced by both a lower feed rate and a higher reaction temperature. The first way, of course, leads to longer addition times, which negatively affect the time-cycle and resulting productivity. Hence a higher temperature presents some advantages: the reaction is faster resulting in a shorter time-cycle and the cooling capacity is increased due to the increased temperature difference between reaction mass and heat carrier. But there is an upper limit for the reaction temperature: At higher temperature levels, secondary reactions may become dominant, higher pressure may be required, and the thermal stability of the reaction mass may be affected.

With the substitution reaction example (Figure 7.10), too low a process temperature, for example 60 °C, leads to a high accumulation and consequently high MTSR that is in a range where the secondary reaction is fast. Thus, the time between cooling failure and runaway is short. At too high a temperature, for example 120 °C, even with a low accumulation, the MTSR is so high that again, the runaway

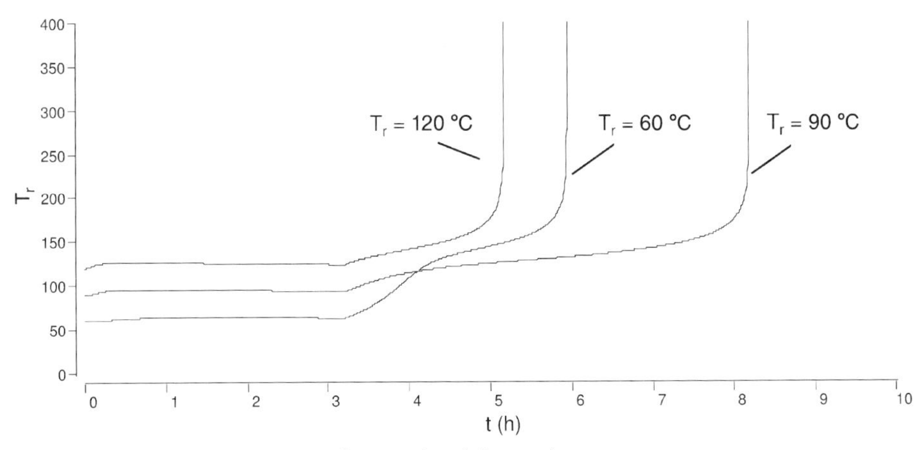

Figure 7.10 Temperature course after a cooling failure at the instant of maximum accumulation with three different process temperatures: 60, 90, and 120 °C.

occurs in a short time. For an intermediate temperature, for example 90 °C, the time to runaway is longer. Hence there is an optimal temperature for a safe semi-batch process [12, 13].

For an irreversible second-order reaction, the optimization of the reaction temperature and feed rate can be performed by using the following equation [14]:

$$Da = \frac{2}{\pi}\frac{1}{(1-X_{st})^2} \tag{7.26}$$

where X_{st} stands for the conversion at the stoichiometric point and is valid for $Da > 6$. From Equations 7.21 and 7.26, an expression for the calculation of the temperature after a cooling failure can be derived [13]:

$$MTSR = T_p + \Delta T_{ad}\sqrt{\frac{2}{\pi \cdot Da}} \tag{7.27}$$

The maximum temperature of synthesis reaction was calculated for the substitution reaction example as a function of the process temperature and with different feed rates corresponding to a feed time of 2, 4, 6, and 8 hours. The straight line (diagonal in Figure 7.11) represents the value for no accumulation, that is, for a fast reaction. This clearly shows that the reactor has to be operated at a sufficiently high temperature to avoid the accumulation of reactant B. But a too high temperature will also result in a runaway due to the high initial level, even if the accumulation is low. In this example, the characteristics of the decomposition reaction

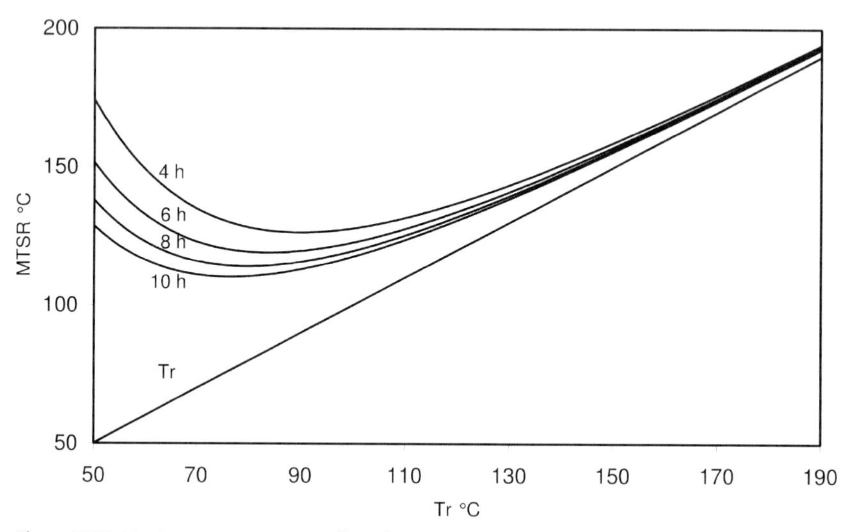

Figure 7.11 Maximum temperature of synthesis reaction occurring at stoichiometric point as a function of the process temperature T_P, with different feed rates, t_{fd} = 2, 4, 6, 8 hours.

require that the MTSR should be limited to 113 °C for a medium probability of triggering the decomposition reaction. This implies that the process should be run with a feed time of 9 to10 hours between 70 and 90 °C.

7.10 Feed Control by Accumulation

In industrial practice, a semi-batch reaction often has to be optimized in order to ensure that a maximum temperature (T_{max}) may not be surpassed, even in the case when a failure occurs. It was shown how the concept of MTSR can be used for that purpose. Apart from traditional changes such as the concentrations, the reaction media, the procedure itself, temperature, and feed rate, there is a further way of achieving the goal: a variable feed rate. If we consider the T_{cf} curve in Figure 7.12, which represents a process optimized as explained in the previous paragraph with process temperature of 85 °C and a constant feed rate during 9 hours, we observe that the maximum allowed temperature is only reached relatively late during the course of the reaction: at the stoichiometric equivalence. Before and after this instant, the temperature would remain below the maximum allowed level, indicating that the accumulation is less than could be tolerated. The gap between the actual and the accepted T_{cf} is "wasted." This fact can be used to improve the process with a variable feed rate.

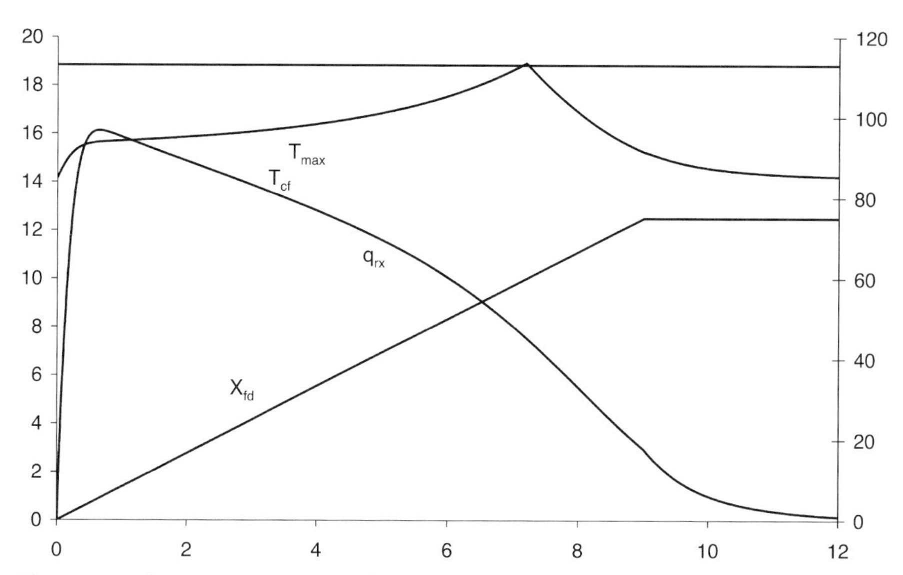

Figure 7.12 Substitution reaction example optimized at a process temperature of 85 °C with a constant feed rate during 9 hours. Heat release rate on left axis and T_{cf} on right axis, time (h).

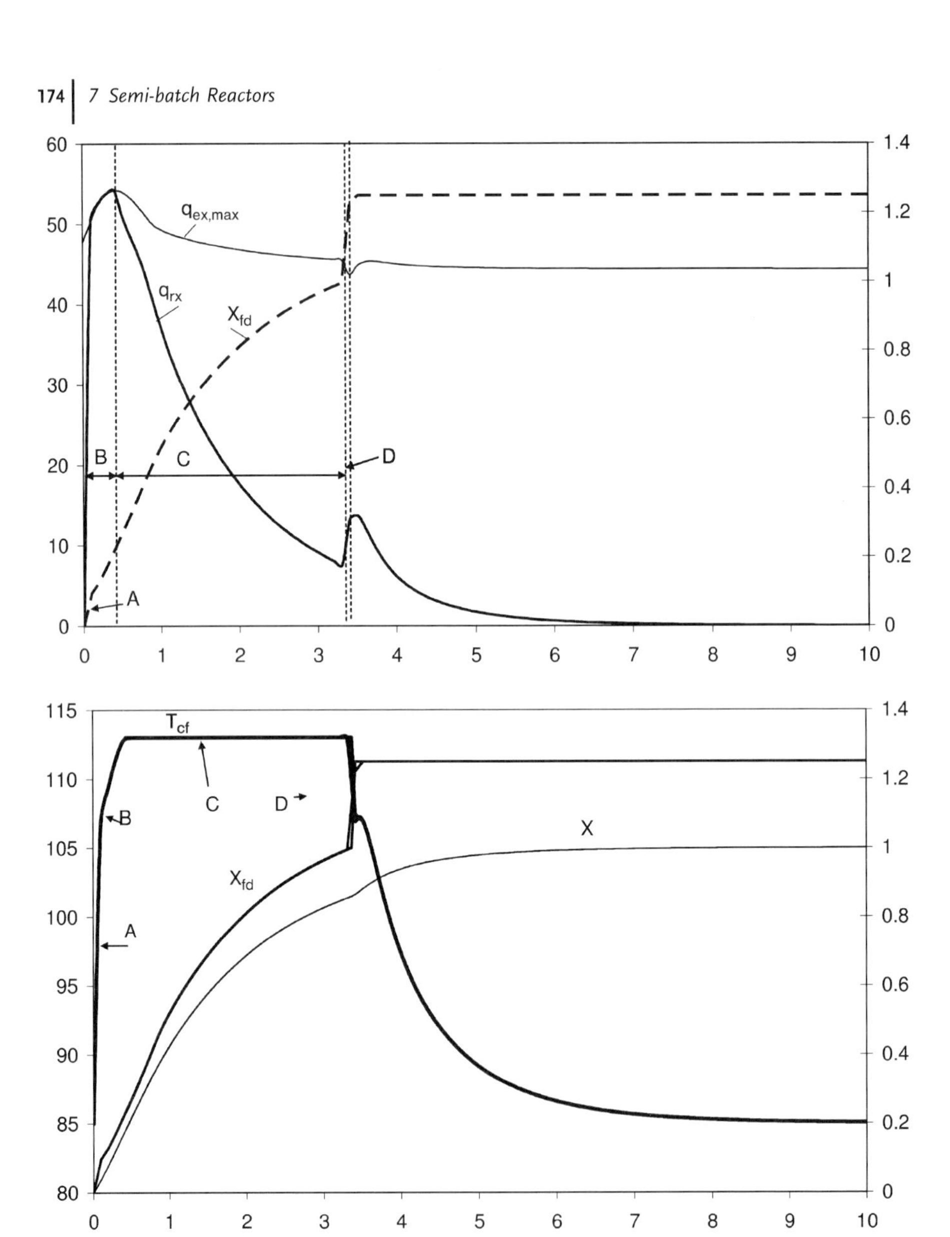

Figure 7.13 Semi-batch reaction with feed rate adapted to the T_{cf}-limitation at 113 °C. Upper diagram: heat release rate and cooling capacity on left scale, feed on right scale. Lower diagram: temperature after cooling failure on left scale, feed and conversion on right scale; time (h).

The method was first proposed by Gygax [12] and was worked out by Ubrich [15–18]. This method applied to the substitution reaction example is illustrated in Figure 7.13. The feed profile is divided into four stages: the first stage (A) consists of building up the maximum allowed accumulation by a fast feed rate until the

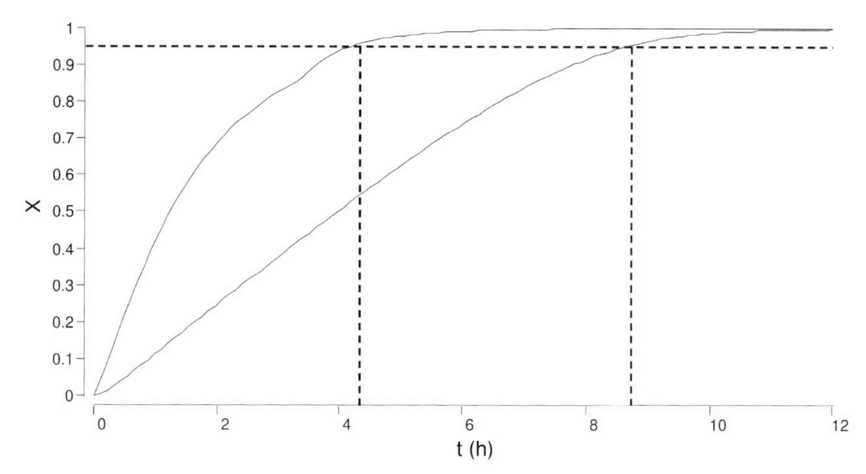

Figure 7.14 Conversion curves obtained with the optimized linear feed and with the accumulation controlled feed; time (h).

limit of the cooling capacity of the reactor is reached. Then, in the second stage (B), the feed is adapted such as not to surpass the cooling capacity until the maximum allowed capacity is reached. In the third stage (C), this accumulation is maintained constant at the maximum allowed level by a relative slow and decreasing feed rate. In the fourth stage (D), beginning at the equimolecular point, the feed can be fast again: in this stage, the accumulation is governed by the initially charged reactant and hence, cannot be influenced by the addition. The reactor then behaves like a batch reactor. Since the concentration of B remains at the maximum allowed, the reaction rate is increased, which leads to shorter cycle time. For the slow reaction example, the conversion of 95% is reached after 8.5 hours with a linear feed and after only 4.2 hours with the accumulation controlled feed (Figure 7.14). This gives a reduction of the reaction time by a factor of two, without altering the safety of the process.

The condition for the practical implementation of such a feed control is the availability of a computer controlled feed system and of an on-line measurement of the accumulation. The later condition can be achieved either by an on-line measurement of the reactant concentration, using analytical methods or indirectly, by using a heat balance of the reactor. The amount of reactant fed to the reactor corresponds to a certain energy of reaction and can be compared to the heat removed from the reaction mass by the heat exchange system. For such a measurement, the required data are the mass flow rate of the cooling medium, its inlet temperature, and its outlet temperature. The feed profile can also be simplified into three constant feed rates, which approximate the ideal profile. This kind of semi-batch process shortens the time-cycle of the process and maintains safe conditions during the whole process time. This procedure was shown to work with different reaction schemes [16, 19, 20], as long as the fed compound B does not enter parallel reactions.

7.11 Exercises

▶ Exercise 7.1

A Grignard reagent is to be prepared at industrial scale at a temperature of 40 °C in THF as a solvent (T_b = 65 °C). The reaction scheme is

$$R-Br+Mg \rightarrow R-Mg-Br$$

In a laboratory experiment in a reaction calorimeter, a bromo-compound was added at a constant feed rate over 1.5 hours. The data obtained during this experiment are summarized as follows:

- Specific heat of reaction Q'_r 450 kJ kg^{-1} (final reaction mass)
- Reaction enthalpy $-\Delta H_r$ 375 kJ mol^{-1} (Bromide)
- Max. heat release rate $-\Delta H_r$ 220 W kg^{-1} at starting of feed (after initiation)
- Specific heat capacity c'_P 1.9 kJ (kg^{-1} K^{-1})
- Maximum accumulation X_{ac} 6% (reached at the end of feed)

At industrial scale, the charge should be 4000 kg, the heat exchange area of the reactor is 10 m^2, and the overall heat transfer coefficient is 400 W m^{-2} K^{-1}. Brine, allowing for an average jacket temperature of –10 °C, will be used as a coolant.

Questions:

1. May the boiling point be reached in the case of cooling failure, if the feed is immediately stopped?
2. What feed time would you recommend at industrial scale?

▶ Exercise 7.2

An exothermal reaction is to be performed in the semi-batch mode at 80 °C in a 16 m^3 water cooled stainless steel reactor with heat transfer coefficient U = 300 W m^{-2} K^{-1}. The reaction is known to be a bimolecular reaction of second order and follows the scheme A + B → P. The industrial process intends to initially charge 15 000 kg of A into the reactor, which is heated to 80 °C. Then 3000 kg of B are fed at constant rate during 2 hours. This represents a stoichiometric excess of 10%.The reaction was performed under these conditions in a reaction calorimeter. The maximum heat release rate of 30 W kg^{-1} was reached after 45 minutes, then the measured power depleted to reach asymptotically zero after 8 hours. The reaction is exothermal with an energy of 250 kJ kg^{-1} of final reaction mass. The specific heat capacity is 1.7 kJ kg^{-1} K^{-1}. After 1.8 hours the conversion is 62% and 65% at end of the feed time. The thermal stability of the final reaction mass imposes a maximum allowed temperature of 125 °C (T_{D24}). The boiling point of the reaction mass (MTT) is 180 °C, its freezing point is 50 °C.

Questions:
1. Is the cooling capacity of the reactor sufficient?
2. Calculate the MTSR and determine the criticality class of this reaction.
3. Does the process appear to be feasible with regard to accumulation?
4. Which recommendations would make for the plant?

▶ Exercise 7.3

A catalytic hydrogenation is performed at constant pressure in a semi-batch reactor. The reaction temperature is 80 °C. Under these conditions, the reaction rate is $10\,\text{mmol}\,\text{l}^{-1}\,\text{s}^{-1}$ and the reaction may be considered to follow a zero-order rate law. The enthalpy of the reaction is $540\,\text{kJ}\,\text{mol}^{-1}$. The charge volume is $5\,\text{m}^3$ and the heat exchange area of the reactor $10\,\text{m}^2$. The specific heat capacity of water is $4.2\,\text{kJ}\,\text{kg}^{-1}\,\text{K}^{-1}$.

Questions:
1. What is the heat release rate of the reaction?
2. What is the required average temperature of the jacket required to maintain a constant reaction temperature at 80 °C? The heat transfer coefficient is $U = 1000\,\text{W}\,\text{m}^{-2}\,\text{K}^{-1}$.
3. What are the other technical means that control the reaction rate?
4. Propose a laboratory experiment to prove the efficiency of the proposed means.

▶ Exercise 7.4

An aromatic di-nitro compound ($M_w = 453.75\,\text{g}\,\text{mol}^{-1}$) is to be reduced by a Béchamp reaction. The reaction is performed at reflux in water as a solvent. Water and iron are initially charged and the nitro compound is fed at a constant rate. The initial charge in the reactor is 300 g water, 320 g iron. The feed consists of 726 g of a 20% w/w suspension of nitro compound (0.32 mol) in water. The reaction should be performed under the same conditions at the industrial scale of 1.74 kmol. In order to check the thermal control of the reaction, an experiment is performed in a reaction calorimeter. The raw heat flow curve (without any correction of the convective cooling by the feed) is rectangular with a heat flow of $350\,\text{W}\,\text{mol}^{-1}$. The heat flow immediately starts as the feed is started and straightaway depletes to zero as it is stopped after 30 minutes.

Questions:
1. What is the heat of reaction?
2. What can be stated concerning the accumulation of unreacted di-nitro compound?
3. Calculate the adiabatic temperature rise of the reaction. The specific heat capacity of the reaction mass is $3\,\text{kJ}\,\text{kg}^{-1}\,\text{K}^{-1}$.

4. What would the consequences of a cooling failure be, if the feed is immediately stopped? The latent heat of evaporation of water is: $\Delta H_v' = 2240\,\mathrm{kJ\,kg^{-1}}$.
5. What is the required cooling power?

References

1 Steinbach, J. (1985) *Untersuchungen zur thermischen Sicherheit des indirekt gekühlten Semibatch-Reaktors*, TU Berlin.

2 Hugo, P. and Steinbach, J. (1985) Praxisorientierte Darstellung der thermischen Sicherheitsgrenzen für den indirekt gekühlten Semibatch-Reaktor. *Chemie Ingenieur Technik*, **57** (9), 780–2.

3 Lerena, P., Wehner, W., Weber, H. and Stoessel, F. (1996) Assessment of hazards linked to accumulation in semi-batch reactors. *Thermochimica Acta*, **289**, 127–42.

4 Regenass, W. (1983) Thermische Methoden zur Bestimmung der Makrokinetik. *Chimia*, **37** (11), 430–7.

5 Gygax, R. (1993) *Thermal Process Safety, Data Assessment, Criteria, Measures*, Vol. 8 (ed. ESCIS), ESCIS, Lucerne.

6 Gygax, R. (1991) Facts finding and basic data, Part II: Desired chemical reactions, in *1st IUPAC-Workshop on Safety in Chemical Production*, Blackwell Scientific Publication, Basel.

7 Stoessel, F. (1993) What is your thermal risk? *Chemical Engineering Progress*, 68–75.

8 Steensma, M. and Westerterp, K.R. (1990) Thermally safe operation of a semibatch reactor for liquid-liquid reactions. Slow reactions. *Industrial and Engineering Chemistry Research*, **29**, 1259–70.

9 Westerterp, K.R., Swaij, W.P.M.v. and Beenackers, A.A.C.M. (1984) *Chemical Reactor Design and Operation*, Wiley-VCH, Weinheim.

10 Westerterp, K.R. (2006) Safety and runaway prevention in batch and semi-batch reactors – a review. *Chemical Engineering Research and Design*, **84** (A7), 543–52.

11 Hoffmann, W. (1989) Reaction calorimetry in safety: the nitration of a 2,6-disubstituted benzonitrile. *Chimia*, **43**, 62–7.

12 Gygax, R. (1988) Chemical reaction engineering for safety. *Chemical Engineering Science*, **43** (8), 1759–71.

13 Hugo, P., Steinbach, J. and Stoessel, F. (1988) Calculation of the maximum temperature in stirred tank reactors in case of breakdown of cooling. *Chemical Engineering Science*, **43** (8), 2147–52.

14 Hugo, P. (1981) Berechnung isothermer Semibatch-Reaktionen. *Chemie Ingenieur Technik*, **53** (2), 107–9.

15 Ubrich, O., Srinivasan, B., Lerena, P., Bonvin, D. and Stoessel, F. (1999) Optimal feed profile for a second order reaction in a semi-batch reactor under safety constraints, Experimental study. *Journal of Loss Prevention in the Process Industries*, **12** (11), 485–93.

16 Ubrich, O. (2000) Improving safety and productivita of isothermal semi-batch reactors by modulating the feed rate, in *Department de Chimie*, EPFL, Lausanne.

17 Stoessel, F. and Ubrich, O. (2001) Safety assessment and optimization of semi-batch reactions by calorimetry. *Journal of Thermal Analysis and Calorimetry*, **64**, 61–74.

18 Ubrich, O., Srinivasan, B., Lerena, P., Bonvin, D. and Stoessel, F. (2001) The use of calorimetry for on-line optimisation of isothermal semi-batch reactors. *Chemical Engineering Science*, **56** (17), 5147–56.

19 Srinivasan, B., Ubrich, O., Bonvin, D. and Stoessel, F. (2001) Optimal feed rate policy for systems with two reactions, in *DYCOPS, 6th IFAC Symposium on Dynamic Control of Process Systems*. International Federation of Automatic Control, 455–460, Cheju Island Corea.

20 Ubrich, O., Srinivasan, B., Stoessel, F. and Bonvin, D. (1999) Optimization of semi-batch reaction system under safety constraint, in *European Control Conference*, European Union Control Association, Karlsruhe.

8
Continuous Reactors

Case History

Sulfonation of p-nitrotoluene (PNT) is performed in a cascade of Continuous Stirred Tank Reactors (CSTR). The process is started by placing a quantity of converted mass in the first stage of the cascade, a 400-liter reactor, and heating to 85 °C with jacket steam (150 °C). PNT melt and Oleum are then dosed in simultaneously (exothermal reaction). When 110 °C is reached, cooling is switched on automatically. On the day of the accident, a rapid increase in pressure took place at 102 °C. The lid of the reactor burst open and the reaction mass, which was decomposing, flowed out like lava, causing considerable damage.

Subsequent investigations revealed the following history and reasons for the accident:

- September 1982: There was a change of the process involving a higher concentration of Oleum. Due to the higher concentration, an additional internal cooling coil was placed in the first reactor. As a result, the anchor-stirring unit was replaced by a smaller turbine stirrer, that was adjusted in height to optimize the circulation of the reaction mass around the coils. This stirrer had a clearance of 43 cm.
- December 1982: For safety reasons, it was decided to revert to the old process, but the modified stirrer and the coil were left in place.
- March 1983: Due to corrosion problems with the additional cooling coil, it was decided to remove the coils, but the turbine stirrer was left in place.
- October 5, 1983: This was the day of the incident. The adjustable stirring unit was positioned such that it became immersed only from 250 litres instead of the original 160 litres. Nobody noticed that the position of the stirring unit had changed. With simultaneous dosing, the light PNT remained unstirred on the surface, and of the heavier Oleum sank to the bottom of the reactor and kept it cool. The temperature sensor fitted near the bottom of the reactor showed the correct temperature for that part, but the jacket had heated the upper part of the reaction mass. The study of the thermal stability of the

Thermal Safety of Chemical Processes: Risk Assessment and Process Design. Francis Stoessel

ISBN: 978-3-527-31712-7

reaction mass showed that at 130–140 °C, the reaction mass decomposes relatively slowly, but the decomposition becomes critical in the temperature range from 170 to 200 °C. At these temperatures, decomposition occurs rapidly and spreads from top to bottom. The undesired overheating and the local reaction of the PNT with the Oleum can explain why high temperatures were reached in the layered reaction mass. After the level had risen sufficiently, the stirring unit became immersed and brought the locally accumulated reactants suddenly to reaction. From the energy potential present, the damage would have been considerably worse if the thermal explosion had taken place in a well-mixed reactor.

The lessons drawn from this incident are:

- A continuous reactor such as the CSTR cascade behaves like a batch reactor at start up.
- The original design, which was in accordance with the process, degenerated due to the successive modifications and the final configuration was no longer in accordance with the process performed in the unit. For continuous reactors, the design takes the reaction kinetics and heat balance into account and modifications of one element should lead to reconsideration of the overall design.

8.1 Introduction

To avoid thermal explosions due to unplanned reactions in continuous reactors, it is essential that any modifications are compatible with the overall design of the reactor. Therefore, the design of continuous reactors must follow well-established rules, reviewed in this chapter. We first consider two types of ideal continuous reactors: the CSTR with full back mixing and the tubular reactor without back mixing, that is, plug flow. Some special reactor types are also considered. For each reactor, basic Chemical Reaction Engineering considerations are followed by a study of the heat balance and their impact on reactor safety. Continuous reactors also present advantages that can be used intentionally during design and make possible reactions that could not be carried out in discontinuous reactors.

8.2 Continuous Stirred Tank Reactors

The continuously operated stirred tank reactor is fed with reactants at the same time as the products are removed by an overflow or a level control system (Figure 8.1). This ensures a constant volume and, consequently with a constant volume flow rate of the feed, a constant space time. We further assume the reactor contents

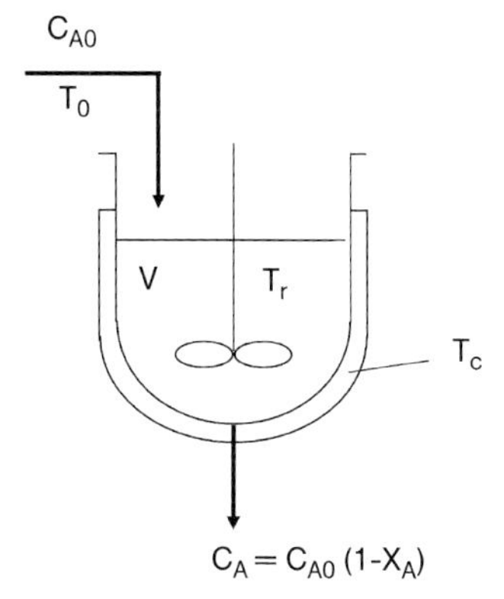

Figure 8.1 Schematic representation of a CSTR.

to be uniform in concentration and temperature, that is, the reactor is ideally stirred. The reactor may or may not be equipped with a temperature control system. These different configurations are analysed in terms of mass and energy balance, and the specific safety aspects are reviewed in the following sections.

8.2.1 Mass Balance

We consider a reaction of type $A \xrightarrow{k} P$; the CSTR (Figure 8.1) is continuously fed with a stream at an initial conversion X_0. Thus, the concentration of the reactant A in the feed stream is C_{A0} and at the outlet of the reactor is at its final value $C_{Af} = C_A = C_{A0}\,(1 - X_A)$, which is also equal to the concentration inside the reactor volume. If the reactor is operated at steady state, the molar flow rate of A, $F_{A,}$ the mass balance can be written for the reactant A:

$$F_{A0} = F_A + (-r_A)\cdot V = F_{A0}(1 - X_A) + (-r_A)\cdot V \Rightarrow F_{A0}\cdot X_A = (-r_A)\cdot V \tag{8.1}$$

Thus, the performance equation that expresses the required space time τ as a function of the initial concentration, the desired conversion, and the reaction rate becomes

$$\tau = \frac{V}{\dot{v}_0} = \frac{V\cdot C_{A0}}{F_{A0}} = \frac{C_{A0}\cdot X_A}{-r_A} \tag{8.2}$$

For a first-order reaction, where $-r_A = k\cdot C_{A0}\cdot(1 - X_A)$, the performance equation of the CSTR becomes

$$\tau = \frac{X_A}{k\cdot(1 - X_A)} \Leftrightarrow X_A = \frac{k\tau}{1 + k\tau} \tag{8.3}$$

The mass balance must be fulfilled together with the heat balance.

8.2.2 Heat Balance

The temperature control of the CSTR can be realized in different ways, such as an adiabatic reaction without cooling system or with jacket cooling. These different modes of operation, and the effect of the operating parameters on the stability of the reactor, are described in the following subsections.

8.2.3 Cooled CSTR

In cases where the isothermal CSTR is cooled by the jacket, the heat balance comprises three terms:

Reaction heat release rate:
$$q_{rx} = -r_A \cdot V \cdot (-\Delta H_r) \cdot X_A = \frac{F_{A0}}{C_{A0}} \cdot \rho \cdot Q'_r \cdot X_A = \dot{m} \cdot Q'_r (X_A - X_{A0}) \tag{8.4}$$

Sensible heat:
$$q_{fd} = \dot{m} \cdot c'_P \cdot (T_r - T_0) \tag{8.5}$$

Cooling by the jacket:
$$q_{ex} = U \cdot A \cdot (T_r - T_c) \tag{8.6}$$

If we assume the initial conversion is zero, the heat balance is

$$U \cdot A \cdot (T_r - T_c) + \dot{m} \cdot c'_P \cdot (T_r - T_0) = \dot{m} \cdot Q'_r \cdot X_A \tag{8.7}$$

This equation, together with the mass balance $F_{A0} \cdot X_A = (-r_A) \cdot V$, calculates the jacket temperature (T_c) required to maintain the reactor temperature at the desired level T_r, while obtaining a conversion X_A [1]. As an example, for a first-order reaction, by combining the mass balance in Equation 8.3 and the heat balances we find:

$$U \cdot A \cdot (T_r - T_c) + \dot{m} \cdot c'_P \cdot (T_r - T_0) = \dot{m} \cdot Q'_r \cdot \frac{k\tau}{1 + k\tau} = q_{rx} \tag{8.8}$$

Since the mass flow rate is related to the volume flow rate, we can write

$$\dot{m} = \frac{\rho V}{\tau} \tag{8.9}$$

Dividing both members by $\rho \cdot V \cdot c'_P$, we find:

$$\frac{U \cdot A}{\rho \cdot V \cdot c'_P} \cdot (T_r - T_c) + \frac{(T_r - T_0)}{\tau} = \Delta T_{ad} \cdot \frac{k}{1 + k\tau} \tag{8.10}$$

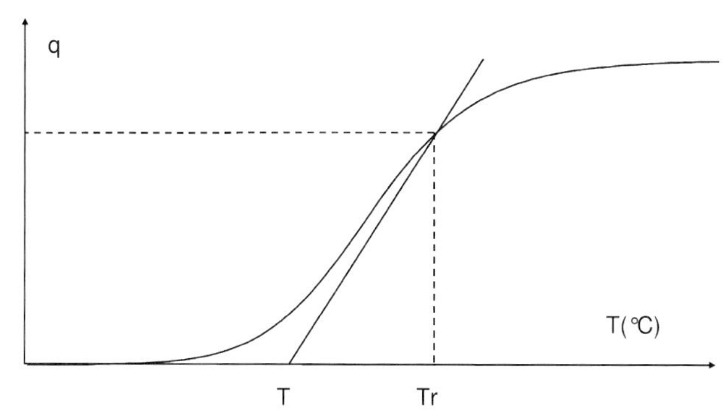

Figure 8.2 Heat balance of a cooled CSTR, with the cooling term (straight line) and the S-shaped heat release rate curve. The working point is at the intercept.

In this equation, the thermal time constant τ_{th} (see Section 9.2.4.1) appears in the cooling term:

$$\frac{(T_r - T_c)}{\tau_{th}} + \frac{(T_r - T_0)}{\tau} = \Delta T_{ad} \cdot \frac{k}{1 + k\tau} \tag{8.11}$$

The left-hand side of the equation represents the cooling term, a linear function of temperature. In the right-hand side of the equation, we find the heat release rate by the reaction, where the rate constant k is an exponential function of the temperature. Thus, the heat release rate curve is S-shaped. The working point of the CSTR is located at the intercept of both curves (Figure 8.2).

8.2.4 Adiabatic CSTR

Without jacket cooling, only three terms remain in the heat balance: the accumulation, the heat release rate of the reaction, and the sensible heat due to the temperature difference between the feed and the reactor contents. Thus, we obtain

$$X_A = \frac{c'_P}{Q'_r} \cdot (T_r - T_0) = \frac{T_r - T_0}{\Delta T_{ad}} = \frac{-r_A \cdot \tau}{C_{A0}} \tag{8.12}$$

In a diagram $X_A = f(T)$, the left member of this equation is a straight line with a slope equal to the inverse of the adiabatic temperature rise. In the same diagram, the mass balance, corresponding to the right member of the equation, gives an S-shaped curve. Since both equations must be simultaneously satisfied, the working point is located at the intercept of both curves (Figure 8.3). Thus the temperature and conversion are determined by the feed temperature (T_0), the thermodynamic parameters, that is, heat of reaction, heat capacity, and reaction kinetics that define the sigmoid curve of the mass balance.

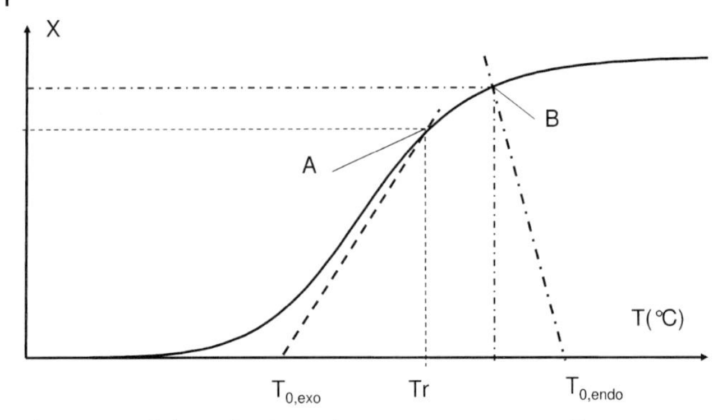

Figure 8.3 Adiabatic CSTR: Working point as intercept of heat and mass balances. Point A is the working point for an exothermal reaction. Point B is the working point for an endothermal reaction.

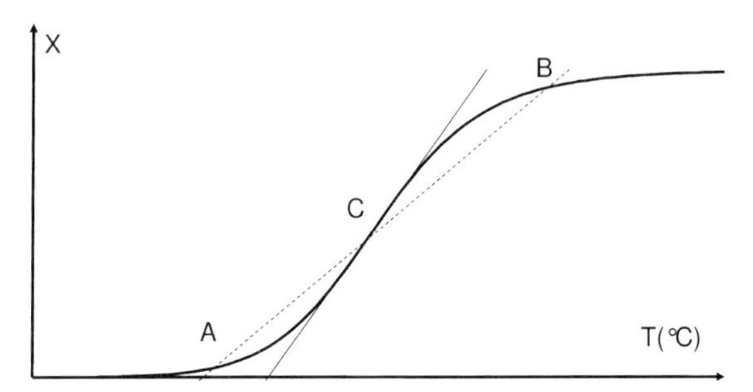

Figure 8.4 Adiabatic CSTR: With the solid straight line, there is no clearly defined working point, which results in an oscillating behavior. With the dashed line, there are multiple solutions (points A, B, C).

When the slope of the heat balance line is close to the slope of the inflexion point of the mass balance curve, there is no longer a well-defined working point (solid line in Figure 8.4) and the reactor may enter cyclic oscillations of temperature and conversion, even for constant working parameters. To avoid these oscillations, the condition must be fulfilled [2]:

$$\left|\frac{dX}{dT}\right|_{\tau} < \frac{1}{\Delta T_{ad}} \tag{8.13}$$

Thus, this equation is a stability condition for the adiabatic CSTR. If this condition is not fulfilled, such as in strongly exothermal reactions, there may also be a situation where there are multiple solutions (dashed line in Figure 8.4). In such a case, a small perturbation of one of the process parameters makes the reactor jump from low conversion to high conversion, or reversely, leading to an instable operation. The stability conditions of the CSTR were studied in detail by

Chemburkar *et al.* [3]. An extensive discussion of the parametric sensitivity of the CSTR is presented by Varma [4]. In the example in Figure 8.4, A is a working point on the cold branch, B is an instable point, and C is a working point on the hot branch. The consequences of this multiplicity are explained in more detail in Section 8.2.6.1.

8.2.5 The Autothermal CSTR

In the CSTR the reaction mass leaves the reactor at reaction temperature. Thus, when an exothermal reaction is carried out in this reactor type, it may be interesting from an energy viewpoint, to recover the heat of the reaction mass for preheating the feed. The reactor outlet then flows through a heat exchanger that heats the feed to the desired temperature. This type of operation is called the autothermal reactor (Figure 8.5). The feed enters the heat exchanger at T_0 and leaves it to enter the reactor at T_{fd}. The reaction mass leaves the reactor at T_r and enters the second side of the heat exchanger to leave it at T_f. Thus, the reactor works without an external heat source, the heat of reaction recovered for heating the reactants.

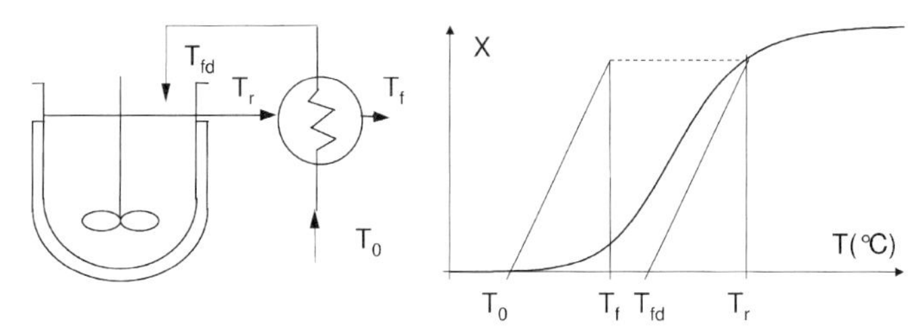

Figure 8.5 Autothermal CSTR.

8.2.6 Safety Aspects

Safety aspects for two different situations are considered. The first is for nominal operating conditions, where the objective is to maintain stable control of the reactor temperature. The second is for deviations of these operating conditions, especially in the case of a cooling failure, where the objective is to design a reactor that behaves safely even under adiabatic conditions.

8.2.6.1 Instabilities at Start- up or Shut Down

When a CSTR is thermally started, that is, the feed temperature is progressively increased from T_1 to T_2 (Figure 8.6), the working point first moves in the cold branch from T_{r1} to T_{r2}. At the feed temperature T_2, two solutions T_{r2} and T'_{r2} are possible (multiplicity). Thus, as the reactor temperature suddenly jumps from T_{r2} to T'_{r2}, it ignites. If the feed temperature continues to increase to T_3, the working

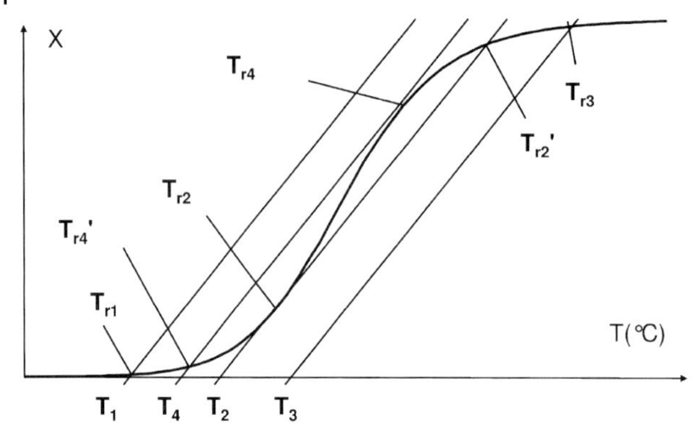

Figure 8.6 Ignition and hysteresis in the CSTR at start-up and shut down.

point moves on the hot branch to T_{r3}. At shut down, the feed temperature progressively decreases from T_3 to T_4 and the working point moves back on the hot branch from T_{r3} to T'_{r4}. At this point, there are two solutions and the reactor temperature suddenly jumps to the cold branch at T'_{r4}: this is the extinction point. When the feed temperature continues decreasing, the working point continues moving on the cold branch until it comes back to T_{r1}. Since ignition occurs at a higher feed temperature (T_2) than the extinction (T_4), there is a hysteresis phenomenon. This must be taken into account when designing the temperature control system of the reactor.

8.2.6.2 Behavior in Case of Cooling Failure

In the case of cooling failure, if the feed remains active, the reactor behaves adiabatically and its temperature and conversion are shifted to the corresponding adiabatic working point. Thus, its behavior can be predicted using the heat and material balance, as developed in Section 8.2.4.

If the feed is suddenly stopped, it behaves as an adiabatic batch reactor with an accumulation corresponding to the non-converted fraction, thus the maximum temperature of the synthesis reaction is

$$MTSR = T_r + (1 - X_A) \cdot \Delta T_{ad} \tag{8.14}$$

Since the conversion is generally high, the accumulation is low and the temperature increase remains low. This is a great advantage of the CSTR for strongly exothermal reactions. Since the MTSR depends on the working temperature (T_r), for a low reactor temperature, the conversion is also low and the accumulation high, if the space time is maintained as constant. Thus, for strongly exothermal reactions, the MTSR curve as a function of the process temperature presents a minimum (Figure 8.7). This corresponds to the optimum reaction temperature from the safety point of view. It was shown [5] that, for second-order reactions, the minimum only appears for a reaction number $B > 5.83$.

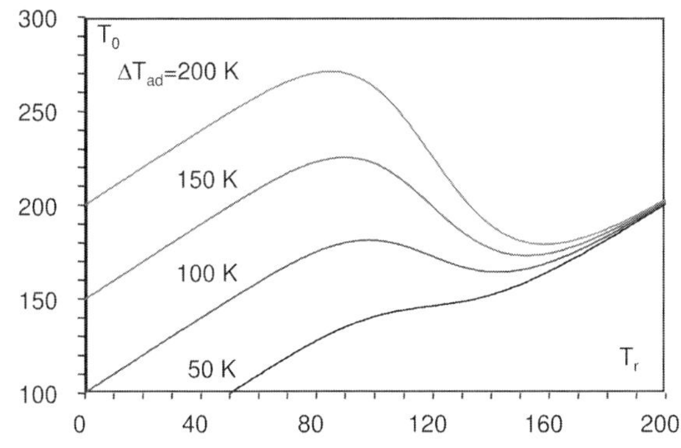

Figure 8.7 MTSR as a function of the reaction temperature in a CSTR with an exothermal reaction.

Worked Example 8.1: Fast Reaction in a CSTR

This is the continuation of Worked Examples 6.3 and 7.1.

A reaction $A \xrightarrow{k} P$ is to be performed in a CSTR. The reaction follows first-order kinetics and at 50 °C, the conversion reaches 99% in 60 seconds. The reaction is to be performed in a CSTR with the same productivity as the semi-batch reactor (Worked Example 7.1). The overall heat transfer coefficient of the reactor is $500\,\mathrm{W\,m^{-2}\,K^{-1}}$. The maximum temperature difference with the cooling system is 50 K.

Data:

$C_A\ (t = 0) = C_{A0} = 1000\,\mathrm{mol\,m^{-3}}$ $\rho = 900\,\mathrm{kg\,m^{-3}}$

$c'_P = 2000\,\mathrm{J\,kg^{-1}\,K^{-1}}$ $-\Delta H_r = 200\,\mathrm{kJ\,mol^{-1}}$

Normalized jacketed reactors:

V (m^3)	0.63	1.0	1.6	2.5
A (m^2)	2.82	4.2	6.6	8.9

Questions:

1. Reactor design: what must be the volume of the CSTR in order to achieve the same performance as the $5\,\mathrm{m^3}$ semi-batch reactor with time-cycle of 2 hours and a conversion of 99%?
2. Isothermal reaction: under what conditions is this strategy possible?
3. Slow reaction: what happens with a 100 times slower reaction?
4. Cooling failure: what happens in case of cooling failure?

Solution:

1. Reactor design:

The semi-batch reactor produces $5\,\mathrm{m^3}$ of product solution every 2 hours. Therefore, in a continuous reactor, the flow rate must be $2.5\,\mathrm{m^3\,h^{-1}}$ or $6.94 \cdot 10^{-4}\,\mathrm{m^3\,s^{-1}}$.

The mass balance is

$$\tau = \frac{V}{\dot{v}_0} = \frac{C_{A0} \cdot X_A}{-r_A}$$

For a first-order reaction:

$$-r_A = k \cdot C_{A0}(1 - X_A)$$

Hence:

$$\tau = \frac{X_A}{k \cdot (1 - X_A)} = \frac{0.99}{0.077 \times 0.01} = 1286\,\text{s}$$

The reactor volume is

$$V = \tau \cdot \dot{v}_0 = 1286\,\text{s} \times 6.94 \cdot 10^{-4}\,\text{m}^3\,\text{s}^{-1} \cong 0.9\,\text{m}^3$$

The $1\,\text{m}^3$ reactor is most suitable, as the volume of the reaction mass will be maintained constant at $0.9\,\text{m}^3$. Then, the heat exchange area is $A = 4.2\,\text{m}^2$.

2. Isothermal reaction; heat balance:

Heat production:

$$q_{rx} = \dot{v}_0 \cdot C_{A0} \cdot X_A \cdot (-\Delta H_R) = 6.94 \cdot 10^{-4}\,\text{m}^3\,\text{s}^{-1} \times 1000\,\text{mol}\,\text{m}^{-3} \times 0.99 \\ \times 200\,\text{kJ}\,\text{mol}^{-1} \cong 139\,\text{kW}$$

Heat exchange:

$$q_{ex} = U \cdot A \cdot \Delta T = 0.5\,\text{kW}\,\text{m}^{-2}\,\text{K}^{-1} \times 4.2\,\text{m}^2 \times 50\,\text{K} = 105\,\text{kW}$$

The difference of 34 kW, which cannot be removed by the jacket, can be compensated for by using the cooling effect of the feed:

$$\frac{q_{RX} - q_{EX}}{\dot{v}_0 \cdot \rho \cdot Cp} = \frac{139\,\text{kW} - 105\,\text{kW}}{6.94 \cdot 10^{-4}\,\text{m}^3/\text{s} \times 900\,\text{kg/m}^3 \times 2\,\text{kJ/(kg} \cdot \text{K)}} \approx 27\,\text{K}$$

Thus, a feed temperature of 23 °C will maintain a reactor temperature of 50 °C.

3. Slow reaction:
For a 100 times slower reaction, the heat exchange becomes fully uncritical, but the conversion of 99% can only be reached with a volume of $90\,\text{m}^3$, which is unrealistic. The situation improves with a higher reactor temperature or with a different combination of reactors: cascade of CSTRs or CSTR followed by a tubular reactor.

4. Cooling failure:
If the feed is stopped immediately in the case of malfunction, the CSTR is uncritical: the non-converted reactant is only 1%, resulting in a ΔT_{ad} of ca. 1 °C only. This result enhances the strength of the CSTR in its behavior after cooling failure. The CSTR is a practicable and elegant solution for the industrial performance of this fast and exothermal reaction. Since the technique is based on a stirred tank, it does not require high investment for it to be realized in a traditional multipurpose plant.

8.3 Tubular Reactors

In a tubular reactor, the reactants are fed in at one end and the products withdrawn from the other. If we consider the reactor operated at steady state, the composition of the fluid varies inside the reactor volume along the flow path. Therefore, the mass balance must be established for a differential element of volume dV. We assume the flow as ideal plug flow, that is, that there is no back mixing along the reactor axis. Hence, this type of reactor is often referred to as Plug Flow Reactor (PFR).

8.3.1 Mass Balance

If we write the mass balance for a reaction of the type $A \xrightarrow{k} P$, at steady state, there is no accumulation term, thus the mass balance becomes [6]

$$\text{input} = \text{output} + \text{disappearance by reaction} \tag{8.15}$$

The input can be written as a molar flow rate F_A, the output is $F_A + dF_A$ (Figure 8.8), and the disappearance by reaction is $(-r_A \cdot dV)$. By substituting these expression into Equation 8.15, we find

$$F_A = F_A + dF_A + (-r_A)dV \tag{8.16}$$

Since $dF_A = d[F_{A0}\ (1 - X_A)] = -F_{A0}dX_A$ we obtain

$$F_{A0}dX_A = (-r_A)dV \tag{8.17}$$

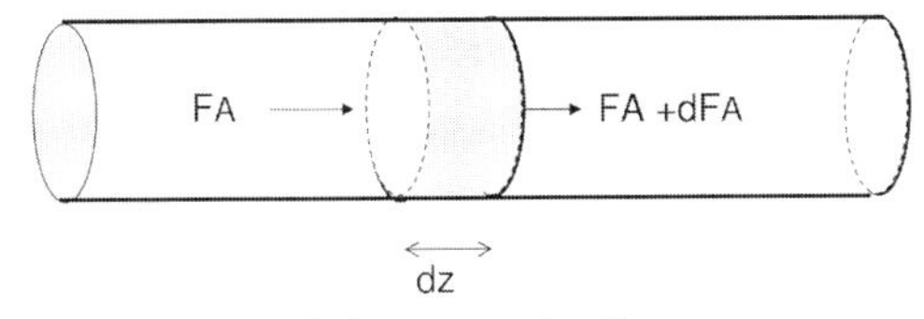

Figure 8.8 Mass balance in a plug flow reactor.

This equation represents the mass balance of A in the differential volume dV. In order to obtain the overall mass balance of the reactor, we must integrate this expression over the reactor volume, by taking into account that the reaction rate is a function of the local concentration:

$$\int_0^V \frac{dV}{F_{A0}} = \int_0^{X_A} \frac{dX_A}{-r_A} \tag{8.18}$$

Thus, since $F_{A0} = \dot{v}_0 \cdot C_{A0}$, the performance equation of the ideal plug flow reactor can be written as

$$\tau = \frac{V}{\dot{v}_0} = C_{A0} \int_0^{X_A} \frac{dX_A}{-r_A} \tag{8.19}$$

The great difference with the CSTR (Equation 8.2) is that here the reaction rate varies within the reactor volume instead of being constant. Hence, the reaction rate is in the integral term. If the initial conversion is not zero, the equation becomes

$$\tau = \frac{V}{\dot{v}_0} = C_{A0} \int_{X_{A0}}^{X_{Af}} \frac{dX_A}{-r_A} \tag{8.20}$$

For constant density, which is valid for liquids, the performance equation can be written as a function of concentration:

$$\tau = \frac{V}{\dot{v}_0} = C_{A0} \int_0^{X_A} \frac{dX_A}{-r_A} = -\int_{C_{A0}}^{C_{Af}} \frac{dC_A}{-r_A} \tag{8.21}$$

This expression can be integrated for different forms of the rate equation.

8.3.2
Heat Balance

The heat balance can be written either globally over the whole reactor volume or locally for a differential element dV. The global heat balance is similar to the CSTR:

Heat production:

$$q_{rx} = \dot{m} \cdot (C'_0 - C'_f) \cdot (-\Delta H_r)$$

Heat exchange across the wall:

$$q_{ex} = U \cdot A \cdot (T_r - T_c)$$

Convective exchange:

$$q_{cx} = \dot{m} \cdot c'_P \cdot (T_f - T_0)$$

The local heat balance written in a differential element dV is:

Heat production:

$$q_{rx} = (-r_A)(-\Delta H_r) \cdot dV = (-\Delta H_r) \cdot C_{A0} \cdot dX_A \cdot dV$$

Heat exchange across the wall:

$$q_{ex} = U \cdot dA \cdot (T_c - T)$$

Convective exchange:

$$q_{cx} = \dot{m} \cdot c'_P \cdot dT + c_w \frac{dT}{dt}$$

Here c_w stands for the heat capacities of the reactor wall and dT the temperature variation in the volume dV.

By ignoring the heat capacity of the wall, the heat balance becomes

$$\dot{m} \cdot c'_P \cdot \frac{dT}{dV} = (-r_A)(-\Delta H_r) + U \cdot (T_c - T) \cdot \frac{dA}{dV} \tag{8.22}$$

Assuming a cylindrical geometry with a tube diameter d_r, we have

$$dV = \frac{\pi d_r^2}{4} \cdot dz \quad \text{and} \quad dA = \pi d_r \cdot dz$$

where z is the length coordinate along the tube axis. At constant density, the mass flow rate can be expressed as

$$\dot{m} = \frac{\rho V}{\tau}$$

and the heat balance becomes [1, 2]

$$\frac{\rho V}{\tau} \cdot c'_P \cdot \frac{dT}{dz} = (-r_A)(-\Delta H_r) + \frac{4U}{d_r} \cdot (T_c - T) \tag{8.23}$$

which can be rearranged as

$$\frac{dT}{dz} = \Delta T_{ad} \frac{-r_A \tau}{C_{A0}} + \frac{4U\tau}{\rho \cdot c'_P \cdot d_r}(T_c - T) \tag{8.24}$$

This equation allows calculating the temperature profile in a polytropic tubular reactor.

8.3.3
Safety Aspects

8.3.3.1 Parametric Sensitivity

The mass balance $\frac{dX_A}{dz} = \frac{-r_A \tau}{C_{A0}}$ must be satisfied simultaneously with the heat balance and we obtain a system of coupled differential equations:

$$\begin{cases} \dfrac{dX_A}{dz} = \dfrac{-r_A \tau}{C_{A0}} \\ \dfrac{dT}{dz} = \underbrace{\Delta T_{ad} \dfrac{dX_A}{dt}}_{a} + \underbrace{\dfrac{4U\tau}{\rho c_P' d}(T_c - T)}_{b} \end{cases} \tag{8.25}$$

These equations calculate the temperature and conversion profiles in a polytropic tubular reactor. The term (a) represents the heat generation rate by the reaction and the term (b) the heat removal rate by the heat exchange system. This equation is similar to Equation 5.2, obtained for the batch reactor. Moreover, since the conversion rate $\frac{dX_A}{dz}$ is a strongly non-linear function of the temperature, the system of differential equations may be parametrically sensitive. In this case, the sensitivity may lead to a local runaway, that is, a hot spot in the tubular

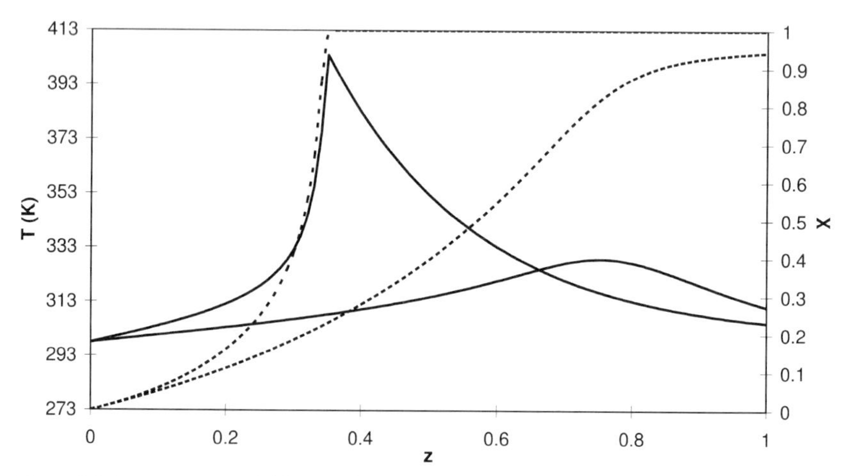

Figure 8.9 Temperature (solid line) and conversion (dashed line) profiles in a tubular reactor for two cooling medium temperatures 293 K and 300 K, the second leading to a hot spot.

Table 8.1 Specific cooling capacity compared for different reactor sizes.

Reactor	A/V m^2/m^3	Cooling capacity $kW\,m^{-3}$
Tubular d_t = 20 mm	200	5000
Tubular d_t = 100 mm	40	1000
Laboratory 0.1 l	100	2500
Kilo lab 2 l	40	1000
Pilot plant 100 l	9	225
Production 1 m^3	3	75
Production 25 m^3	1	25

The cooling capacities are calculated for a heat exchange coefficient of $500\,W\,m^{-2}\,K^{-1}$ and a temperature difference of 10 K with the cooling medium.

reactor (Figure 8.9). Thus, the temperature control of a tubular reactor may be difficult. Nevertheless, the great advantage compared to the batch reactor is that the specific heat exchange area is by far greater, which leads to a comparatively high cooling capacity. This point is discussed in the next section.

8.3.3.2 Heat Exchange Capacities of Tubular Reactors

The geometry confers the tubular reactors with a high specific heat exchange area compared to stirred tank reactors. This allows tubular reactors to have a specific cooling capacity, in that cooling capacity per unit of volume is comparable to laboratory stirred tank reactors (Table 8.1). Thus, tubular reactors perform strongly exothermal reactions in a safe way. A further advantage is that the reactor volume is small compared to that of batch or semi-batch reactors, which also reduces the overall energy potential exposed to reaction conditions at the same time. Moreover, a tubular reactor can be constructed to resist high pressures, with less investment than would be the case for a stirred tank reactor. This may allow a fail-safe construction. Nevertheless, continuous operation is only possible for fast reactions, in order to obtain a short residence time. This must not be a drawback, since a reaction can be at a higher temperature in order to accelerate it and obtain a high conversion in a short residence time.

8.3.3.3 Passive Safety Aspects of Tubular Reactors

The small dimensions of the tubular reactor, compared to the stirred tank reactor, represent a further advantage. The full heat capacity of the reactor cannot be neglected before the heat capacity of the reaction mass. Thus, in the case of cooling failure, the heat released by the reaction not only serves to increase the temperature of the reaction mass, but also to increase the temperature of the reactor wall. This is a kind of "thermal dilution" of the reaction mass by the reactor itself, meaning that if there is cooling failure, the adiabatic temperature increase is strongly reduced. This can be compared to the adiabacity coefficient used in adiabatic calorimeters, the major difference being the objective to realize a high adiabacity coefficient.

Worked Example 8.2: Fast Reaction in PFR

This example is a continuation of Worked Examples 6.3, 7.1, and 8.1.

A reaction $A \xrightarrow{k} P$ is to be performed in a PFR. The reaction follows first-order kinetics, and at 50 °C in the batch mode, the conversion reaches 99% in 60 seconds. Pure plug flow behavior is assumed. The flow velocity should be $1\,m\,s^{-1}$ and the overall heat transfer coefficient $1000\,W\,m^{-2}\,K^{-1}$. (Why is it higher than in stirred tank reactors?). The maximum temperature difference with the cooling system is 50 K.

Data:

$C_A\ (t=0) = C_{A0} = 1000\,mol\,m^{-3}$ $\quad \rho = 900\,kg\,m^{-3}$

$c'_P = 2000\,J\,kg^{-1}\,K^{-1}$ $\quad -\Delta H_r = 200\,kJ\,mol^{-1}$

Questions:

1. Reactor design: calculate the required tube length and diameters required to achieve the same performance as the $5\,m^3$ semi-batch reactor with time-cycle of 2 hours and a conversion of 99%.
2. Isothermal reaction: under which conditions is this strategy possible?
3. Slow reaction: what happens with a 100 times slower reaction?
4. Cooling failure: what happens in case of cooling failure?

Solution:

1. Reactor design:

The semi-batch reactor produces $5\,m^3$ of product solution every 2 hours. Therefore, in a continuous reactor, the flow rate must be $2.5\,m^3\,h^{-1}$ or $6.94 \cdot 10^{-4}\,m^3\,s^{-1}$. In order to achieve a space time $\tau = 1$ minute, necessary to reach a conversion of 99%, the volume is

$$V = \tau \cdot \dot{v}_0 = 0.042\,m^3$$

With a flow rate of $1\,m\,s^{-1}$, the length is 60 m. Hence the diameter is

$$d_r = \sqrt{\frac{4V}{\pi}} \cong 0.03\,m$$

2. Isothermal reaction: The overall heat balance gives:

Heat production:

$$q_{rx} = \dot{v}_0 \cdot C_{A0} \cdot (-\Delta H_R)$$

$$q_{rx} = 6.94 \cdot 10^{-4}\,m^3\,s^{-1} \times 1000\,mol\,l^{-1} \times 200\,kJ\,mol^{-1} = 139\,kW$$

Worked Example 8.3: Adiabacity Factor for a Tubular Reactor

An exothermal reaction with an adiabatic temperature rise of 100 K is to be performed in a tubular reactor with internal diameter of 30 mm, wall thickness of 2 mm, and surrounding jacket of thickness 30 mm containing water. Calculate the effective temperature rise that would occur if the reactor suddenly lost the utilities. In this situation, the reactant flow is stopped and there is no water flow in the jacket.

The reaction mass is a typical organic liquid with density of 900 $kg\,m^{-3}$ and specific heat capacity of 1800 $J\,kg^{-1}\,K^{-1}$. The wall has density of 8000 $kg\,m^{-3}$ and specific heat capacity of 500 $J\,kg^{-1}\,K^{-1}$. Water has a density of 1000 $kg\,m^{-3}$ and a specific heat capacity of 4180 $J\,kg^{-1}\,K^{-1}$.

Solution:
The calculation is made for a unit length of 1 m. The volume of the reaction mass is

$$V = \pi r_1^2 l = 3.14 \times 0.015^2\,\mathrm{m}^2 \times 1\,\mathrm{m} = 7.069 \cdot 10^{-4}\,\mathrm{m}^3$$

The heat capacity is

$$c_{P,r} = \rho V c_P' = 900\,\mathrm{kg\,m^{-3}} \times 7.069 \cdot 10^{-4}\,\mathrm{m}^3 \times 1800\,\mathrm{J\,kg^{-1}\,K^{-1}} = 1145\,\mathrm{J\,K^{-1}}$$

In a similar way, the heat capacity of the reactor wall is

$$c_{P,w} = \pi(r_2^2 - r_1^2) l \cdot \rho \cdot c_P'$$

$$c_{P,w} = 3.14 \times (0.017^2 - 0.015^2)\,\mathrm{m}^2 \times 1\,\mathrm{m} \times 8000\,\mathrm{kg\,m^{-3}} \times 500\,\mathrm{J\,kg^{-1}\,K^{-1}} = 804\,\mathrm{J\,K^{-1}}$$

For the water layer we find $c_{P,c} = 25\,213\,J \cdot K^{-1}$.

The adiabacity coefficient is

$$\Phi = \frac{c_{P,r} + c_{P,w} + c_{P,c}}{c_{P,r}} = \frac{1145 + 804 + 25213}{1145} = 23.7$$

Thus the temperature rise is only

$$\Delta T = \frac{\Delta T_{ad}}{\Phi} = \frac{100\,\mathrm{K}}{23.7} = 42\,\mathrm{K}$$

which may not cause any harm to the reactor.

8.4
Other Continuous Reactor Types

8.4.1.1 Cascade of CSTRs and Recycle Reactor

With regard to flow conditions, the great difference between the continuous stirred tank reactor and the tubular reactor is full back mixing in the CSTR and pure plug flow, that is, no back mixing in the tubular reactor. From the safety point of view, this results in a low accumulation in the CSTR, whereas the accumulation decreases along the reactor length in the tubular reactor. Intermediate flow characteristics are achieved in two ways, either by placing several CSTRs in series, building a so-called cascade reactor, or by redirecting some of the outlet of a plug flow reactor to its entrance, building a so-called recycle reactor.

The safety aspects of the cascade reactor are the same as for the CSTR, whereas the focus should be on the first stage, where usually the greatest conversion increase takes place, that is, the heat release is also greatest. Moreover, at this stage the conversion is lowest, implying the highest degree of accumulation.

The recycling reactor behaves similarly to the plug flow reactor, with one major difference, that the conversion range in the reactor is narrower than in the true plug flow reactor (Figure 8.10). This reactor type is also named the differential reactor. The performance equation is [1, 2, 6, 7]

$$\tau = \frac{V}{\dot{v}_0} = C_{A0}(1+R)\int_{\frac{R}{1+R}X_f}^{X_f} \frac{dX}{-r_A} \tag{8.26}$$

The recycling ratio R is the ratio of recycled flow rate to the flow rate leaving the reactor. It adjusts the degree of back mixing and consequently the accumulation of non-converted reactants. On the other hand, the geometry of the reactor is the same as for the tubular reactor. Therefore, it allows the same high specific heat exchange area. Moreover, this geometry also confers the recycle reactor with the same high heat capacity, providing a possible “thermal dilution.” From a safety viewpoint, this brings together the advantages of both reactor types: the low accumulation of the CSTR, the high cooling capacity and the high heat capacity of the tubular reactor. Therefore, this type of reactor presents many advantages for performing strongly exothermal reactions in a safe way.

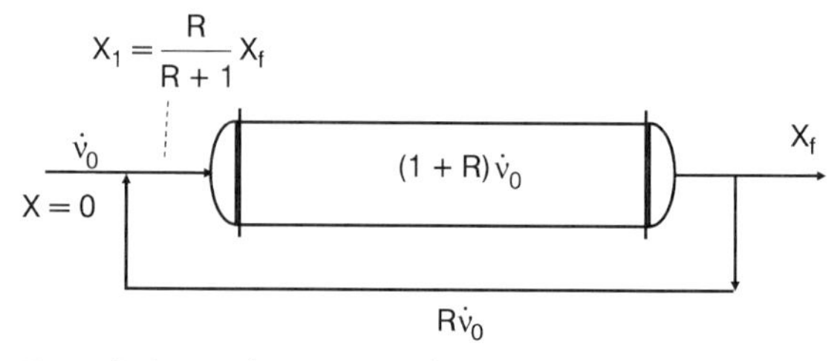

Figure 8.10 Recycling reactor. The conversion at reactor entrance is $X_1 = R/(1 + R)X_f$.

Table 8.3 Specific heat exchange area and cooling capacity of different reactor types. The values are calculated for a heat transfer coefficient of 1000 W m^{-2} K^{-1} and a ΔT of 10 K.

Reactor	Dimension	A/V m^2/m^3	q_{ex} kW m^{-3}
Stirred tank bench scale	2 liters	40	400
Stirred tank production	1 m^3	4	40
Tubular diameter 100 mm	100 mm	40	400
Tubular diameter 10 mm	10 mm	400	4 000
Milli-reactor, diameter	1 mm	4 000	40 000
Micro-reactor, diameter[a)]	0.1 mm	40 000	400 000

a) For micro-reactors, the heat transfer coefficient may be even higher by one order of magnitude.

8.4.1.2 Micro Reactors

Pushing the geometry of tubular reactors to the extreme leads to the micro reactor. The high surface to volume ratio provides a highly efficient heat transfer. The small volume of the micro-channel means that only small amounts of potentially dangerous chemicals are used, which also represents an extreme in terms of process intensification (see Section 10.3). The small radius of the micro-channel results in a short time for radial diffusion and thus good mass transfer. Hence, a micro reactor can be designed as a plug-flow reactor operating continuously at steady state. The reactants are mixed at the beginning of the reaction micro-channel, which allows good control of the reaction progress as well as of the thermal effects.

The increase in the specific heat exchange area is illustrated in Table 8.3. The values in this table were calculated assuming a constant heat transfer coefficient of 1000 W m^{-2} K^{-1}. In fact, the effect is even greater since the heat transfer coefficient increases for small tube dimensions. For example, for a typical set of physical properties resulting in a heat transfer coefficient of 500 W m^{-2} K^{-1} in a 10 m^3 stirred tank, a tubular reactor of 10 mm diameter will present a heat transfer coefficient of 1000 W m^{-2} K^{-1}. A micro reactor with a tube diameter of 0.1 mm will present a heat transfer coefficient as high as 20 000 W m^{-2} K^{-1}. Thus, the effects of the dimensions on the safety characteristics are even greater.

In Section 8.3.3, the safety performance of the tubular reactor was compared to the stirred tank reactor. This comparison can now be extended to a micro reactor. For this, we take a 10 m^3 stirred tank vessel, a tubular reactor with 10 mm tube diameter length 1 m, and a micro reactor with 0.1 mm tube diameter and length 1 cm. We compare the following criteria that are important for the reactor safety:

- The thermal time constant, which gives an indication of temperature control dynamics, that is, of the response time of the reactor to a change in the temperature set point (see Section 9.2.4.1):

9
Technical Aspects of Reactor Safety

Case History "Process Transfer"

A Vielsmeier reaction was performed over several years in a reactor equipped with a water circulating jacket. The reaction proceeds in two steps: first the Vielsmeier complex is formed at 35 °C by slow addition of phosphor-oxy-chloride ($POCl_3$) to the reactants. In a second step, the reaction mass is heated to 94 °C within 1 hour and maintained at this temperature for a given time. This process was then transferred to another plant, where the first batch led to a runaway reaction during the heating phase. The temperature was not stabilized at the required 94 °C, continuing to rise, leading to a gas release to the environment. The PVC-ventilation line also melted due to the high gas temperature, causing an eruption of the reaction mass, resulting in heavy damage to the plant.

The analysis of the incident showed that this reactor was equipped with an indirect heating-cooling system with oil circulation and computerized temperature control. In this plant, to obtain a "nervous" temperature control, the control algorithm of the cascade controller (see Section 9.2.4.3) was adjusted to have an on-off behaviour, by calculating the set temperature of the jacket proportional to the squared difference between the actual and set values of the reaction medium temperature as

$$T_{c,set} = T_c + G \cdot (T_{r,set} - T_r)^2 \cdot \frac{T_{r,set} - T_r}{|T_{r,set} - T_r|}$$

The temperature curves of the initial reactor and the incident batch are shown in Figure 9.1. In the initial reactor, where the water circulation operated at atmospheric pressure, the jacket temperature was physically limited to 100 °C. Further, the control algorithm made the jacket start to cool at a reactor temperature of around 90 °C. In the new reactor, the jacket temperature reaches higher temperatures, which begin decreasing only as the reactor temperature reaches 94 °C. Since it takes time for the jacket to decrease below the reactor temperature, the cooling effect only starts as the reactor temperature 97 °C.

Thermal Safety of Chemical Processes: Risk Assessment and Process Design. Francis Stoessel

ISBN: 978-3-527-31712-7

Then the reaction was so fast that the cooling capacity was not sufficient to compensate for the heat released by the reaction. The reaction was no longer controllable and runaway was inevitable.

Thus changing the parameters and the properties of the temperature control system was sufficient to cause an incontrollable reaction course. If the temperature of the jacket had simply been limited to 100 °C, the incident would have been avoided (Figure 9.2). This case history shows how important the dynamics of the temperature control system are.

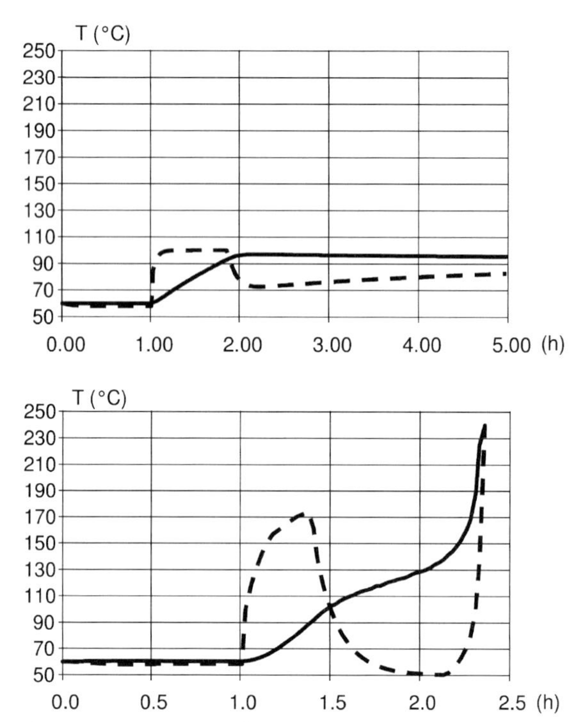

Figure 9.1 Comparison of temperature curves in the initial reactor (upper) and in the new reactor (lower). Dashed line: jacket temperature; solid line: reactor temperature.

9.1 Introduction

In order to control the reaction course and so avoid a runaway incident, it is essential to understand how heating-cooling systems of reactors work and what their performance is. These topics are reviewed in this chapter, where different heating and cooling systems are reviewed from the particular implications on process safety. In the first section, the different heating and cooling techniques

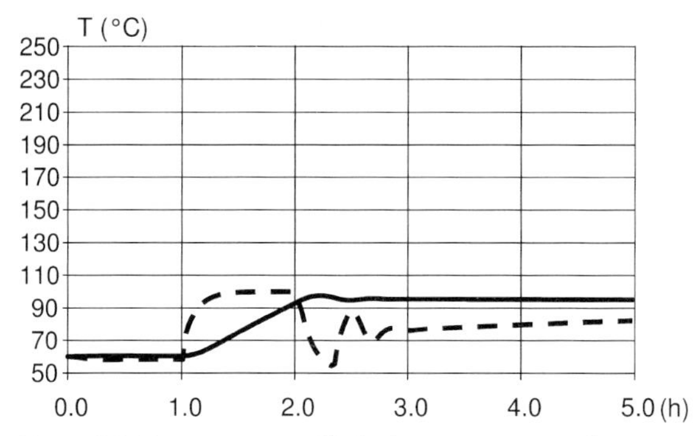

Figure 9.2 Temperature profile in the new reactor with a limitation of jacket temperature at 100 °C. Dashed line: jacket temperature; solid line: reactor temperature.

are presented with their particular features. A novel technique, for taking the dynamics of industrial reactors into account during process development, is presented. The second section is devoted to heat exchange across the reactor wall and the last section to evaporative cooling.

9.2 Temperature Control of Industrial Reactors

9.2.1 Technical Heat Carriers

9.2.1.1 Steam Heating

The most common heat carrier for heating industrial reactors is steam, providing an efficient and simple means. The efficiency of steam is due to its high latent heat of condensation ($\Delta H_v = 2260\,\text{kJ}\,\text{kg}^{-1}$ at 100 °C). For saturated steam, the temperature can be controlled by its pressure. Some values are presented in Table 9.1. The pressure and latent heat of evaporation corresponding to a given temperature may easily be estimated using Regnault law:

$$P[bar] = \left[\frac{T[^\circ\text{C}]}{100}\right]^4$$
$$\Delta H'_V[\text{kJ}\cdot\text{kg}^{-1}] = 2537 - 2.89\cdot T[^\circ\text{C}] \qquad (9.1)$$

The temperature is controlled by a pressure control valve, which provides a technically simple system. The condensate drains from the jacket via a purge, which continues to maintain the required pressure in the system when the condensate is drained off (Figure 9.3). The drawback is that for high temperatures, the valves and piping system become heavy and expensive, for example, 240 °C requires a pressure of over 30 bar. Another important point is that when the

Table 9.1 Temperature as a function of pressure for saturated steam.

Temperature (°C)	100	125	150	175	200
Pressure (bar a)	1	2.3	4.8	8.9	15.5

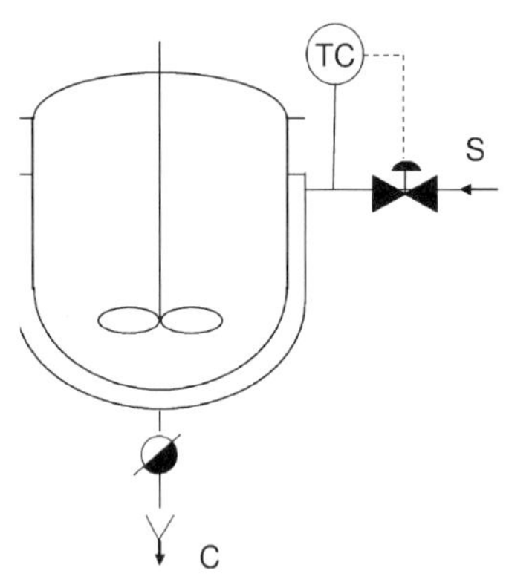

Figure 9.3 Reactor with steam heated jacket (S: steam, C: condensate).

temperature of the reactor contents surpasses the dew temperature, practically no heat exchange takes place between the reaction mass and the jacket: steam does not cool.

9.2.1.2 Hot Water Heating

Heating, using hot water circulation, may be used either open to the atmosphere, in this case limited to 100 °C, or under pressure, in which case the limitations are the same as for steam. The circulating water may be heated either by direct steam injection into the circulation (Figure 9.4), or indirectly by a heat exchanger. This system can easily switch from heating to cooling by closing the steam valve and opening the cold-water inlet valve. Opposite to steam heating, in case the reactor temperature surpasses the jacket temperature, the heat flow reverses, that is, the jacket cools the reactor. Thus, for safety reasons, this system is preferred to direct steam heating into the jacket, since it provides passive safety.

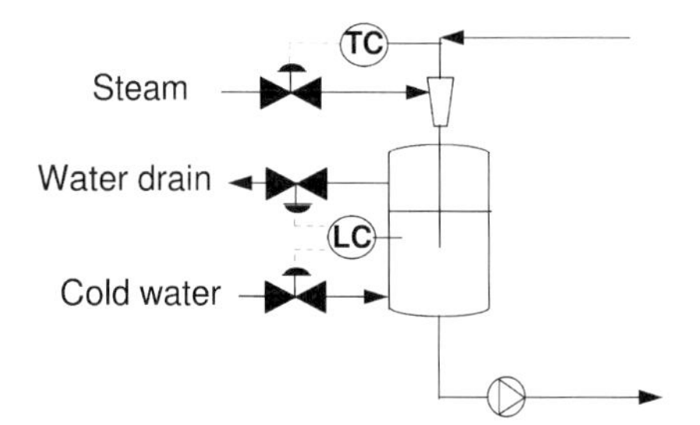

Figure 9.4 Hot water circulation with direct steam injection.

9.2.1.3 Other Heating Media

In industrial systems, other thermal fluids are used in heat transport, such as mineral oils, organic liquids (i.e. Dowtherm, an eutectic mixture of diphenyl and diphenyloxide), Marlotherm, or silicon oils. These liquids are selected, essentially for their stability and physical transport properties, which confer on them good heat transport capabilities over a broad temperature range, also reaching below ambient temperatures. Temperatures above 200 °C are reached for certain fluids if the circulation system is protected against oxidation by, for example, nitrogen. They are used in a closed loop system (see double circuit below) that allows heating as well as cooling with the same fluid. These fluids may also be used in electrical heating systems. They can be useful where water needs to be excluded, as in the case when alkaline metals are processed. In certain cases, molten salts are used in high temperature systems, but these are special cases.

9.2.1.4 Electrical Heating

Electrical heating can be achieved by using a resistor in a protection tube, directly immersed into the reaction mass. In these systems, overheating due to a high surface temperature of the tube may be a hazard. Therefore, in most cases, electrical heating is achieved indirectly using a secondary circulation system with a heat carrier, as described in Section 9.2.1.3. The heat carrier then flows through the electrical heater and / or a cooled heat exchanger, allowing a smooth transition from heating to cooling, or reversely cooling to heating. The main hazard with electrical heaters is that they may achieve high temperatures. Thus, the maximum temperature must be limited by appropriate technical means.

9.2.1.5 Cooling with Ice

In the past, ice was often used as a coolant, poured directly into the medium that needed to be cooled. This is obviously only possible when the chemistry is compatible with water, since the medium soon becomes diluted by molten ice. Ice acts by releasing it latent heat by melting $\Delta H'_{melt} = 320\,\text{kJ}\cdot\text{kg}^{-1}$. The amount of ice used must be taken into account during this process. Ice cooling may be interesting as an emergency coolant.

9.2.1.6 Other Heat Carriers for Cooling

The most common cooling medium is water, which may be used for temperatures from ca 5 °C to above 100 °C, if the circulation system is closed and works under pressure. Other fluids are also used for cooling, for example, brine, a solution of sodium chloride in water, which may reach –20 °C, or calcium chloride in water for temperatures down to –40 °C. These fluids have a major drawback, causing corrosion due to the presence of chloride ions. Other mixtures of alcohols with water, or polyethylene glycols, sometimes with water, are often used in industrial systems. With ethylene glycol and water at 1:1, a temperature of –36 °C can be achieved. The heat carrier is cooled by a refrigerating system using freons (CFCs), now replaced by more environmentally friendly fluids or ammonia. In this case, the heat exchanger is an evaporator. With such cooling systems, great care must be used to avoid the decreased wall temperatures below the solidification temperature of the reaction mass. If this should happen, the consequences would formation of a viscous film or even a solid on the wall, impinging the heat transfer and thus the cooling capacity. Too much cooling may well achieve these adverse effects.

9.2.2
Heating and Cooling Techniques

Different technical solutions are used in the temperature control of industrial reactors. The heat carriers mentioned in Section 9.2.1 may be used by different technical means: the direct way whereby the heat carrier is directly mixed with the reaction mass, internal or external coils, jacket, simple circuits, and indirect systems with a double circulating system. These techniques with their advantages and drawbacks, in terms or process safety, are reviewed in the following sections.

9.2.2.1 Direct Heating and Cooling

Direct heating means that steam is directly condensed in the reactor contents. For cooling, either cold water or ice is directly mixed with the reactor contents. This principle is simple to use and efficient since there is no heat transfer across the wall. Moreover, a certain agitation effect is also attained by this injection. Nevertheless, the reactor contents are diluted by water (or condensed steam), which implies that water needs to be compatible with the reaction or it may also lead to contamination of the reaction mass with impurities. Direct cooling can be advantageous in emergency cooling (see Section 9.4), but is seldom used as an operational heating or cooling technique.

9.2.2.2 Indirect Heating and Cooling of Stirred Tank Reactors

The temperature control of stirred tank reactors can be attained by heat exchange across the reactor wall, where a jacket or external coils are used. The jacket is mainly used on glass-lined vessels, whereas external welded half coils are used on stainless steel reactors. The jacket generally presents a more important surface coverage, but the circulation of the heat carrier inside the jacket is less precisely-defined than with external coils. This generally leads to lower overall heat transfer coefficients for jackets compared to coils. Nevertheless, techniques using injectors and baffles that correct this behavior are available on the market. The coils may allow higher fluid pressures than a jacket, which is essential in cases where high temperatures must be achieved with steam heating. The main limitation of this technique is that the heat exchange area is defined by the reactor geometry, for example, a $1\,m^3$ reactor has a heat-exchange area of only $4\,m^2$ and cannot be extended.

In order to provide a higher heat exchange area, internal coils may also be used. This technique may double the heat exchange area. Even if this technique is by itself very simple to apply, it also presents important drawbacks: The internal coil takes up a significant part of the reaction volume, it may be sensitive to corrosion, and it makes cleaning operations more difficult. Finally, it is more difficult to achieve good internal circulation of the reaction mass, that is, the agitator design must be carefully prepared for that purpose (see case history in Chapter 8).

Another technique is to add an external loop with a circulation pump that passes the reaction mixture over an external heat exchanger. In this case, the heat exchange area can be significantly increased for a given reactor volume, since it can be designed independently from the reactor geometry. By using this technique, high specific cooling capacities can be achieved. Nevertheless, the technique is more complex and reserved for strongly exothermal reactions. Another safety aspect must be considered with this type of reactor, in case of pump failure or plugging in the line, whereby a heat confinement situation may result (see Chapter 13).

9.2.2.3 **Single Heat Carrier Circulation Systems**

In this configuration, which is the simplest for indirect heating and cooling, steam is injected into the jacket or external coil for heating, and cold water for cooling. Since steam condenses while delivering its latent heat of condensation, the condensate must be removed from the jacket in order to avoid accumulation. This is achieved by a purge that keeps the jacket closed under steam saturation pressure, while condensate is released to a drain that recycles the treated water to the boiler (Figures 9.5 and 9.6). Thus during the heating period, the jacket is under pressure and contains steam admitted from the top, with the condensate flowing to the bottom where it is removed. During cooling, the water is injected from the bottom of the jacket and drained from the top in order to avoid any air plug. Therefore, when switching from heating to cooling, or from cooling to heating, several intermediate operations have to be carried out: the steam pressure must be relieved before the cold water is admitted. When switching from cooling to heating, the water contained in the jacket must first be drained off before steam is admitted. These operations are performed by automatic valves, as described in Table 9.2.

Single circuit systems can also be used with two different cooling media, water for the range between, say 20 °C and 100 °C and brine for lower temperatures. In such cases, the valve system becomes more complex (Figure 9.6). Care must be taken to avoid brine loss to the water system, and to totally purge the water before

Table 9.2 Valve switching for a single circuit heating cooling system.

Function valve	1	2	3	4	5
Heating	Controlled	Closed	Closed	Closed	Open
Purge	Closed	Open	Open	Closed	Closed
Cooling	Closed	Open	Closed	Controlled	Closed

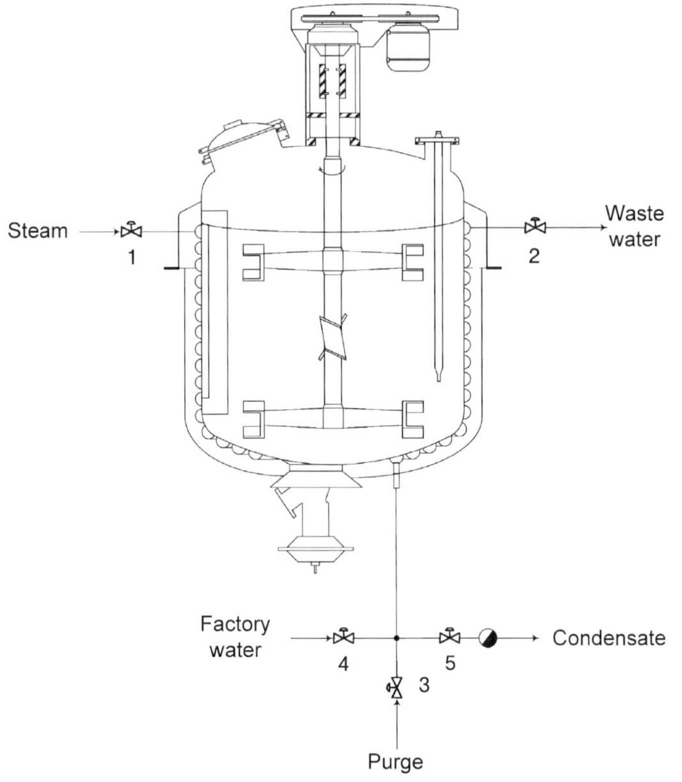

Figure 9.5 Heating cooling system with single circuit. The valves are as flows: (1) steam inlet, (2) cooling water outlet, (3) purge, (4) cold water inlet, (5) condensate outlet.

brine is admitted to avoid icing. This requires additional intermediate operations, with purges with compressed air or nitrogen. The major problem with these systems is that time is required when switching from heating to cooling, or cooling to heating. This can become critical from a safety point of view, for example, in case an exothermal reaction is performed, as in a polytropic batch operation (see Section 6.6). When reaching the process temperature after a heating period, the reactor must be immediately cooled, in order to avoid a temperature overshoot that could result in runaway. Thus, when designing a batch process for a reactor with the type of heating cooling system described here, great care must be taken to provide process conditions with sufficient time between heating and cooling phases.

A more flexible arrangement is to use a single circuit with pressurized water and steam injection into this circulation (Figure 9.7). Such a system can achieve temperatures up to 200 °C with 16 bar of steam. When cooling is required, the steam injection is shut off and cold water injected into the circuit. An expansion vessel separates steam from water and maintains a constant water level in the

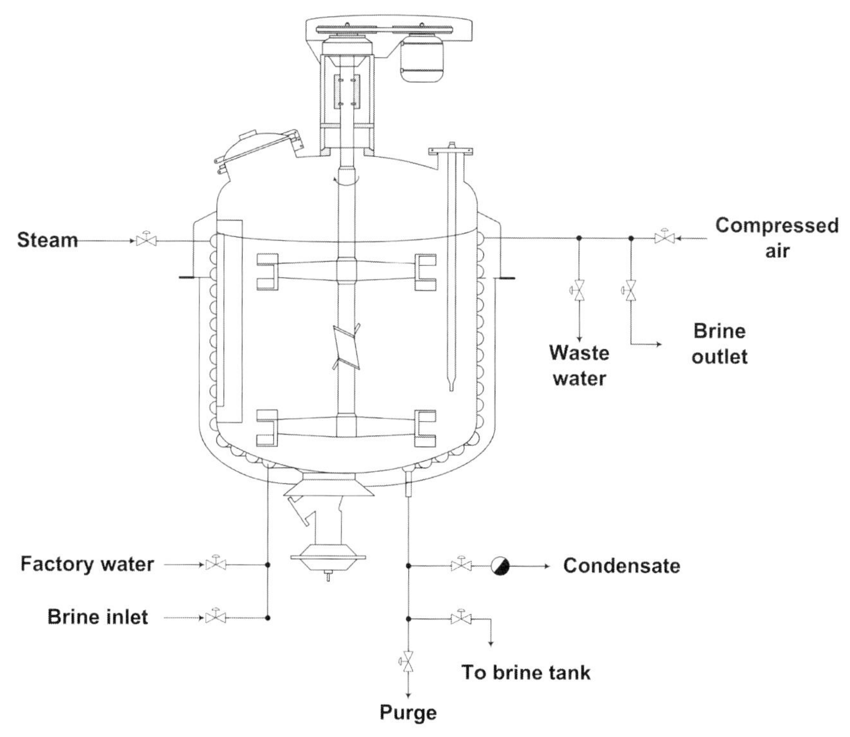

Figure 9.6 Single-circuit heating cooling system with steam, water, and brine.

system. This allows a fast and smooth transition between heating and cooling phases and gives more flexible temperature control.

9.2.2.4 Secondary Circulation Loop Temperature Control Systems

This technique consists of circulating a heat carrier, such as organic heat transfer oil (see Sections 9.2.1.3 and 9.2.1.6), through the reactor jacket or coils, and through different heat exchangers (Figure 9.8). There are at least two heat exchangers, one heated by steam or electricity, and one cooled by water. Often a third heat exchanger, cooled by brine, is used. The temperature control of the circulating heat carrier acts by controlling the position of the different control valves, allowing the heat carrier to flow through the hot heat exchanger when heating is required, and inversely through the cold heat exchanger when cooling is required. This gives a smooth transition between heating and cooling, with no idle time between heating and cooling. Moreover, a precise and flexible temperature control is achieved over a broad temperature range. This type of system is also useful when the reaction medium is not compatible with water. In such cases, an inert heat carrier is chosen, giving great safety, even in the case of reactor wall breakthrough. Nevertheless, the investment is more important for this type of equipment than for single circuit systems.

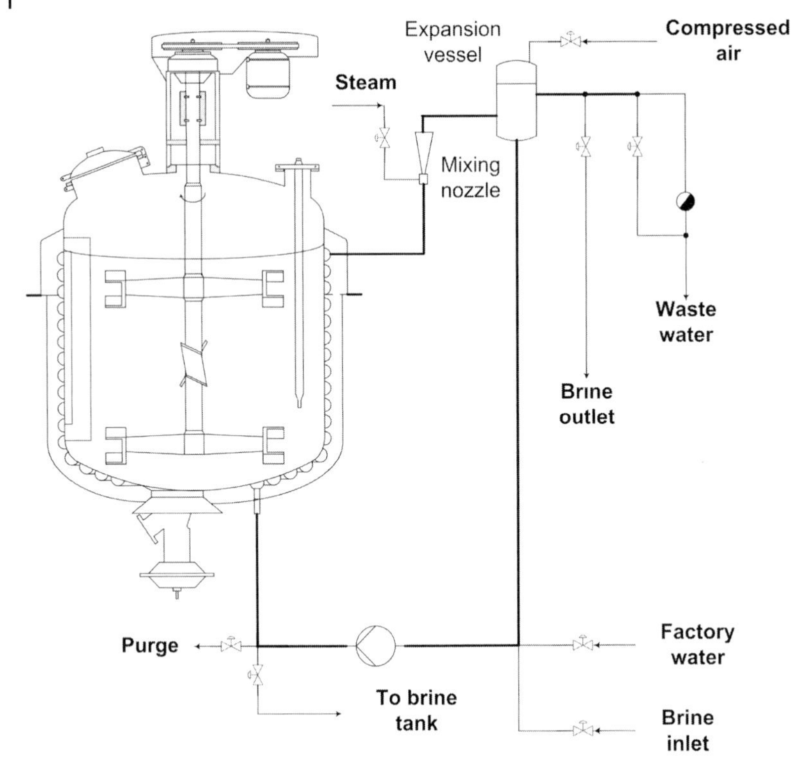

Figure 9.7 Single-circuit heating cooling system with pressurized water circulation.

9.2.3 Temperature Control Strategies

9.2.3.1 Isoperibolic Temperature Control

This is the simplest system for temperature control of a reactor: only the jacket temperature is controlled and maintained constant, leaving the reaction medium following its temperature course as a result of the heat balance between the heat flow across the wall and the heat release rate due to the reaction (Figure 9.9). This simplicity has a price in terms of reaction control, as analysed in Sections 6.7 and 7.6. Isoperibolic temperature control can be achieved with a single heat carrier circuit, as well as with the more sophisticated secondary circulation loop.

9.2.3.2 Isothermal Control

Performing a reaction under isothermal conditions is somewhat more complex. It requires two temperature probes, one for the measurement of the reaction mass temperature and a second for the jacket temperature. Depending on the internal reactor temperature, the jacket temperature is adjustable. The simplest method is to use a single heat carrier circuit to act either on the flow rate of cooling water or on the steam valve. With a secondary heat carrier circulation loop, the temperature controller acts directly on the heating and cooling valves by using a conventional

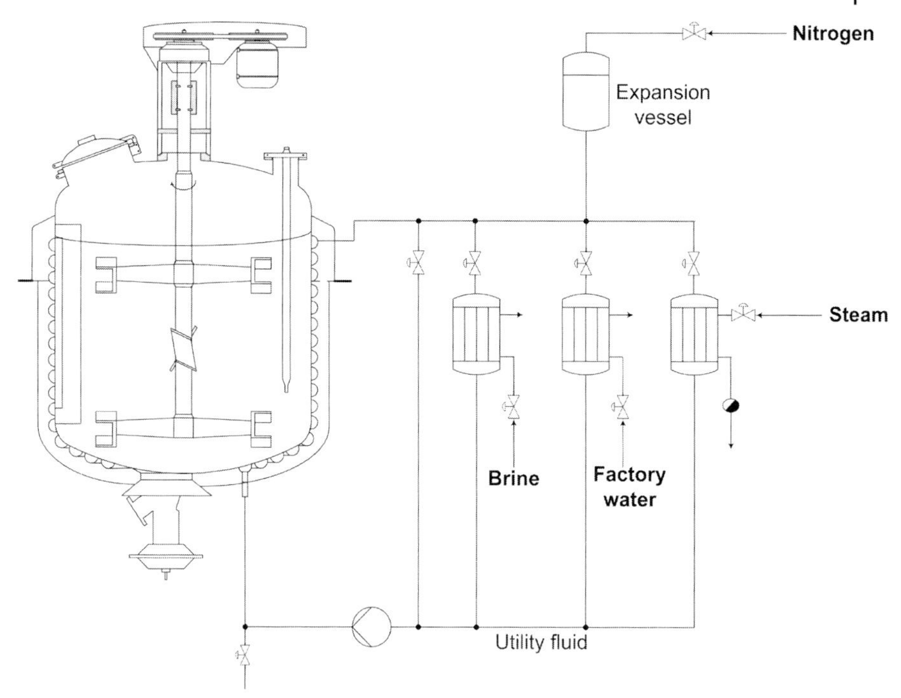

Figure 9.8 Heating cooling system with secondary circulation loop.

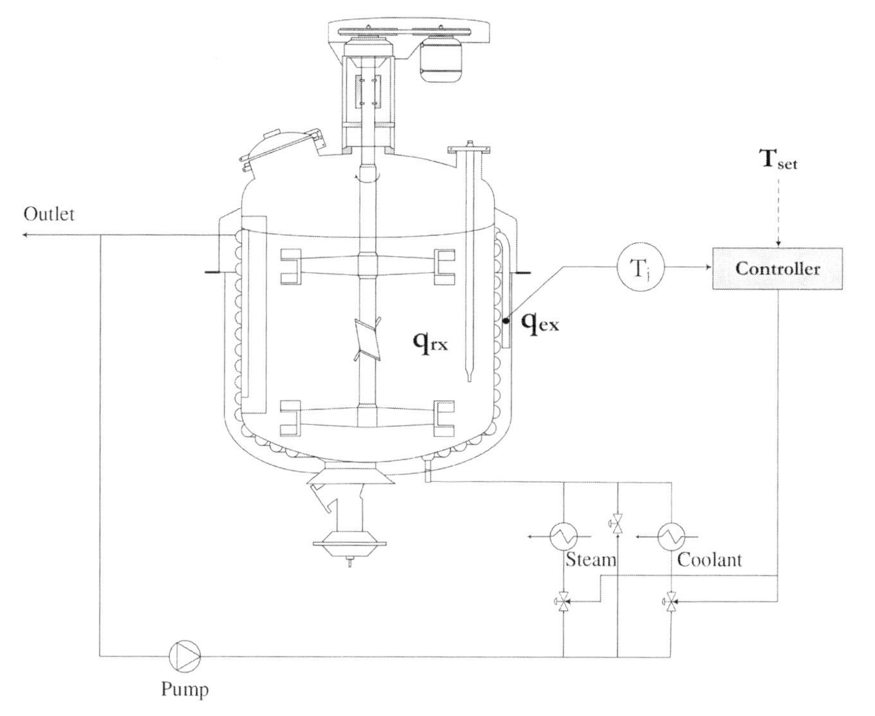

Figure 9.9 Isoperibolic temperature control with a secondary circulation loop.

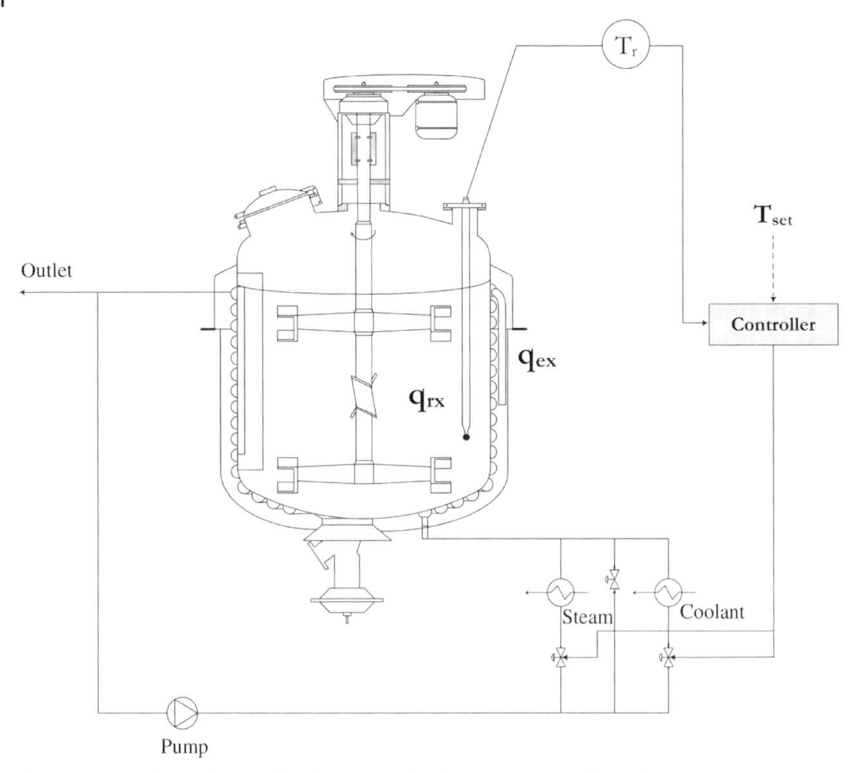

Figure 9.10 Isothermal control with a secondary heat carrier circulation loop.

P, I, D system (Figure 9.10). This type of temperature control requires careful tuning of the control parameters, in order to avoid oscillations, which may lead to loss of control of reactor temperatures in cases where an exothermal reaction is carried out. The main advantage of the isothermal control is to give a smooth and reproducible reaction course, as long as the controller is well tuned.

9.2.3.3 **Isothermal Control at Reflux**

This type of temperature control is simple to achieve. As for isoperibolic control, only the jacket temperature is controlled and the reaction is performed at boiling point, that is, at a constant temperature, which is physically limited (Figure 9.11). If a temperature below boiling point is required, a vacuum may be applied to the system. Besides its simplicity, the main advantage of this temperature control strategy is that the main heat exchange takes place in the reflux condenser, where high heat exchange capacities can be achieved (see Section 9.4). This strategy is often used when a volatile product has to be eliminated during the reaction, for example, azeotropic water elimination. This provides a safe way of controlling the reaction temperature, but special care is required in systems where the boiling point varies with conversion, Here the temperature no longer remains constant, but may increase if volatiles are converted to higher boiling compounds.

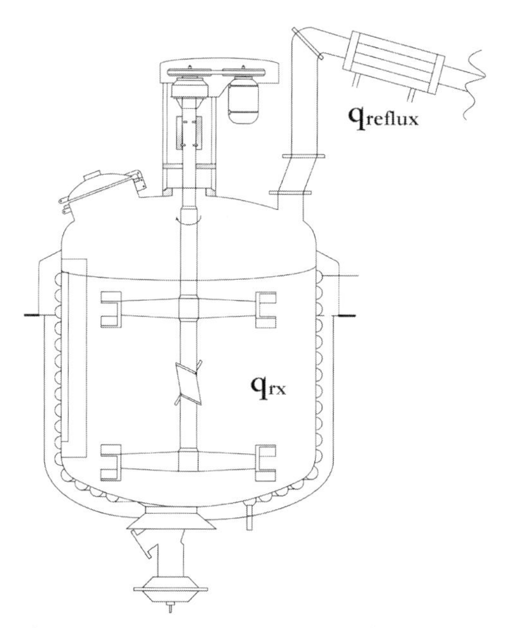

Figure 9.11 Reactor temperature control at reflux.

9.2.3.4 Non Isothermal Temperature Control

This system is the most complex, but also the most versatile. In fact, with this type of system, all the previous modes are accessible without further modification. The temperature set point corresponds to a predefined function of time (Figure 9.12). Polytrophic conditions can be achieved (see Section 6.6). The reactor is heated up at a temperature lower than that of the reaction and is then run under adiabatic conditions, Finally, cooling is started to stabilize the temperature at the desired level. By doing so, energy is saved because it is the heat of reaction that attains the process temperature. Moreover, for batch reactions, the cooling capacity is not oversized, since the low temperature at the beginning of the reaction diminishes the heat production rate. Other control strategies are possible, such as the ramped reactor, where the temperature varies with time (see Section 7.7).

With all these control strategies, but perhaps mostly with the latter, the dynamics of the temperature control system plays an important role. This is analysed in the next section.

9.2.4 Dynamic Aspects of Heat Exchange Systems

9.2.4.1 Thermal Time Constant

If we consider a stirred tank filled with a mass (M) of a liquid with a specific heat capacity (c'_p) and where no reaction takes place, a simplified heat balance can be written with only the heat accumulation and the heat exchange terms:

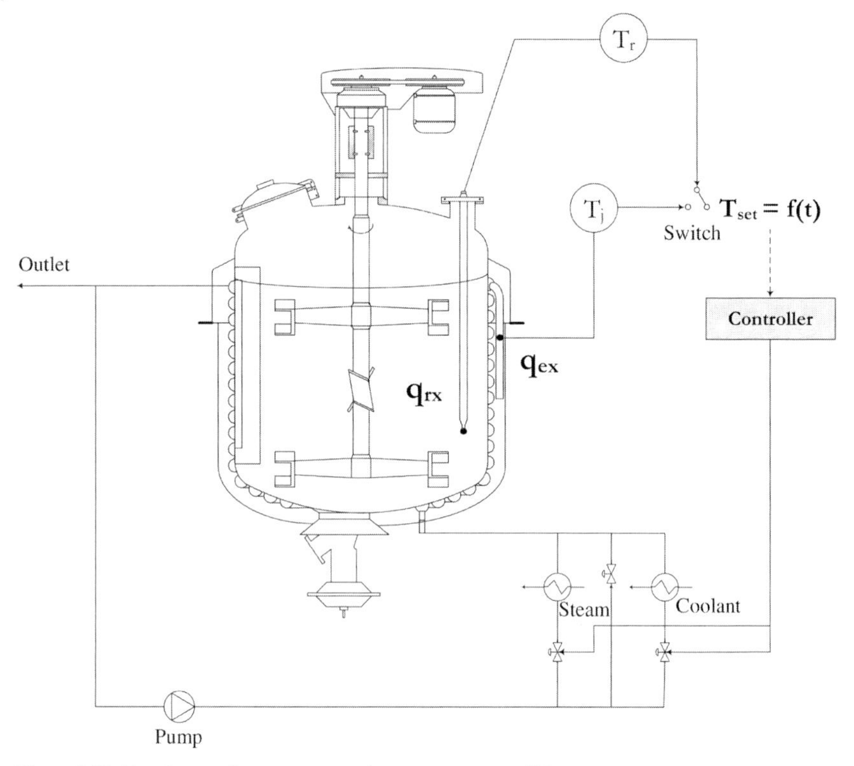

Figure 9.12 Reaction under programmed temperature conditions.

$$q_{ac} = q_{ex} \tag{9.2}$$

If we also consider that the reactor is heated to a constant heat carrier temperature (T_C), the heat balance in Equation 9.2 becomes

$$M \cdot c_P' \cdot \frac{dT}{dt} = U \cdot A \cdot (T_c - T) \tag{9.3}$$

Obviously, this equation is also valid for cooling with ($T_c < T$), the derivative of the temperature becoming negative. Equation 9.3 can be rewritten as a function of the temperature gradient across the reactor wall:

$$(\Delta T = T_c - T) \quad \text{so that} \quad d(\Delta T) = dT$$

as

$$\frac{M \cdot c_P'}{U \cdot A} \cdot \frac{d(\Delta T)}{dt} = \Delta T \tag{9.4}$$

Table 9.3 Time constants of normalized stainless steel reactors
The reactors are considered to be filled to their nominal volume with water (ρ = 1000 kg m^{-3} and $c_P' = 4{,}2$ kJ kg^{-1}K^{-1}) and a heat transfer coefficient of 500 W m^{-2} K^{-1}.

V (m^3)	0.1	0.25	0.4	0.63	1	1.6	2.5	4	6.3	10	16	25
A (m^2)	0.63	1.1	1.63	2.05	2.93	4.2	6	7.4	10	13.5	20	24
A/V (m^{-1})	6.3	4.4	4.1	1.7	2.9	2.6	2.4	1.9	1.6	1.3	1.2	1
τ (h^{-1})	0.37	0.53	0.57	0.72	0.8	0.89	0.97	1.26	1.47	1.73	1.87	2.43

Equation 9.4 is a simple first-order differential equation, simply expressing that the rate of variation of the reactor temperature is proportional to the temperature gradient between reactor contents and cooling medium. The ratio $\frac{M \cdot c_P'}{U \cdot A}$ that appears in Equation 9.4, is called the thermal time constant of the reactor:

$$\tau_c = \frac{M \cdot c_P'}{UA} \tag{9.5}$$

The thermal time constant of a reactor characterizes the dynamics of the evolution of the reactor temperature. In fact, since it contains the ratio of the mass proportional to volume with the dimension L^3 to the heat exchange area with the dimension L^2, it varies non-linearly with the reactor scale, as is explained in Section 2.4. Some values of the time constant obtained with normalized stainless steel reactors [1] are summarized in Table 9.3. The variation by a factor of about 7, over the range considered here, is critical during scale-up. The heating or cooling times are often expressed as the half-life, the time required for the temperature difference to be divided by two:

$$t_{1/2} = \ln 2 \cdot \tau_c \cong 0.693 \cdot \tau_c \tag{9.6}$$

The thermal time constant is only one aspect of reactor dynamics. In practice, the heat carrier temperature cannot be adjusted instantaneously at industrial scale, as it has its own dynamics, depending on the equipment and the temperature control algorithm. These aspects of the dynamics of the heat exchange and temperature control systems are considered in the next sections.

9.2.4.2 Heating and Cooling Time

The thermal time constant, defined in Equation 9.5, is useful for calculating heating and cooling times, which often take up a considerable amount of the cycle time of batch and semi-batch reactions. We start from Equation 9.4 that, after variable separation, combines with Equation 9.5, to become

$$\frac{d(\Delta T)}{\Delta T} = \frac{-dt}{\tau_c} \tag{9.7}$$

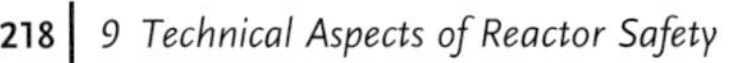

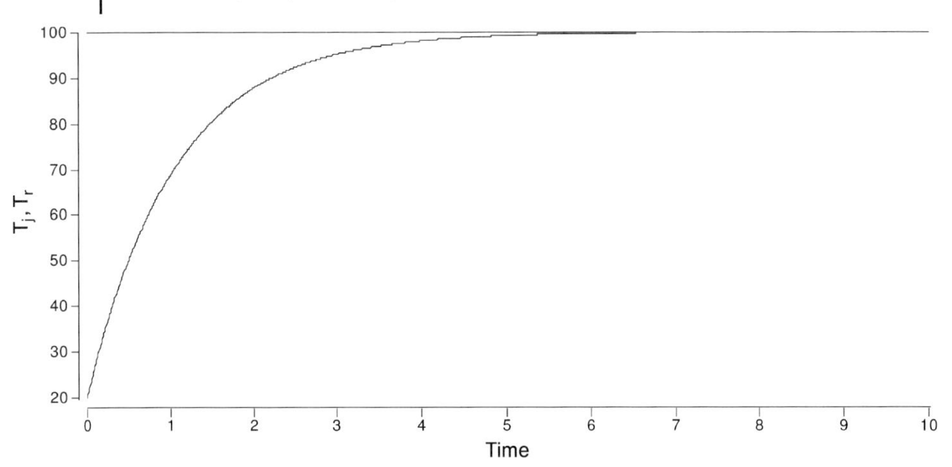

Figure 9.13 Heating curve of a reactor heated with a constant heat carrier temperature of 100 °C.

With the initial conditions:

$$t = 0 \leftrightarrow \Delta T = \Delta T_0 = T_c - T_0 \tag{9.8}$$

it can be integrated to give

$$\frac{\Delta T}{\Delta T_0} = e^{-t/\tau_c} \tag{9.9}$$

This expresses that the reactor contents temperature approaches the heat carrier temperature asymptotically, following an exponential law (Figure 9.13).

From Equation 9.9, different practical useful forms can be derived:

- Calculation of the reactor contents temperature (T) as a function of time, when heated with a heat carrier at a constant temperature (T_C):

$$T = T_c + (T_0 - T_c)e^{-t/\tau_c} \tag{9.10}$$

- Calculation of the time required for the reactor contents to reach a temperature (T), starting from (T_0) when heated with a constant heat carrier temperature (T_c):

$$t = \tau_c \cdot \ln\frac{\Delta T_0}{\Delta T} = \tau_c \cdot \ln\frac{T_0 - T_c}{T - T_c} \tag{9.11}$$

- Required heat carrier temperature (T_C) to reach the temperature (T), starting from (T_0) within a given time (t):

$$T_c = \frac{T - T_0 \cdot e^{-t/\tau_c}}{1 - e^{-t/\tau_c}} \tag{9.12}$$

Obviously all these expressions are also valid for cooling, then ($T_c < T$). They are useful in process design, for calculating cycle times, where heating and cooling phases are often time-consuming. Of course, their use requires knowledge of the overall heat transfer coefficient U. Its determination is described in Section 9.3.

9.2.4.3 Cascade Controller

In order to achieve an accurate control of the internal reactor temperature, a cascade controller can be used. In this type of controller, temperature control is managed by two controllers arranged in cascade, that is, in two nested loops (Figure 9.14). The external loop, called the master, controls the temperature of the reaction mixture by delivering a set value to the slave, the inner loop, which controls the temperature of the heat carrier (T_c).

The set point of the heat carrier temperature is calculated proportionally to the deviation of the reactor temperature from its set point:

$$T_{c,set} = T_{r,set} + G(T_{r,set} - T_r) \tag{9.13}$$

The constant G is the gain of the cascade. This is an important parameter for tuning the dynamics of the temperature control system: too low a gain results in slow temperature control where the set point may be surpassed, causing a hazardous situation (see Section 7.8.3). On the other hand, too high a gain results in oscillations that may cause loss of control of the reactor temperature.

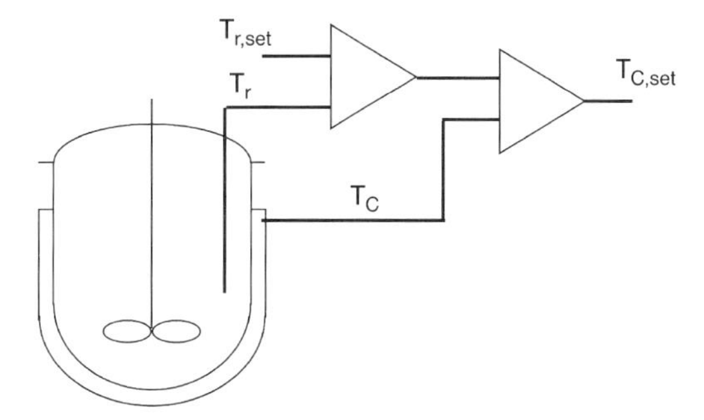

Figure 9.14 Principles of a cascade controller. The master controller controls the reactor temperature (T_r); the slave controls the cooling system temperature (T_c).

9.3 Heat Exchange Across the Wall

9.3.1 Two Film Theory

In a reactor working under normal operating conditions, meaning the heat exchange system is working as designed, the mechanism of heat transfer is forced convection [2]. The reaction mixture is agitated or flows through a tube, and the heat carrier also flows through the coil or jacket. In the immediate neighborhood of the wall, the accommodation of the fluid results in a film with a slow flow rate, increasing the resistance to heat transfer. This phenomenon occurs on both sides of the reactor wall. This gave rise to the two films model, that is, the overall resistance to heat transfer consists of three resistances in series: the resistance of the internal film, the resistance of the wall itself with conductive heat transfer, and the resistance of the external film.

$$\frac{1}{U} = \underbrace{\frac{1}{h_r}}_{\substack{\text{depends on}\\ \text{reaction mass}}} + \underbrace{\frac{d}{\lambda} + \frac{1}{h_c}}_{\substack{\text{depends}\\ \text{on reactor}}} = \frac{1}{h_r} + \frac{1}{\varphi} \tag{9.14}$$

The first term depends entirely on the physical properties of the reactor contents and degree of agitation. It represents resistance to heat transfer of the internal film and of eventual deposits at the wall, which may determine the overall heat transfer [3]. Therefore, the reactor should be regularly cleaned with a high pressure cleaner. Both last terms depend on the reactor itself and on the heat exchange system, that is, reactor wall, fouling in the jacket, and external liquid film. They are often grouped under one term: the equipment heat transfer coefficient (φ) [4, 5].

9.3.2
The Internal Film Coefficient of a Stirred Tank

For description of the heat transfer coefficient of the internal film, there are several correlations available. The most popular of them is presented by [2]

$$Nu = C^{te} \cdot Re^{2/3} \cdot Pr^{1/3} \cdot \left(\frac{\mu}{\mu_w}\right)^{0.14} \tag{9.15}$$

The last term in Equation 9.15 represents the ratio of the viscosity of the reaction mass at reaction temperature (bulk) to its viscosity at wall temperature. It accounts for the changes of the heat transfer coefficient, when switching from heating to cooling. This produces an inversion of the temperature gradient and therefore affects the viscosity of the product close to the reactor wall. If reactions are in solvents, it can usually be ignored, but may become important in the case of polymers. The viscosity of the reaction mass is often important and its temperature dependence may give this term a value that can no longer be ignored. This expression is valid for Newtonian fluids, therefore its validity with polymers or suspensions must be verified. In this correlation, the dimensionless numbers are defined in

$$Nu = \frac{h_r \cdot d_r}{\lambda} \qquad Re = \frac{n \cdot d_s^2 \cdot \rho}{\mu} \qquad Pr = \frac{\mu \cdot c_P'}{\lambda} \tag{9.16}$$

Here the Reynolds number is expressed for a stirred tank where the flow rate corresponds to the tip speed ($n \cdot d_s$) of the agitator.

9.3.3
Determination of the Internal Film Coefficient

By grouping the terms, the equation can be solved for h_r, the heat transfer coefficient of the reaction mass. This can be written as a function of the technical data of the reactor and the physico-chemical properties of the reaction mass:

$$h_r = \underbrace{C^{te} \frac{n^{2/3} d_s^{4/3}}{d_r g^{1/3}}}_{\substack{\text{technical data} \\ \text{of the reactor}}} \underbrace{\sqrt[3]{\frac{\rho^2 \lambda^2 c'_P g}{\mu}}}_{\substack{\text{physical-chemical data} \\ \text{of the reaction mass}}} = z \cdot \gamma \tag{9.17}$$

Thus for a given reaction mass, the heat transfer coefficient of the internal film can be influenced by the stirrer speed and its diameter. The value of the equipment constant (z) can be calculated using the geometric characteristics of the reactor. The value of material constant for heat transfer (γ) can either be calculated from the physical properties of the reactor contents–as far as they are known–or measured by the method of the Wilson plot in a reaction calorimeter [4, 5]. This parameter is independent of the geometry or size of the reactor. Thus, it can be determined at laboratory scale and used at industrial scale. The Wilson plot determines the overall heat transfer coefficient as a function of the agitator revolution speed in a reaction calorimeter:

$$\frac{1}{U} = \frac{1}{z\gamma} + \frac{1}{\varphi} = f(n^{-2/3}) \tag{9.18}$$

The Wilson plot (Figure 9.15) verifies the correlation in Equation 9.18, that is, if the measures fit on a straight line, a validation is built into the method. The

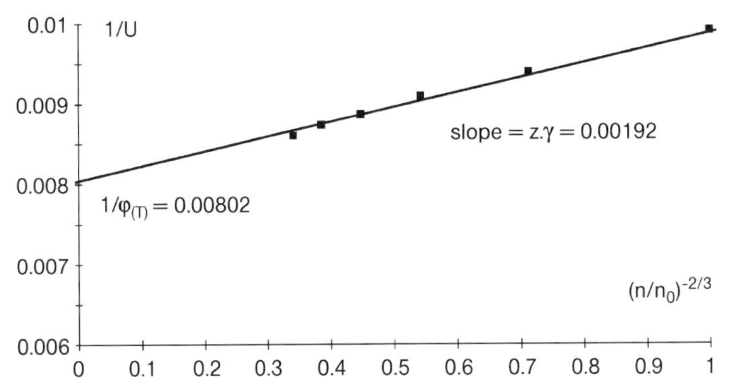

Figure 9.15 Wilson plot obtained for toluene in a reaction calorimeter, $1/U$ as a function of the stirrer speed to power $-2/3$. The reference stirrer speed n_0 is taken as $1\,s^{-1}$.

Table 9.4 Typical values of the agitator constant for Equation 9.17.

Agitator	Constant
Plate stirrer	0.36
Rushton turbine	0.54
Rushton turbine with pitched blades	0.53
Propeller	0.54
Anchor	0.36
Impeller	0.33
Intermig (Ekato)	0.54

intercept with the ordinate represents the reciprocal heat transfer coefficient of the equipment, φ, representing the wall and external cooling system of the calorimeter. The slope is the product of z by γ, which determines either one of these parameters. In the first stage, the equipment constant (z) is determined by a calibration performed using a solvent with known physical properties. In a second stage, γ is determined during the actual measurement with the reaction mixture.

The value of z, characterizing the internal part of the equipment factor, can be calculated using the geometric characteristics of the reactor. Some typical values of the agitator constant are given in Table 9.4 [2].

9.3.4 The Resistance of the Equipment to Heat Transfer

The resistance of the reactor wall $\frac{d}{\lambda}$ and external film h_c can be determined in a cooling experiment using the production reactor filled with a known mass M of a substance with known physical properties. The temperature of the contents T_r of the reactor and the average temperature of the cooling system T_c are recorded during this experiment. A heat balance is calculated between two instants, t_1 and t_2 (Figure 9.16):

Heat removed from the system:

$$Q = M \cdot c'_P \cdot (T_{r1} - T_{r2}) \tag{9.19}$$

Average cooling power:

$$q_{ex} = U \cdot A \cdot \overline{\Delta T} \tag{9.20}$$

Average temperature difference:

$$\overline{\Delta T} = \frac{1}{2} \cdot [(T_{r1} - T_{c1}) + (T_{r2} - T_{c2})] \quad \text{or} \quad \overline{\Delta T} = \frac{(T_{r1} - T_{c1}) - (T_{r2} - T_{c2})}{\ln(T_{r1} - T_{c1}) - \ln(T_{r2} - T_{c2})} \tag{9.21}$$

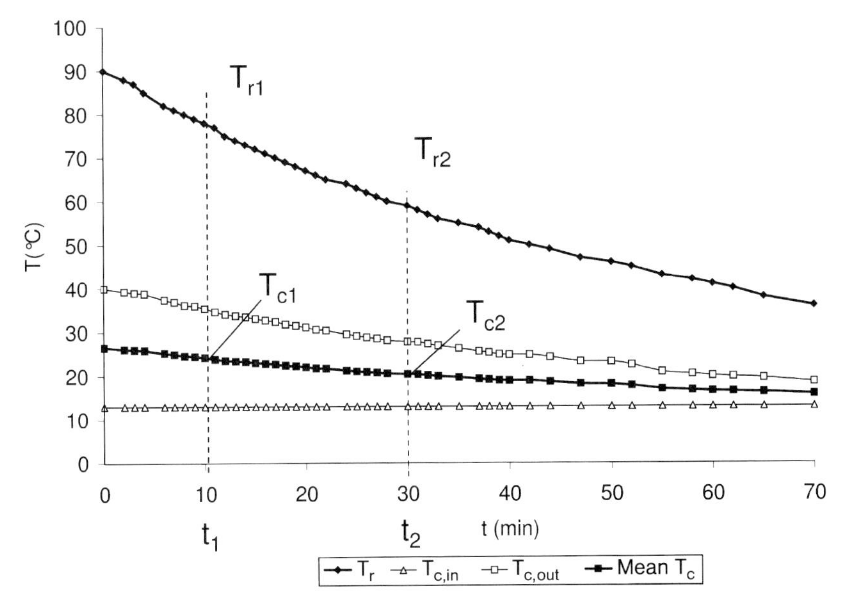

Figure 9.16 Cooling experiment in a full-scale reactor. The heat balance is calculated between the two instants, t_1 and t_2.

Heat Balance:

$$Q = q_{ex} \cdot (t_2 - t_1) \tag{9.22}$$

Placing (9.19) and (9.20) into (9.22) and solving for U gives

$$U = \frac{M \cdot c'_P \cdot (T_{r1} - T_{r2})}{A \cdot \overline{\Delta T} \cdot (t_1 - t_2)} \tag{9.23}$$

A more accurate method determines the thermal time constant from a plot of the natural logarithm of temperature difference between the reactor contents (T_r) and the heat carrier (T_c) as a function of time. This is an application of Equation 9.11:

$$\ln\left(\frac{\Delta T}{\Delta T_0}\right) = -\frac{t}{\tau_c} \tag{9.24}$$

This gives a linear plot where the slope is the inverse of the thermal time constant. An example of such a linear fit is represented in Worked Example 9.1. Since the mass (M), the specific heat capacity of the contents (c'_P), as well as the heat exchange area of the reactor (A) are known, the only unknown is the overall heat transfer coefficient (U). As during heating and cooling experiments, the reactor contains a known substance with known physical properties h_r, which can be calculated from Equation 9.17. Knowing the overall heat transfer coefficient U, from Equations 9.23 or 9.24, the only unknown that remains is the equipment heat transfer φ, determined from

$$\frac{1}{\varphi}=\frac{1}{U}-\frac{1}{h_r} \tag{9.25}$$

For the calculation of the heat transfer coefficient of the external film, some models are also available [2], describing the hydraulics of the flow in the jacket or in the half-welded coils. The results depend strongly on the technical design of the equipment. Thus, the direct experimental determination is mostly preferred.

9.3.5 Practical Determination of Heat Transfer Coefficients

The determination of the overall heat transfer coefficient, across the wall of a stirred vessel, is based on Equation 9.18. Thus, two steps are required:

1. The heat transfer coefficient of the internal film is determined from:
   a) the equipment constant z, calculated from the geometric and technical data of the reactor;
   b) the material constant for heat transfer γ calculated from the physical properties or determined in a reaction calorimeter using the Wilson plot.
2. The equipment heat transfer coefficient determined from a cooling (or heating) experiment performed in the industrial reactor containing a known amount of a compound with known physical properties.

Some typical values of heat transfer coefficients are given in Table 9.5. The values provided for h_r without stirrer and h_c without flow, show the influence of failure of the stirrer or of the cooling system on the heat transfer.

Worked Example 9.1: Determination of a Heat Transfer Coefficient

A 2.5 m^3 stainless steel stirred tank reactor is to be used for a reaction with a batch volume of 2 m^3 performed at 65 °C. The heat transfer coefficient of the reaction mass is determined in a reaction calorimeter by the Wilson plot as $\gamma = 1600\,W\,m^{-2}\,K^{-1}$. The reactor is equipped with an anchor stirrer operated at 45 rpm. Water, used as a coolant, enters the jacket at 13 °C. With a contents volume of 2 m^3, the heat exchange area is 4.6 m^2. The internal diameter of the reactor is 1.6 m. The stirrer diameter is 1.53 m. A cooling experiment was carried out in the temperature range around 70 °C, with the vessel containing 2000 kg water. The results are represented in Figure 9.16.

The physical properties of water at 70 °C are [2]:

$\rho = 978\,\mathrm{kg\,m^{-3}}$
$c_p' = 4.19\,\mathrm{kJ\,kg^{-1}\,K^{-1}}$;
$\lambda = 0.662\,\mathrm{W\,m^{-1}\,K^{-1}}$
$\mu = 0.4\,\mathrm{mPa \cdot s}$

The material constant for heat transfer of water at 70 °C is

$$\gamma = \sqrt[3]{\frac{\rho^2 \lambda^2 c_P' g}{\mu}} = \sqrt[3]{\frac{978^2 \times 0.662^2 \times 4.19 \times 9.81}{0.4 \cdot 10^{-3}}} \cong 35000\ \mathrm{Wm^{-2}\,K^{-1}}$$

The equipment constant is

$$z = C^{te} \frac{n^{2/3} d_s^{4/3}}{d_r g^{1/3}} = 0.36 \times \frac{\left(\frac{45}{60}\right)^{2/3} \times 1.53^{4/3}}{1.6 \times 9.81^{1/3}} = 0.153$$

Thus the heat transfer coefficient of the internal water film is

$$h_r = z \cdot \gamma = 0.153 \times 35000 = 5355\,\mathrm{Wm^{-2}\,K^{-1}}.$$

The cooling experiment can be evaluated as above, determining the thermal time constant from the plot of the natural logarithm of the temperature difference between T_r and T_c as a function of time. For this, the average coolant temperature is calculated as an arithmetical mean between the jacket inlet and outlet temperatures (Figure 9.17). The slope given by the linear regression is $0.167\,\mathrm{min^{-1}}$, which corresponds to a time constant of 59.5 minutes. From the known mass and properties of the contents, the overall heat transfer coefficient is calculated as

$$U = \frac{M \cdot c_P'}{\tau \cdot A} = \frac{2000\,\mathrm{kg} \times 4190\,\mathrm{J\,kg^{-1}\,K^{-1}}}{59.5\,\mathrm{min} \times 60\,\mathrm{s \cdot min^{-1}} \times 4.6\,\mathrm{m^2}} = 510\,\mathrm{Wm^{-2}\,K^{-1}}$$

This overall heat transfer coefficient is valid for the reactor containing water. The average cooling capacity around 70 °C during this experiment was 105 kW or $52\,\mathrm{W \cdot kg^{-1}}$. In order to calculate the heat transfer coefficient with the reaction mass, the equipment heat transfer, that is, reactor wall and external film, must be determined:

$$\frac{1}{\varphi} = \frac{1}{U} - \frac{1}{h_r} = \frac{1}{510} - \frac{1}{5355} \Rightarrow \varphi = 564\,\mathrm{Wm^{-2}\,K^{-1}}$$

With the reaction mass, the heat transfer coefficient becomes

$$\frac{1}{U} = \frac{1}{z\gamma} + \frac{1}{\varphi} = \frac{1}{0.153 \times 1600} + \frac{1}{564} \Rightarrow U \cong 170\,\mathrm{W\,m^{-2}\,K^{-1}}$$

This heat transfer coefficient gives the reactor a cooling capacity of approximately 35 kW or $18\,\mathrm{W\,kg^{-1}}$ under reaction conditions. This was calculated with a cooling medium temperature of 25 °C. In this example, it is worth noting that with water in the reactor, the main resistance to heat transfer is in the equipment, whereas with the reaction mass, it is in the internal film. This is not surprising since water has excellent heat transfer capabilities, $\gamma = 35\,000\,\mathrm{W\,m^{-2}\,K^{-1}}$, and the reaction mass rather poor, $\gamma = 1600\,\mathrm{W\,m^{-2}\,K^{-1}}$. Thus, it is strongly recommended to estimate the heat transfer coefficient of a reactor for the reaction mass to be processed, before running a reaction.

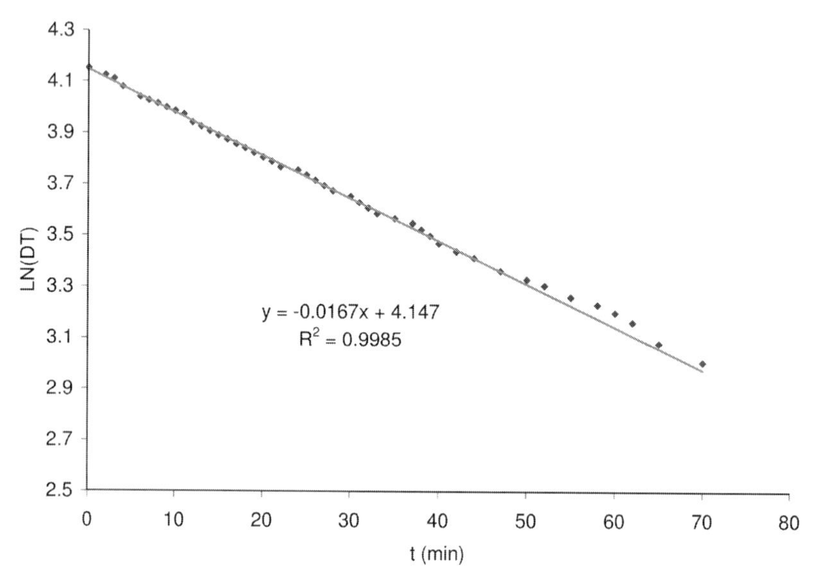

Figure 9.17 Example of linearized cooling curve recorded with a 2.5 m³ reactor filled with 2000 kg water. The linear regression was performed only on the data points before 50 minutes.

9.4 Evaporative Cooling

Cooling by solvent evaporation is an efficient method. On one hand, it is independent of the heat transfer at the reactor wall and, on the other hand, the condenser can be dimensioned independently of the reactor geometry. This reaches relatively high specific cooling powers. In case a reaction cannot be performed at boiling temperature, it is possible to work under partial vacuum to decrease the boiling

Table 9.5 Factors influencing the heat transfer with some typical values of heat transfer coefficients in an agitated reactor.

Type	Influencing factors	Typical values $W\ m^{-2} K^{-1}$	
Internal film	Stirrer: speed and type	Water	1000
h_r forced convection	Reaction mass C_p,λ,ρ,η	Toluene	300
	Physical data especially ρ = f(T)	Glycerol	50
h_r natural convection		Water	100
(stirrer failure)		Gases	10
Polymer deposit	Thermal conductivity λ	With d = 1 mm	
	Thickness of deposit	PE	300
		PVC, PS	170
Reactor wall λ/d	Construction	With d = 10 mm	
	Wall thickness (d)	Iron	4800
	Construction material	Stainless steel	1600
	Coating		
		Glass	100
		Glass lined	800
Fouling at external wall	Thermal conductivity λ	With d = 0.1 mm	
	Thickness of deposit	Gelatine	3000
		Scale	5000
External film h_c	Jacket	Water	
	Construction, flow rate	with flow	1000
	Heat carrier, physical	no flow	100
	properties, phase change	condensation	3000
	Welded half coil	Water	
	Construction, flow rate	with flow	2000
	Physical properties	no flow	200

point and to work at reflux. Evaporative cooling can be used as the main cooling system for a reactor under normal operating conditions, as well as for emergency cooling, if the boiling point is reached during temperature increase following cooling system failure. In such a case, delay in boiling must be avoided, for example, by stirring or ensuring a regular bubble nucleation by gas injection (nitrogen). Nevertheless, this is only possible provided the condenser is equipped with an independent cooling system and the equipment has been designed for this purpose.

Some technical aspects and limitations must be considered in the design of such cooling systems:

- mass of solvent being evaporated: this is a function of the energy release of the reaction or decomposition reaction.
- boiling rate of the solvent: depending on the instantaneous heat release rate of the reaction, this point will govern the whole design of the reflux system.
- flooding of the vapor tube: when the condensate flows counter-current to the rising vapor, flooding may occur.

- swelling of the reaction mass: the apparent volume varies due to the presence of bubbles in the reaction mass.
- Cooling capacity of the condenser: this is a standard engineering task, which will not be treated here.

These different points are reviewed in detail in the following subsections.

9.4.1 Amount of Solvent Evaporated

If the boiling point is reached during runaway (i.e. scenario classes 3 and 4), a possible secondary effect of the solvent evaporation is the formation of an explosive vapor cloud, which in turn, if ignited, can lead to a severe room explosion. In some cases, there is enough solvent present in the reaction mixture to compensate for the energy release, allowing the temperature to be stabilized at boiling point. This is only possible if the solvent can be distilled off in a safe way, to a catch pot or a treatment system. The thermal stability of the concentrated reaction mixture must also be ensured. In general, it is preferable to reflux the condensed solvent to the reactor. This avoids concentration of the reaction mass and allows additional cooling with the sensible heat of the cold condensate.

The amount of solvent evaporated can easily be calculated using the energies of the reaction and of the decomposition, as follows (see also Section 2.2.2.3):

$$M_v = \frac{Q_r}{\Delta H_v'} = \frac{M_r \cdot Q_r'}{\Delta H_v'} \tag{9.26}$$

All these considerations are purely static aspects. Only the amount of vapor was calculated and there was no concern about the dynamic aspects, especially the rate of evaporation. This is considered in the next section.

9.4.2 Vapor Flow Rate and Rate of Evaporation

This second aspect is important when the capacity of a distillation system must be assessed, or when such a system must be designed. The capacity is sufficient if all the vapor produced by an exothermal reaction can be conducted from the reactor to the condenser, where it must be entirely condensed:

$$\dot{m}_v = \frac{q_r' \cdot M_r}{\Delta H_v'} \tag{9.27}$$

As a first approximation, under pressure conditions close to atmospheric, the vapor can be considered as an ideal gas, thus its density is given by

$$\rho_G = \frac{P \cdot M_W}{R \cdot T} \tag{9.28}$$

Thus, the vapor velocity can be calculated from the cross-section of the vapor tube:

$$u = \frac{q_{RX}}{\Delta H_V \rho_G S} \tag{9.29}$$

$$u = \frac{4 \cdot R}{\pi} \cdot \frac{q'_R \cdot M_R \cdot T}{\Delta H'_V \cdot d^2 \cdot P \cdot M_W} = 106 \cdot \frac{q'_R \cdot M_R \cdot T}{\Delta H'_V \cdot d^2 \cdot P \cdot M_W} \tag{9.30}$$

The vapor velocity is essential information for the assessment of reactor safety at boiling point. This is particularly the case when cooling by evaporation, as normal operating conditions or if the boiling point is reached after a failure.

9.4.3
Flooding of the Vapor Tube

When the vapor flow rises in the liquid and cold condensate flows down in counter currents, waves build up at the liquid surface, which may even form bridges (Figure 9.18) across the tube, causing flooding. When the vapor tube diameter is too narrow for a given vapor release rate, the high vapor velocity results in a pressure increase in the reactor, leading to an increase in boiling temperature and further acceleration of the reaction. The consequence will be a runaway reaction until the rupture of weak equipment parts allows pressure relief. In order to avoid this type of reaction course, it is important to know the maximum vapor velocity

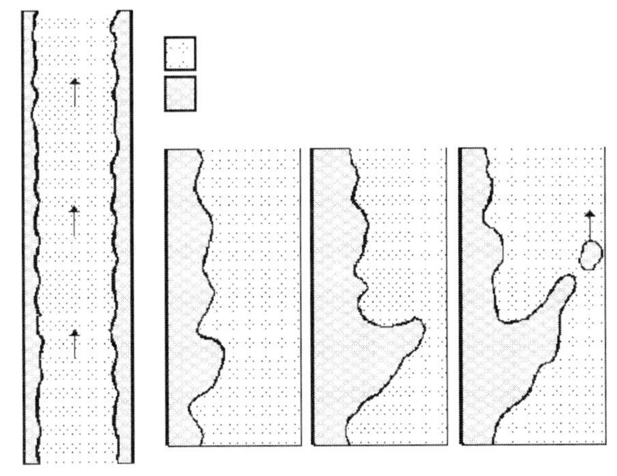

Figure 9.18 Bridging the vapor tube due to the condensate flow counter current with the vapor.

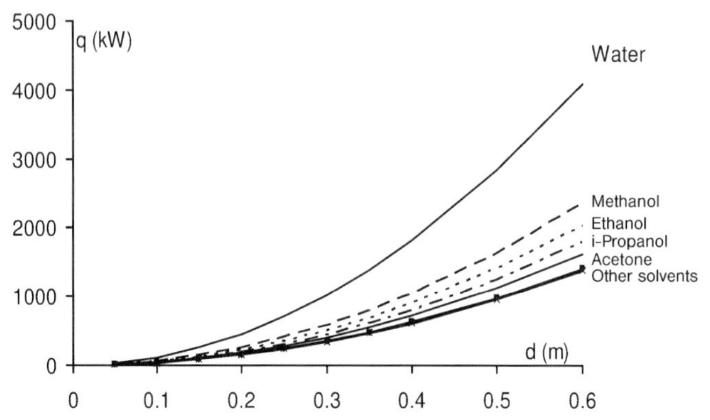

Figure 9.19 Maximum heat release rate with respect to flooding as a function of the vapor tube diameter for different solvents.

admissible in a given tube and consequently the maximum admissible heat release rate for the reaction. To predict if flooding will occur in existing equipment, an empirical correlation was established experimentally [6]. The experimental study was performed in the laboratory, at the pilot plant, and at the industrial scale with various organic solvents and water for tubes with an inside diameter between 6 and 141 mm. The maximum allowable heat release rate is obviously a function of the latent heat of evaporation and of the tube cross-section:

$$q_{\max} = (4.52 \cdot \Delta H_v + 3.37 \cdot 10^6)S \tag{9.31}$$

with ΔH_v the latent heat of evaporation expressed in J kg^{-1} and S the cross-section of the vapor tube in m^2. The vapor limit superficial velocity, $u_{G\max}$ can be calculated from the physico-chemical properties of the solvent:

$$u_{G,\max} = \frac{(4.52 \cdot \Delta H'_v + 3.37 \cdot 10^6)}{\Delta H'_v \cdot \rho_G} \tag{9.32}$$

This expression calculates the maximum admissible heat release rate compatible with evaporation cooling (Figure 9.19). Calculations performed for different common solvents show that the limiting velocity is characteristic for the different solvent classes. These calculations are shown in Table 9.6. The vapor velocities calculate the required vapor tube diameter for a given heat release rate as a function of the nature of the solvent, or inversely, the maximum admissible heat release rate for a given equipment.

9.4.4 Swelling of the Reaction Mass

During boiling of a reacting mass, vapor bubbles form in the liquid phase and rise to the gas–liquid interface. During the time they travel to the surface, they occupy a certain volume in the liquid, which results in an apparent volume increase of

Table 9.6 Maximum vapor velocity of different solvents in a tube with liquid in counter current. The calculations were performed at atmospheric pressure (1013 mbar).

Solvent	Water	Methanol	Ethanol	Acetone	Dichloro-methane	Chloro-benzene	Toluene	M-xylene
ΔHv kJ kg^{-1}	2260	1100	846	523	329	325	356	343
T_b °C	100	65	78	56	40	132	111	139
M_w g mol^{-1}	18	32	46	58	85	112	92	106
ρ_G kg m^{-3}	0.59	1.15	1.60	2.15	3.31	3.37	2.92	3.13
u_{gmax} m s^{-1}	10.2	6.6	5.3	5.1	4.5	4.4	4.8	4.6

the liquid in the reactor. This is called swelling and sometimes results in a two-phase flow into the vapor tube. The apparent volume increase or swelling of the reaction mass can be estimated using the Wilson correlation, first established for air in water [7, 8]. This correlation is easy to use and is experimentally found [6] to describe the swelling of a liquid by vapor bubbles, with enough accuracy:

$$\alpha = K \cdot \left(\frac{\rho_G}{\rho_L - \rho_G} \right)^{0.17} \cdot (d_h^*)^{-0.1} \cdot (u_G^*)^a \tag{9.33}$$

with

$$\alpha = \frac{h_b - h_0}{h_b}$$

$$d_h^* = \frac{d_h}{\sqrt{\frac{\sigma}{g \cdot (\rho_L - \rho_G)}}}$$

$$u_G^* = \frac{u_G}{\sqrt{g \sqrt{\frac{\sigma}{g \cdot (\rho_L - \rho_G)}}}}$$

The coefficient K and α depend on the value of the limit velocity u_G^*

if $u_G^* < 2$, $K = 0.68$ and $\alpha = 0.62$
if $u_G^* \geq 2$, $K = 0.88$ and $\alpha = 0.40$

Thus, by knowing the maximum admissible level increase, depending on the degree of filling of a reactor, it is possible to calculate the maximum admissible heat release rate at boiling (Figure 9.20).

9.4.5
Practical Procedure for the Assessment of Reactor Safety at the Boiling Point

A practical and systematic procedure for the assessment of the safety of chemical reactions at boiling point is presented in Figure 9.21. This procedure allows one

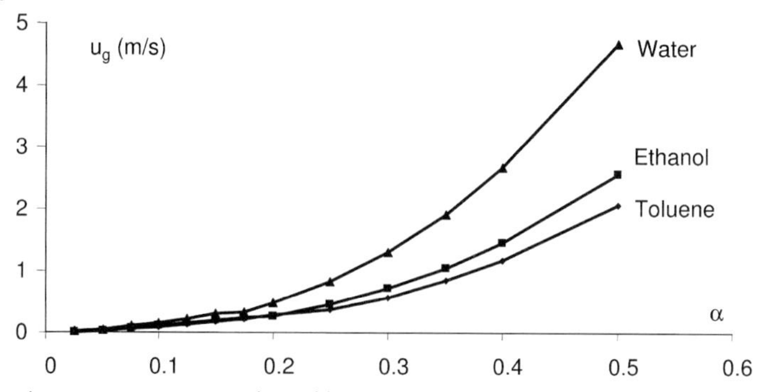

Figure 9.20 Maximum admissible vapor velocity at the gas–liquid interface as a function of the allowed relative volume increase for different solvents.

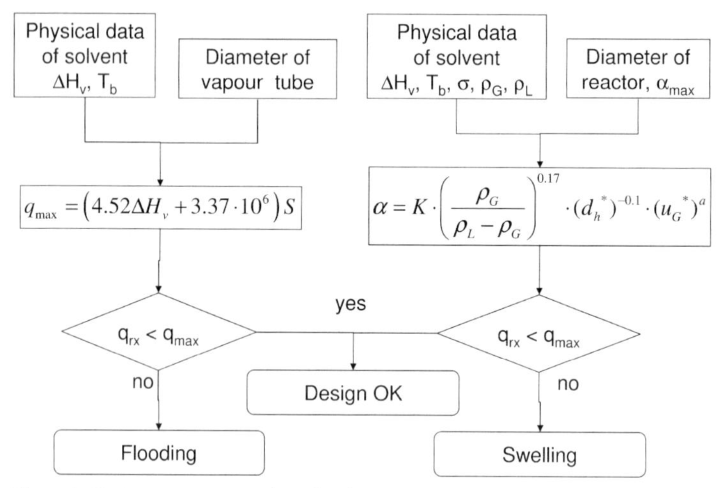

Figure 9.21 Systematic procedure for the assessment of the behavior of a reaction mass at boiling point.

to predict the behavior of a reactor at boiling point. The assessment may be performed in two different situations:

1. The heat release rate at boiling point is known, and the equipment has to be designed to cope with this requirement
2. The maximum admissible heat release rate has to be determined for existing equipment (rating).

As an example, for a $6.3\,m^3$ stirred tank filled with $6.3\,m^3$ acetone, the maximum heat release rate at boiling is:

- $35\,W\,kg^{-1}$ with a 200 mm vapor tube; the diameter of the vapor tube is limiting
- $68\,W\,kg^{-1}$ with a 300 mm vapor tube; the swelling of the reaction mass is limiting.

Such considerations allow adapting the equipment or the process, that is, the degree of filling of the reactor, to the safety requirements. This kind of measure often allows running processes under safe conditions, whereas a standard assessment would consider them as critical. They are based both on easily accessible physico-chemical properties of the boiling solvent and on geometric data of the reactor.

9.5 Dynamics of the Temperature Control System and Process Design

9.5.1 Background

The thermal characteristics of a reaction, including its heat production rate, the necessary cooling power, and the reactant accumulation, are fundamental for safe reactor operation and process design. A successful scale-up is achieved, only when the different characteristic time constants of the process, such as reaction kinetics, thermal dynamics of the reactor, and its mixing characteristics are in good agreement [9]. If we focus on the reaction kinetics and thermal dynamics, that is, we consider that the reaction rate is slow compared to the mixing rate, in principle, there are two ways to predict the behavior of the industrial reactors:

1. determine the reaction kinetics and use numerical simulation, or
2. determine the thermal dynamics of the reactor and perform an experimental simulation at laboratory scale.

This second approach is preferred because it avoids the tedious determination of the reaction kinetics. This is replaced by the determination of the reactor dynamics.

The background of this approach is that an industrial chemical reactor not only behaves according to the kinetics of the reaction, but also to the dynamics of its temperature control system. This is essentially for two reasons. First the heat transfer area to volume ratio decreases as the size of the reactor increases, leading to serious limitations of the heat transfer capacity of large reactors. The second deals with problems due to the thermal inertia (long time constant) of the jacket wall [10]. Moreover, reaction enthalpies, kinetic parameters, and hence product selectivity and global safety are known to be temperature-dependent. Therefore, only the combination of both reaction kinetics and reactor dynamics allows describing and predicting the behavior of an industrial reactor with respect to productivity, selectivity, and safety. Since, in the fine chemicals and pharmaceutical industries, the process development consists more of adapting a process to an existing plant, perhaps with few modifications, than building a new plant for a given process, a specific approach is required that takes the plant equipment characteristics into account.

Such an approach, based on the experimental simulation of an industrial reactor at laboratory scale, was proposed by Zufferey [11, 12]: the scale down approach. In order to simulate the thermal behavior of full-scale equipment at laboratory scale, it is necessary to combine process dynamics and calorimetric techniques.

9.5.2
Modeling the Dynamic Behavior of Industrial Reactors

The characteristics of industrial reactors are identified in a series of heating and cooling experiments, as described in Section 9.3.4, building a dynamic model of the reactor:

$$(M_r \cdot c'_{P,r} + c_w)\frac{dT_r}{dt} = q_{ex} + q_{st} + q_{loss} \tag{9.34}$$

Here c_w represents the heat capacity of the equipment that is to be identified in the experiments, together with the heat losses. The stirrer power is calculated according to Equation 2.24. The heat exchange term is

$$q_{ex} = U \cdot A \cdot (T_c - T) \tag{9.35}$$

where the overall heat transfer coefficient U is developed using Equation 9.14, in which the external film heat transfer coefficient is expressed as a linear function of temperature:

$$h_{e(T)} = p_1 T + p_2 \tag{9.36}$$

The dynamics of the jacket is described, using the implemented temperature control algorithm. In most cases, a so-called P-Band controller is used, that requires full cooling or heating capacity when the temperature is far from the set point, and a proportional control when the temperature lies within a certain range close to the set point. Such control systems can be described as two first-order dynamic systems using only one time constant. All the model parameters are identified using a least squares fitting with the experimental data and can be stored in a database comprising the different reactors available in a plant. This simulates the thermal behavior of a given plant reactor that could be used for full-scale production.

9.5.3
Experimental Simulation of Industrial Reactors

The principle of the scale-down methodology, using a reaction calorimeter, is as follows:

- observe on-line the instantaneous heat production rate of a chemical reaction in the reaction calorimeter;
- use this value in a numerical simulation model of the plant reactor dynamics;
- deduce the evolution of the jacket and reaction mixture temperatures of the plant reactor model if this chemical reaction took place in it;
- force the reaction calorimeter to track this temperature evolution and hence not to behave ideally anymore;
- repeat the first four points during the entire course of the chemical reaction.

By doing so, the reaction calorimeter mimics the behavior of the industrial reactor that optimizes the process without any information, neither about the stoichiometry nor about the reaction. The results of such an experiment can be evaluated by classical analytics or in process control.

As an example, a series parallel reaction of the type:

$$\begin{cases} A + B \xrightarrow{k_1} P \\ P + B \xrightarrow{k_2} S \end{cases} \tag{9.37}$$

should be performed in a $4\,m^3$ batch reactor with thermal initiation. The reactants are charged at 30 °C and the reactor should be heated to 90 °C with $15\,°C\,min^{-1}$. The kinetic parameters are unknown. This reaction was studied at laboratory scale in a reaction calorimeter and gave selectivity in the desired product P of 95 %. The same process, run in the reaction calorimeter simulating the $4\,m^3$ plant reactor, gave selectivity of only 82 % (Figure 9.22). This difference is due to the temperature

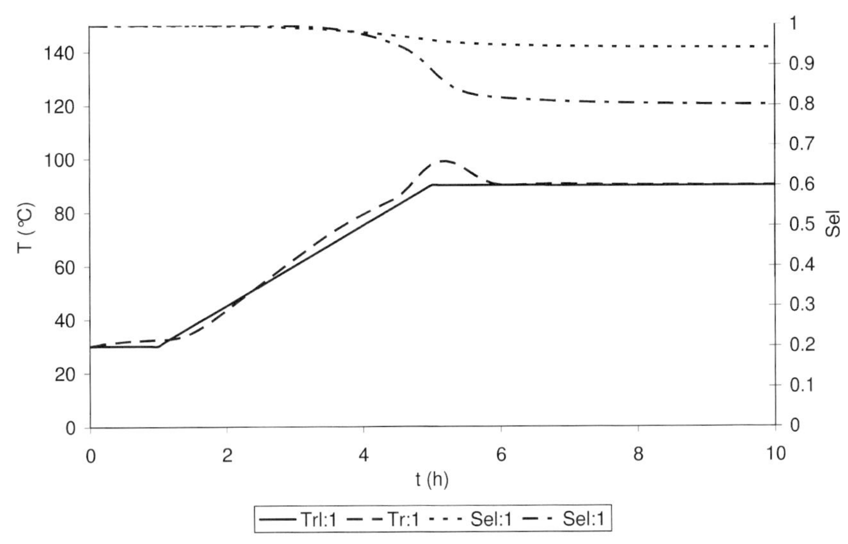

Figure 9.22 Example of series parallel reaction in a batch reactor. Temperature on left scale, selectivity on right scale.

control dynamics that lead to an overshoot of the desired temperature by 10 °C and higher temperatures during the heating phase. It is worth noting that these results were obtained in a reaction calorimeter at laboratory scale. By changing the process conditions to a slower temperature ramp of $10\,°C\,min^{-1}$ and a final set point of 85 °C only, selectivity of 89% can be obtained. Thus, the scale-down method predicts the final product distribution at laboratory scale, avoiding expensive full-scale experiments, and without the explicit knowledge of the reaction kinetics. This method is very powerful in the case of non-isothermal processes, as here or for isoperibolic processes.

9.6 Exercises

▶ Exercise 9.1

Consider a reaction in a $16\,m^3$ reactor at 100 °C. At this temperature, using a feed time of at least one hour, the reaction is feed-controlled. The feed rate must be adapted to the cooling capacity of the vessel. There are 15 000 kg of reaction mass in the vessel, the specific heat of reaction is $200\,kJ\,kg^{-1}$ final reaction mass. During the reaction, the heat exchange area remains constant at $20\,m^2$. Ambient pressure is 1013 mbar.

The cooling capacity of the reactor was determined by means of a cooling experiment, where the vessel was filled with 16 000 liters of 96% sulfuric acid. The melting point of the reaction mass is 65 °C. Therefore, to avoid crystallization at the wall, the wall temperature was kept above 70 °C, and the cooling experiment was performed using a coolant inlet temperature of 70 °C. Two points of the resulting cooling curve are shown in Table 9.7.

The material constant for heat transfer of the reaction mass at 100 °C was determined in the reaction calorimeter: $\gamma = 6700\,W\,m^{-2}\,K^{-1}$ at 100 °C. The relevant physical properties of 96% sulfuric acid are as follows:

- dynamic viscosity: 4.2 mPa s
- specific heat capacity: $1640\,J\,kg^{-1}\,K^{-1}$
- specific weight $1740\,kg\,m^{-3}$
- thermal conductivity $0.375\,W\,m^{-1}\,K^{-1}$

Table 9.7 Two points extracted from the cooling experiment.

Time	0.5 h	1.5 h
Temperature of reaction mass (°C)	107	91
Coolant inlet temperature (°C)	70	70
Coolant outlet temperature (°C)	78	74

The data of the reaction vessel

- internal vessel diameter: 2.80 m
- diameter of stirrer: 1.40 m
- stirrer speed: 45 rpm
- stirrer constant: 0.36

Calulate:

1. the overall heat transfer coefficient during the cooling experiment (U);
2. the reactor part of the heat transfer coefficient, that is the contribution from its jacket and cooling system (φ);
3. the overall heat transfer coefficient (U);
4. the shortest admissible feed time. Assume a mean coolant temperature of 75 °C.

► Exercise 9.2

A reaction should be performed at reflux. The maximum heat release rate of the reaction is $100\,W\,kg^{-1}$ reaction mass. The $4\,m^3$-reactor with a diameter of 2 m contains 4000 kg of reaction mass; the vapor tube has a diameter of 200 mm.

Questions:
Assume a heat exchange area of the reactor of $6\,m^2$ and an overall heat exchange coefficient of $500\,W\,m^{-2}\,K^{-1}$.

1. Is the cooling capacity by the reactor jacket sufficient to remove the heat of reaction?
2. Calculate the mean temperature of the cooling medium.
3. Can the reaction heat be removed by reflux cooling using: (i) water; (ii) toluene, as solvent?

The relevant physical properties of water and toluene are summarized in Table 9.8.

Hint: Consider three elements.

First element: Flooding of vapor tube Calculate the necessary cooling capacity by the reactor jacket and compare it to the limit for flooding of the heat release rate,

Table 9.8 Physical properties of water and toluene.

Property	Water	Toluene
Molecular weight ($g\,mol^{-1}$)	18	92
Specific heat of evaporation ($kJ\,kg^{-1}$)	2400	356
Boiling point (°C)	100	110

or calculate the velocity of the solvent vapor and compare it to the vapor velocity limit for flooding.

Second element: Swelling of the reaction mass. Swelling is determined by the flow of vapor across the surface of the reaction mass. A vapor velocity below $0.1\,m\,s^{-1}$ is uncritical, except for vessels filled to their maximum level.

Third element: Capacity of the reflux condenser. For the reflux condenser, assume a heat transfer coefficient of $1000\,W\,m^2\,K$. The mean coolant temperature is 30 °C. Calculate the necessary heat exchange area of the condenser.

▶ Exercise 9.3

The neutralization of a reaction mixture by caustic soda releases $120\,kJ\,l^{-1}$ of final reaction mixture. The batch size is 1200 liters of final reaction mass in a glass-lined stirring vessel equipped with a jacket for cooling with water entering the jacket at 17 °C. The maximum temperature for the water leaving the system is 30 °C.

Questions:
1. How fast can the addition of caustic soda be (shortest addition time)?
2. What other measures should be taken to ensure a safe process?

Data:
Heat exchange area: $A = 5m^2$ (assumed to be constant)
Heat transfer coefficient of internal film $h_r = 1000\,W\,m^{-2}\,K^{-1}$
Heat transfer coefficient of the external film (cooling water/wall)
$h_e = 1500\,W\,m^{-2}\,K^{-1}$
Thermal conductivity of glass: $\lambda = 0.5\,W\,m^{-1}\,K^{-1}$; thickness 2 mm
Thermal conductivity of steel: $\lambda = 50\,W\,m^{-1}\,K^{-1}$; thickness 5 mm

▶ Exercise 9.4

An epoxy compound is to be condensed with an amine at 40 °C. The reaction is fast and will be performed in a $4\,m^3$ reactor cooled with water. The process is semi-batch operation, the solvent 800 kg isopropanol, 240 kg of the epoxy compound are initially charged, and 90 kg of amine are added at a constant rate during 45 minutes, while the temperature is not allowed to surpass 40 °C.

Questions:
1. Is the cooling capacity sufficient?
2. What other solution could you suggest, knowing the temperature must be kept at a maximum of 40 °C and the addition time at a maximum of 45 minutes?

Data:
Specific heat of reaction $Q_r' = 130\,kJ\,kg^{-1}$ (exothermal)
Specific heat capacity $c_p' = 2.1\,kJ\,kg^{-1}\,K^{-1}$

Overall heat transfer coefficient $U = 310\,\mathrm{W\,m^{-2}\,K^{-1}}$
Heat exchange area (assumed to be constant) $A = 5.5\,\mathrm{m^2}$
Mean cooling water temperature $T_c = 20\,°\mathrm{C}$
Latent heat of evaporation (Isopropanol) $\Delta H_v = 700\,\mathrm{kJ\,kg^{-1}}$

► Exercise 9.5

A fast exothermal reaction is to be performed in a semi-batch reactor. In order to control the temperature course of the reaction, one of the reactants is added at a constant rate, producing a constant heat flow. The reactor is cooled with water from a river (at 15 °C in winter). The cooling water should not be rejected at a temperature higher than 30 °C.

Questions:
1. What is the shortest addition time allowing to maintain a reaction temperature of 50 °C, if the added reactant is heated to 50 °C before addition?
2. What will the required cooling water mass flow rate be?
3. Same questions if the added reactant is at room temperature of 25 °C?
4. Same questions in summer with a cooling water temperature of 25 °C.

Data:
Specific heat of reaction $Q'_r = 100\,\mathrm{kJ\,kg^{-1}}$ (exothermal)
Charge: 3000 kg initially charged, 2000 kg added
Heat exchange area (average during addition) A = $6\,\mathrm{m^2}$
Overall heat transfer coefficient $U = 400\,\mathrm{W\,m^{-2}\,K^{-1}}$
Specific heat capacities: Water $c'_p = 4.2\,\mathrm{kJ\,kg^{-1}\,K^{-1}}$
Added reactant $c'_p = 1.8\,\mathrm{kJ\,kg^{-1}\,K^{-1}}$

► Exercise 9.6

An aromatic nitro compound is to be reduced to the corresponding aniline by catalytic hydrogenation. The reaction scheme is

$$\mathrm{Ar - NO_2 + 3H_2 \xrightarrow{cat} Ar - NH_2 + 2H_2O}$$

Under the usual operating conditions, 100 °C and 20 bar g hydrogen pressure, the reaction follows an apparent zero-order, indicating that the reaction is operated in the mass transfer control regime. The reaction is performed as semi-batch, the aromatic nitro compound being initially charged together with a solvent, and hydrogen is fed continuously through a feed control valve. Thus, the reaction rate can be controlled by the hydrogen feed. The feed controller has two functions:

1. limit the pressure at max. 20 bar g,
2. limit the hydrogen feed rate such that the temperature remains constant.

Questions:

1. What is the cooling capacity of the reactor?
2. What is the maximum allowed hydrogen feed rate (in $m^3 h^{-1}$ under standard conditions with 1 mol = 22.4 litre)
3. What other safety issues must be considered with this reaction?

Data:

Charge: 3.5 kmol nitro compound
Heat exchange area: $7.5\,m^2$
Reaction enthalpy: $-\Delta H_r = 560\,kJ\,mol^{-1}$
Overall heat transfer coefficient $U = 500\,W\,m^{-2}\,K^{-1}$
Cooling water inlet temperature: 20 °C
Cooling water maximum outlet temperature: 40 °C

References

1 DIN28131 (1979) *Rührer für Rührbehälter, Formen und Hauptabmessungen*, Beuth, Berlin, Germany.

2 VDI (1984) *VDI-Wärmeatlas, Berechnungsblätter für den Wärmeübergang*, VDI-Verlag, Düsseldorf.

3 Stoessel, F. (ed.) (2005) Safety of polymerIzation processes, in *Handbook of Polymer Reaction Engineering* (eds T. Meyer and J. Keurentjes), Vol. 2, Wiley-VCH, Weinheim.

4 Bourne, J.R., Buerli, M. and Regenass, W. (1981) Heat Transfer and Power Measurement in stirred tanks using heat flow calorimetry. *Chemical Engineering Science*, **36**, 347–54.

5 Choudhury, S. and Utiger, L. (1990) Wärmetransport in Rührkesseln: Scale-up. *Methoden. Chemie Ingenieur Technik*, **62** (2), 154–5.

6 Wiss, J. (1993) A systematic procedure for the assessment of the thermal safety and for the design of chemical processes at the boiling point. *Chimia*, **47** (11), 417–23.

7 Wilson, J.F., Grenda, R.J. and Patterson, J.F. (1961) Steam volume fraction in a bubbling two-phase mixture. *Transactions of the American Nuclear Society*, **4** (Session 37), 356–7.

8 Wilson, J.F., Grenda, R.J. and Patterson, J.F. (1962) The velocity of rising steam in a bubbling two-phase mixture. *Transactions of the American Nuclear Society*, **5** (Session 25), 151–2.

9 Machado, R. (2005) Practical mixing concepts for scale-up from lab reactors, in *European RXE-User Forum, Mettler Toledo, Lucerne.*

10 Toulouse, C., Cezerac, J., Cabassud, M., Lann, M.V.L. and Casamatta, G. (1996) Optimization and scale-up of batch chemical reactors: impact of safety constraints. *Chemical Engineering Science*, **51** (10), 2243–52.

11 Zufferey, B. (2006) *Scale-down Approach: Chemical Process Optimisation Using Reaction Calorimetry for the Experimental Simulation of Industrial Reactors Dynamics*, EPFL, n°3464, Lausanne.

12 Zufferey, B., Stoessel, F. and Groth, V. (2007) *Method for Simulating a process plant at laboratory scale*, European Patent Office, Pat. Nr. EP 1764662A1.

10
Risk Reducing Measures

Case History

During the risk analysis of an exothermal batch process, runaway was identified as the major risk of the process. Therefore, great efforts were made to design a series of measures for avoiding such an event, and so a three protection levels system was designed (Figure 10.1). As the first level, the cooling system was improved by installing a back-up pump in the cooling circuit. In addition, a third pump was installed, powered by an emergency power supply network independent of other utilities. As the second level, since the reaction was catalytic, it was decided to install an inhibitor injection system. This consisted of a small vessel containing the inhibitor placed above the reactor and under nitrogen pressure of 5 bar g. The inhibitor injection was triggered by a temperature alarm, which opened an automatic valve. An additional hand-operated valve was placed in parallel, to allow for a manual injection. As the third level, a bursting disk was installed that lead to a catch tank.

However, this three-level system failed. The first level failed due to an error in design of the electrical connections to the pumps. The two main pumps were powered by the main power system and the third pump was powered by the emergency network. However, the control systems of all three pumps were powered by the emergency network. On the day of the incident, maintenance work on the emergency power system was planned. Since this should not have affected the main power system, it was decided to start a batch despite the missing emergency power system, as nobody was aware of the fact that the pump control systems would be out of use. As the emergency power system was disconnected, all three cooling medium circulation pumps stopped. The temperature of the batch increased rapidly and the inhibition system was triggered, but the inhibitor failed to inject into the reactor. An operator opened a hand valve, but this also had no effect.

The reason for this system failure was that the importance of a reliable temperature measurement for triggering the inhibitor injection was duly recognized during the risk analysis and an especially thick tube was installed for the temperature probe, providing a high mechanical resistance ... but also a high time constant for the thermometer! Hence, as the temperature probe reached

Thermal Safety of Chemical Processes: Risk Assessment and Process Design. Francis Stoessel

ISBN: 978-3-527-31712-7

the alarm level, the actual temperature of the reaction mass was higher than designed, causing a higher vapor pressure above 5 bar g at the time it was triggered (Figure 10.2). The inhibitor could not flow into the reactor. At the third protection level, the pressure relief system failed because it was not designed for two-phase flow. The reaction mass was discharged to the environment, causing heavy spillage with toxic material.

This example shows that the design of protection systems must be careful and "thought through," to be efficient in any emergency. Further, it also enhances the fact that technical measures may fail. Absolute reliability can never be guaranteed.

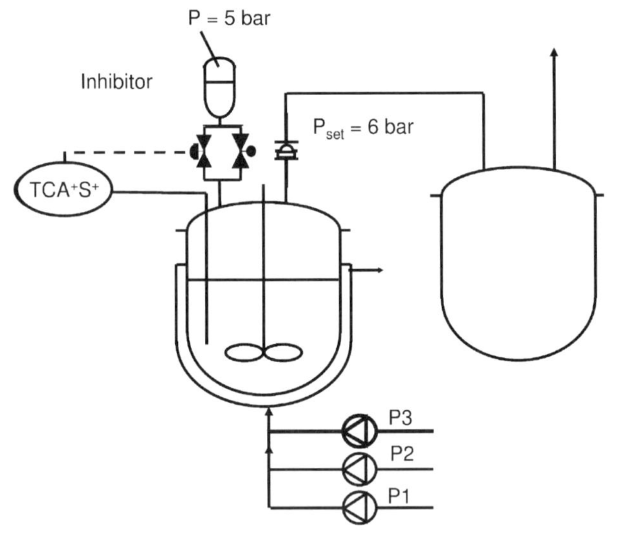

Figure 10.1 Three protection levels for a batch reactor.

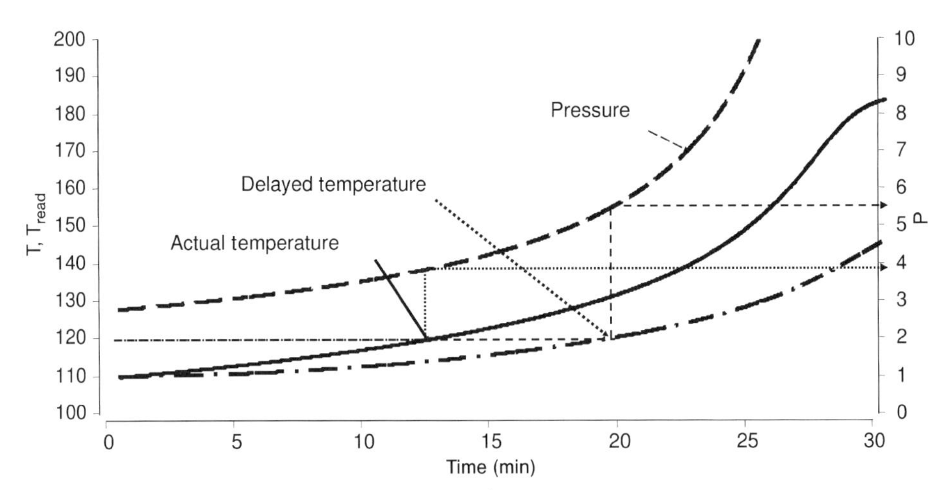

Figure 10.2 Temperature pressure course during a runaway.

10.1 Introduction

To allow for emergencies that could occur due to a sudden technical failure, the strategy of protection of reactors against runaway is of primary importance. This chapter is devoted to the presentation of typical protection measures. The first section gives some hints on the strategy of choice. The sections that follow present various measures that can be applied: first to avoid runaway occurring by eliminating measures, then stopping a developing runaway by preventive measures, and finally mitigating the consequences by emergency measures. A separate section is devoted to the design of protection measures, based on the data defining the cooling failure scenario.

10.2 Strategies of Choice

A discussion about risk reducing measures is presented in Section 1.3.1.7. When reducing risks linked to a runaway reaction, the same principles apply. Thus, the first priority should be implementing measures that avoid runaways. As as second priority, preventive measures should then be applied, in case a runaway has been triggered. As a last resort, emergency measures should mitigate the consequences of a runaway that cannot be avoided. Of course, the priority is to "avoid the problem rather than to solve it" [1].

The international standard IEC 61511 [2] gives advice on the design of safety instrumented systems (SIS) and presents a "layer concept" to achieve reliability of protection systems. These principles can be applied to the protection of chemical reactors [3]. Figure 10.3 represents this layer of protection principles. The first layer is the process itself, meaning that it should be designed in such a way that it cannot give rise to a runaway reaction. Some concepts for achieving this objective are reviewed in Section 10.3.

These different categories of measures are reviewed, with a focus on runaway reactions, in the next sections.

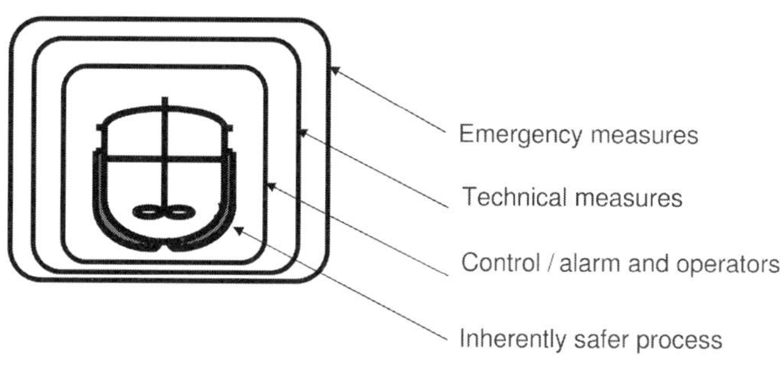

Figure 10.3 Layer design of protection systems for a batch reactor.

10.3 Eliminating Measures

Eliminating the risk of a runaway reaction means reducing the severity (see Section 3.3.2). If we follow the evaluation criteria for such severity, we see that the energy released, by the desired as well as secondary reactions, should be reduced to a level leading to an adiabatic temperature rise below 50 K. This temperature increase would lead to a smooth temperature increase and not a runaway reaction. Different possibilities exist for reducing the energy release:

The first possibility is to reduce the adiabatic temperature rise by dilution. Even if this strategy is efficient in terms of risk reduction, it is not economic since it reduces productivity. Moreover, it may also cause environmental problems, if large amounts of solvent must be handled.

The same goal may be achieved by using a semi-batch reactor, which by its limitation of the accumulation of non-converted reactants can significantly reduce the runaway potential of a reaction, as was extensively showed in Chapter 7.

Another far more fundamental approach is to reduce the absolute energy released by the reaction. This is achieved in different ways according to the principles of the design of inherently safer processes [1, 4–7]. Kletz, who promoted these ideas, gave some principles to follow for the reduction of severity.

The first principle is substitution, consisting of choosing an appropriate synthesis route, which avoids the use of hazardous materials, instable intermediates, or highly energetic compounds. This implies that decisions involving process safety aspects are made at very early stages of process development. In this frame, the methods of screening for the energetic aspects of reactions or compounds by micro-calorimetric methods, for example, DSC or Calvet Calorimetry (see Section 4.3.2), may be very useful. Thus, it is advantageous to take safety and environmental aspects into account from the beginning of process development, following the principles of integrated process development [7]. This practice also requires methods that can be used in early development stages [8–10], at the time when process knowledge is small, but degrees of freedom left for even radical changes, are the greatest.

The second principle is process intensification, consisting of limiting the quantity of hazardous material by scale reduction, that is, the absolute energy potentially released is reduced accordingly. This objective can be achieved by using continuous processes [11] that generally use smaller reactor volume (see Chapter 8). The extreme in scale reduction is to use micro reactors that are able to maintain a reaction mass under isothermal conditions, even with a high heat release rate [12, 13]. In general, smaller reactors are also easier to protect or to build with resistance to high pressure, that helps to makes them fail safe. On the other hand, in case of material release, since the amount is reduced, it is easier to contain, which avoids large-scale consequences.

The third principle is the principle of attenuation, which consists of using a hazardous material in a safer form. As an example, the use of diphosgene instead of phosgene for a chlorination reaction follows this principle. Diphosgene is less volatile than phosgene and so is easier to handle, improving the process safety.

These measures have one point in common, in that they do not limit the effect of failures by adding protective equipment, but instead by process design or changing the process conditions. It is a huge advantage in not needing to rely on protective equipment, but instead to make the process intrinsically safe.

Kletz also considers that the technical design of the process has a positive impact on its safety. Here too he quotes several principles to improve process safety by technical design:

- simplification, as complex plants provide more opportunities for human error and contain more equipment that can go wrong,
- avoiding cumulative or domino effects,
- making the state of equipment, such as open or shut, clear,
- designing equipment that is able to withstand incorrect installation or operation,
- making equipment easy to control,
- making incorrect assembly difficult or impossible.

10.4 Technical Preventing Measures

In this section, the focus is on technical measures, with the objective of avoiding a runaway being triggered, before it takes a non-controllable course.

10.4.1 Control of Feed

In semi-batch or continuous operations, the feed rate controls the reaction course. Hence, it plays an important role in the safety of the process. With exothermic reactions, it is important to be able to limit the feed rate by technical means. Some methods are:

- Feed by portions: this method, presented in Section 7.8.1, is obviously only applicable to discontinuous processes as semi-batch. It reduces the amount of reactant present in the reactor, that is, the accumulation and therefore the energy that may be released by the reaction in case of loss of control. The amount allowed in one portion can be determined in such a way that the maximum temperature of the synthesis reaction (MTSR) does not reach a critical level as the maximum temperature for technical reasons (MTT) or the temperature at which secondary reactions become critical (T_{D24}). The difficulty is to ensure that an added portion has reacted away, before adding the next portion. Generally, the feed control is performed by the operator, but can also be automated.

- Feed tank with control valve and gravimetric flow: the desired flow rate is ensured by the appropriate opening of the valve, that is, the actuator of a control loop using the weight of the reactor or of the feed tank, the level in the feed tank, or a flow meter as input. The maximal feed rate can be limited by the clearance of the valve or by a calibrated orifice.
- Feed tank with centrifugal pump: the centrifugal pump replaces the gravity, positioning the feed tank, even below the reactor level. Since a centrifugal pump is not volumetric, it is necessary to provide an additional control valve to limit the flow rate. The flow control strategies are the same as described above.
- Feed tank and metering pump: the flow rate through such a pump can be controlled by a stroke adjusting mechanism or a variable speed drive acting on the stoke frequency. Control can be achieved by a fixed adjustment or through a flow meter.

It is state of the art to interlock the feed to a reactor with its temperature interrupting the feed, either for too high or too low temperatures (avoiding accumulation). For stirred reactors the feed must also be interlocked with the stirrer in order to avoid accumulation due to lack of mixing.

In cases where it is required, the maximum amount of reactant contained in the feed tank can also be limited, for example, by an overflow.

10.4.2 Emergency Cooling

Emergency cooling replaces the normal cooling system in case of failure. This requires an independent source of cooling medium, generally cold water, which flows through the reactor jacket or through cooling coils. The cooling medium must also be able to flow in case of utilities failure, especially electrical power, often a common cause of cooling failure. The time factor is very important for emergency cooling, which must be applied before the heat release rate of the reaction that is to be controlled is higher than the cooling capacity of the system. The concept of Time to no Return (TNoR), presented in Section 2.5.5, is very useful in this respect.

For emergency cooling, it is critical that the temperature does not fall below the solidification point of the reaction mass. Otherwise a crust would form, resulting in reduced heat transfer, which again may favor a runaway situation. The remediation may then have worse effects than the initial failure.

Agitation of the reaction mass may also be critical in such a situation: without agitation, cooling being provided by natural convection only, leads to a considerable reduction of the heat transfer coefficients (see Section 9.3.5). Generally, by natural convection, the heat transfer coefficient is reduced to 10% of its value with stirring [14]. Nevertheless, this is only valid when natural convection is established, that is, for smaller vessels and contents with moderate viscosity (see Section

13.3.3). Large reaction masses behave almost adiabatically, even if cooled from the outside. Here the injection of nitrogen into the bottom of the reaction mass has proved helpful for emergency mixing. However, this method must be tested under practical conditions.

10.4.3 Quenching and Flooding

Some reactions can be stopped by adding a suitable component. This is sometimes possible for catalytic reactions where a "catalyst killer" can be added in small amounts. For pH-sensitive reactions, a pH modification may also slow down or even stop the reaction. In these cases, the addition of only a small amount of a compound will suffice. Agitation is a critical factor, especially to ensure that a small amount of an inhibitor must be dispersed homogeneously in a large volume of reaction mixture. In order to achieve a fast and homogenous dispersion, often the vessel containing the inhibitor is pressurized, for example, nitrogen and nozzles are used to spray it into the reaction mass.

For other reactions a greater amount of an inert and cold material is required to flood the reaction mass. Flooding has two effects, dilution and cooling, by lowering the concentration and / or the temperature to slow down or stop the reaction. If flooding is to be at a temperature above the boiling point of the fluid, a pressure relief system must be provided. For flooding, the critical factors are the amount, the rate of addition, and the temperature of the quenching material. Obviously, the required empty volume must also be available in the reactor, which may limit the batch size and consequently the productivity.

Calorimetric methods are of great help in designing such measures, since they measure the heat of mixing, which is often important. They also verify that the resulting mixture is thermally stable.

Worked Example 10.1: Emergency Flooding

A reaction should be stopped by flooding with a cold solvent. The amount of solvent needs to be sufficient to cool the reaction mass to a thermally stable level. To test this theory, flooding was tested in a Calvet calorimeter (Figure 10.4). The experiment showed that the dilution is endothermal with a heat release of $-18\,\text{kJ}\,\text{kg}^{-1}$ of mixture (reaction mass and solvent). The reaction mass (2230 kg) has a specific heat capacity of $1.7\,\text{kJ}\,\text{kg}^{-1}\,\text{K}^{-1}$ and a temperature of 100 °C. The dilution is with 1000 kg of a solvent at 30 °C, with a specific heat capacity of $2.6\,\text{kJ}\,\text{kg}^{-1}\,\text{K}^{-1}$. The resulting mixing temperature (T_m) can be calculated from a heat balance:

$$T_m = \frac{M_{r1}\cdot c_{p1}\cdot T_{r1} + M_{r2}\cdot c_{p2}\cdot T_{r2} + (M_{r1}+M_{r2})\cdot Q'_r}{M_{r1}\cdot c_{p1} + M_{r2}\cdot c_{p2}}$$

$$T_m = \frac{2230 \times 1.7 \times 100 + 1000 \times 2.6 \times 30 + 3230 \times (-18)}{2230 \times 1.7 + 1000 \cdot 2.6} \cong 62°C$$

To assess the thermal stability of the quenched mixture at 62 °C, a reference heat release rate of 2 W kg^{-1} can be read at a temperature of 180 °C (Figure 10.4, bottom). With a conservative activation energy of 50 kJ mol^{-1}, the decomposition becomes uncritical below T_{D24} = 145 °C, that is, the quenched reaction mass can be considered stable at 62 °C, even if the potential (520 kJ kg^{-1}) is still high.

10.4.4 Dumping

This measure is similar to quenching, except that the reaction mass is not retained in the reactor, but transferred into a vessel containing the inhibitor or diluting compound. This vessel must be prepared to receive the reaction mass at any instant during the process. The assessment of the suitability of dumping as a preventive measure against runaway is the same as for quenching. The advantage of dumping is that the reaction mass may be transferred to a safer place, thus protecting the plant where the reactor is located.

The transfer line is critical for the success of this measure: plugging or closed valves must be totally avoided. It must be designed to permit an emergency transfer, even if there is a breakdown in utilities. The presence of diluting agent or quenching fluid in the receiving vessel must be assured. This measure is particularly suitable in cases where the final reaction mass must normally be transferred for workup.

10.4.5 Controlled Depressurization

This measure is different from emergency pressure relief. A controlled depressurization is activated in the early stages of the runaway, while the temperature increase rate and the heat release rate are slow.

If a runaway is detected at such an early stage, a controlled depressurization of the reactor may be considered. As an example, during an amination reaction, a 4 m^3 reactor could be cooled from 200 °C to 100 °C within 10 minutes without external cooling, just by using a controlled depressurization allowing evaporative cooling. Obviously, the scrubber and the reflux condenser must be designed to work with independent utilities.

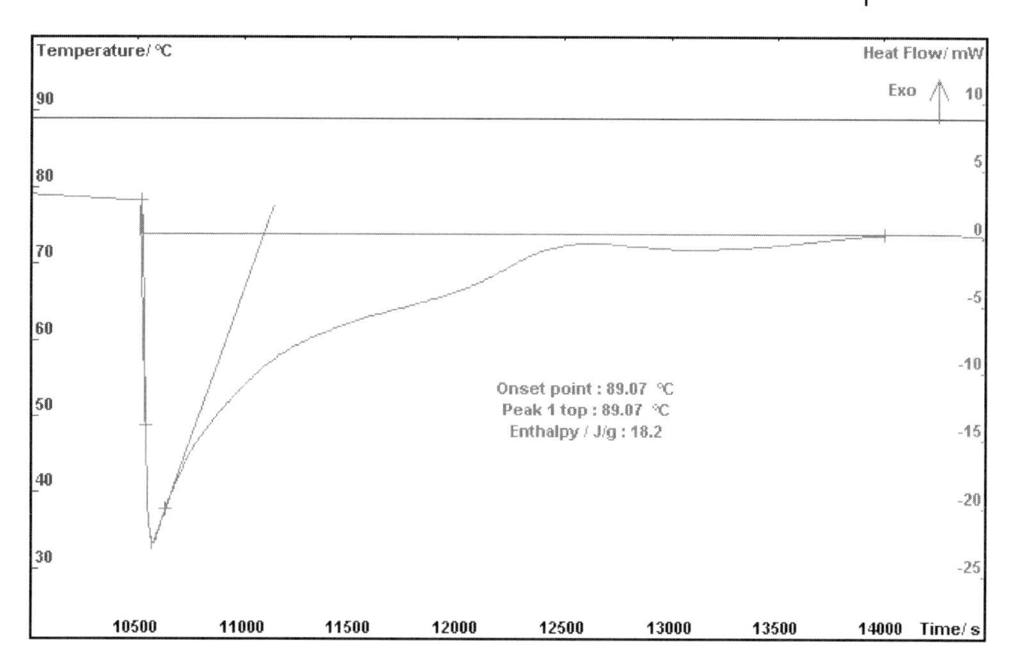

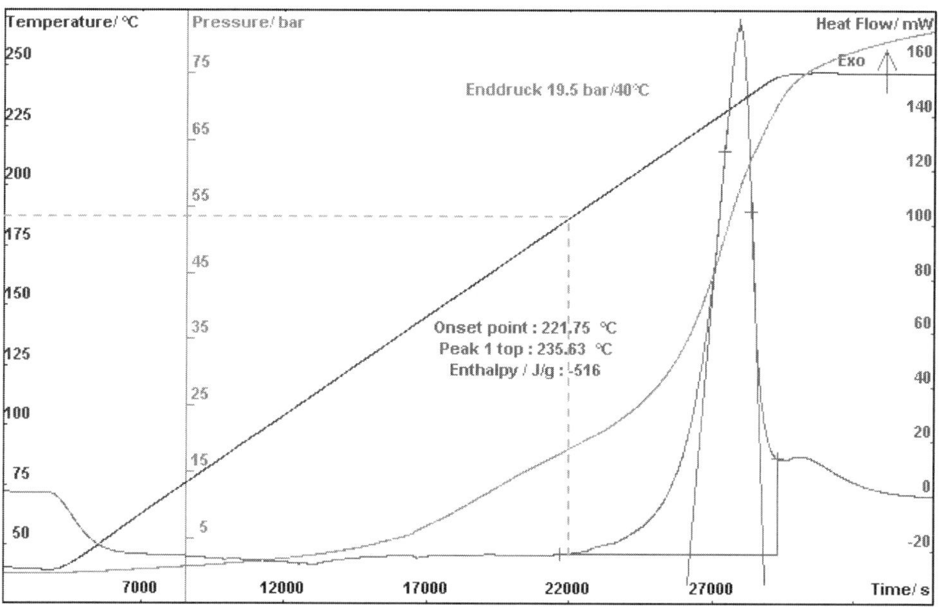

Figure 10.4 Example of thermograms obtained during flooding of a reaction mass with a cold solvent (top). The thermal stability is checked by a heating ramp after flooding (bottom).

Worked Example 10.2: Controlled Depressurization

This is the continuation of Worked Example 3.1. If there is loss of control of an amination reaction, the temperature could reach 323 °C (MTSR), but the maximum allowed working pressure of 100 bar g would be reached at 249 °C (MTT). Thus, the question is: "If the reaction can be controlled by depressurizing the reactor before the safety valve opens, that is, before 240 °C is reached, what would the vapor release rate be?" To answer this question, information about the reaction kinetics is required. The only information is that at 180 °C, a conversion of 90% is reached after 8 hours. If we consider the reaction to follow a first-order rate equation, justified by the fact that ammonia is in large excess, we can calculate the rate constant at 180 °C:

$$\frac{dX}{dt} = k(1-X) \Rightarrow k = \frac{-\ln(1-X)}{t} = \frac{-\ln(0.1)}{8} \cong 0.288\,\mathrm{h^{-1}} = 8 \cdot 10^{-5}\,\mathrm{s^{-1}}$$

Thus the heat release rate is

$$q_{rx} = k(1-X) \cdot Q_{rx} = 8 \cdot 10^{-5} \times 1 \times 175 \times 2000 = 28\,\mathrm{kW}$$

This is calculated at 180 °C for the charge of 2 kmol and for a conversion of zero, which is conservative. It would be reasonable to interrupt the runaway at its very beginning, for example, at 190 °C. If we consider that the reaction rate doubles for a temperature increase of 10 K, the heat release rate would be 56 kW at 190 °C. The latent heat of evaporation can be estimated from the given Clausius–Clapeyron expression:

$$\ln(P) = 11.46 - \frac{3385}{T} \Rightarrow \Delta H_v = 3385 \times 8.314 \cong 28\,\mathrm{kJ\,mol^{-1}}$$

Thus, the vapor release rate is

$$\dot{N}_{NH_3} = \frac{56\,\mathrm{kW}}{28\,\mathrm{kJ\,mol^{-1}}} = 2\,\mathrm{mol\,s^{-1}}$$

with a molar volume of

$$0.0224\,\mathrm{m^3\,mol^{-1}} \times \frac{463}{273} = 0.038\,\mathrm{m^3\,mol^{-1}}$$

At atmospheric pressure and 190 °C, the volume flow rate is

$$\dot{v} = 0.076\,\mathrm{m^3\,s^{-1}}$$

and in a pipe with a diameter of 0.1 m, the vapor velocity is

$$u = 0.076 \times \sqrt{\frac{4}{\pi \times 0.1^2}} = 0.86\,\text{m}\,\text{s}^{-1}$$

which is a low velocity allowing for scrubbing without any problems, provided the scrubber works in the case of failure. Thus, using a controlled depressurization is a technically feasible measure. The depressurization must be slow enough to avoid two-phase flow that would entrain reaction mass with the vapor. In this estimation, we considered that the vapor is only ammonia, in fact, water will also be evaporated, but since the latent heat of evaporation of water is higher than for ammonia, the estimation remains on the safe side. We also neglected the reactant depletion due to conversion (zero-order approximation) that also decreases the heat release rate.

10.4.6 Alarm Systems

In the examples given above, we see how important an early intervention is in case of runaway. Whatever the measure considered, the sooner it becomes active the better. An exothermal reaction is obviously easier to control at its beginning, before the heat release rate becomes too great. This is true for emergency cooling as well as for controlled depressurization. Thus, the idea arose to detect a runaway situation by an alarm system. The first attempt in this direction stems from Hub [15, 16], who proposed evaluating the second time derivatives of the reactor temperature and the first derivative of the temperature difference between reactor and jacket, giving a criteria for a runaway:

$$\frac{d^2T_r}{dt^2} > 0 \quad \text{and} \quad \frac{d(T_r - T_c)}{dt} > 0 \tag{10.1}$$

One important difficulty is that the noise of the temperature signal in the industrial environment is amplified by the derivation, which is at the cost of the accuracy of the method. These criteria give an alarm between 20 and 60 minutes before runaway [16], which is a short time, that is, the runaway is well-developed when the alarm is activated.

A more sophisticated method to construct an early warning detection system (EWDS) is proposed by the research group led by Stozzi and Zaldivar, in the frame of an European research project [17–21]. This approach uses a divergence criterion applied on the state space describing a temperature trajectory. By doing so, it is possible to use only one process variable, generally the temperature, but pressure was tested too. Despite its complex mathematical background, this approach is very promising, and easy to implement in an industrial environment. Nevertheless, an alarm always detects a runaway but cannot prevent it. Moreover, it assumes that measures are provided that will be able to stop the commencing runaway.

10.4.7
Time Factor

Time plays a primary role in the effectiveness of a measure. The following steps must be taken from the instant a failure occurs until regaining control over the process (Figure 10.5):

- When a failure or a malfunction occurs, first it needs to be detected. The detection time is influenced by the choice of appropriate alarm settings or by the use of more sophisticated alarm systems, as described in Section 10.4.6. Most important, is choosing the appropriate parameter which must be monitored to detect a malfunction. The design of alarms, interlocks, and control strategies is an important part of process design and should always follow the principles of simplicity in the concept of inherently safer processes (see Section 10.3).
- Once the alarm is switched on, there is some time before remedial measures can be applied. Quenching or dumping requires considerable time for the transfer of the quenching fluid or of the whole reaction mass. An emergency cooling must be switched on and the cooling medium must flow at the required temperature with the required flow rate, to initiate this cooling.
- The measure must become effective: There is also some time between the instant measures being applied and their effects on the process. Here the process dynamics again play a determining role.

This time factor must be estimated for the effective design of safety measures and compared with the Time to Maximum Rate (TMR_{ad}), giving the upper limit of the time frame. In fact, by applying Van't Hoff rule, the reaction rate doubles for a temperature increase of 10 K. If a temperature alarm is typically set at 10 K

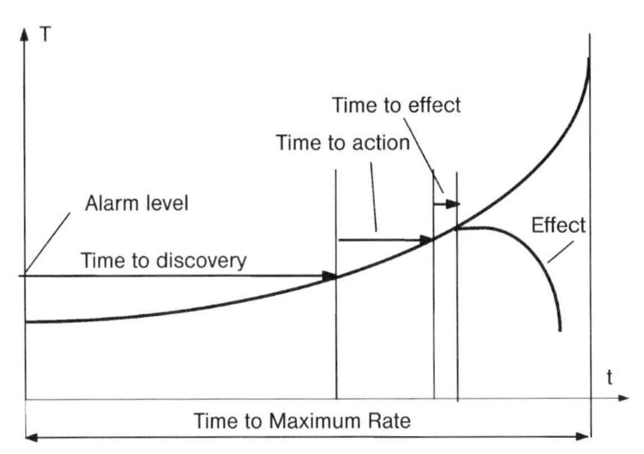

Figure 10.5 Typical temperature and pressure course during pressure relief. Left: tempered system, Right: gassy system.

above the process temperature, half of the TMR_{ad} will have already been spent to activate the alarm, and the measure needs to become effective as soon as possible before the runaway develops. Thus, the time left for the measure to become effective is rather short. This also explains the time criteria (8 and 24 hours) used to assess the probability of triggering a runaway (see Section 3.3.3).

10.5 Emergency Measures

Emergency measures should only be used as a last resort, when runaway cannot be prevented. Thus, they should only be considered after all other approaches have been tried and found to be unsuccessful.

10.5.1 Emergency Pressure Relief

Emergency pressure relief consists of protecting a reactor against overpressure, by stopping the pressure increase by opening a vent line to allow gases or vapor to escape. The design of venting lines, for reactions with thermal potential, is a complex matter, so is not described in detail here. There have been examples of where pressure relief was able to protect reactors from an explosion, but also cases where a reactor exploded, even with an open manhole (see case history in Chapter 11). This measure only applies to reaction systems where the pressure increases significantly for small temperature increases above the normal operating level. In addition, the discharge line must end in a catch tank or a scrubber to avoid spillage of possibly toxic or flammable material.

An essential point, when considering emergency relief, is that as the relief system suddenly opens, reaction mass may be entrained by the gas or vapor, leading to two-phase flow that decreases the relief capacity of the system. Thus the

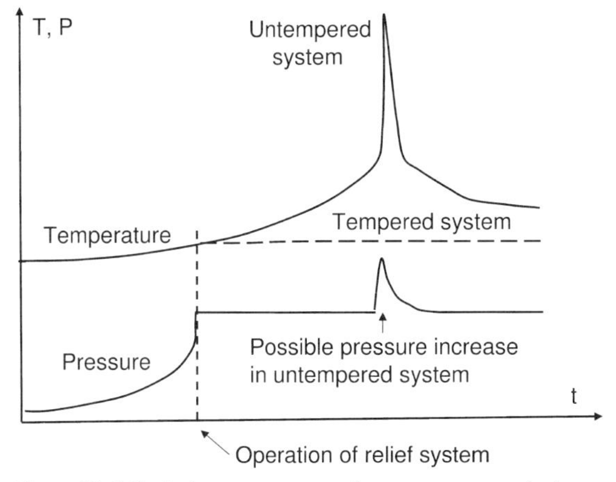

Figure 10.6 Typical temperature and pressure course during pressure relief.

methods developed by the Design Institute of Emergency Relief systems (DIERS) should be used [22–24].

The design of emergency pressure relief systems usually proceeds in three steps, described in the next subsections.

10.5.1.1 Definition of the Relief Scenario

The key factor of success in the design of emergency pressure relief systems lies in a good understanding of the behavior of the reaction under relief conditions. The first point in this context is the cause of pressure increase. This may be the vapor pressure of the reaction mass, the so-called tempered system. Pressure increase may also be due to gas release by a reaction, the so-called gassy system. There are also cases where the pressure stems from both vapor pressure and gas release, the so-called hybrid system, which may or may not be tempered.

A tempered system behaves differently from a gassy system. In the case of a tempered system, vapor is generated during relief that brings an endothermal contribution to the heat balance. Consequently, the temperature increase slows down or even stops during relief. In the case of a gassy system, this endothermal effect does not exist and the temperature continues to rise, accelerating the reaction further until the maximum rate is reached. Thus, a major difference is that for a gassy system, the relief system must be designed to cope with the maximum reaction rate, whereas for a tempered system the sizing may be done for fairly lower reaction rates.

One must be aware of the fact that when the set pressure of the protection device is reached, the pressure increase does not stop immediately, but continues to increase to the maximum pressure before decreasing. These two pressure levels, set pressure and maximum pressure, have to be defined during the design procedure. The design can be for two different scenarios, the physical scenario where no chemical reaction is involved and the chemical scenario where a chemical reaction determines the behavior of the system.

There are many different physical scenarios to be considered. In the first category, we find scenarios that result from gas compression, such as by liquid transfer into a closed reactor or gas inlet from a line connected to the reactor. With such scenarios, two-phase flow is unlikely to occur. Other common physical scenarios are linked to unwanted heating of the reactor contents, either by fire or by inadvertently heating of the reactor by the heating system. In this case, two-phase flow may occur.

For chemical scenarios, the kinetic behavior of the reaction, the temperature and pressure increase rate must be known under runaway conditions in the interval between set pressure and maximum pressure. This implies a good knowledge of the thermo-chemical properties of the reaction mass. The required data are traditionally obtained from adiabatic calorimetric experiments [22, 25, 26]. Nevertheless, other calorimetric methods, especially dynamic DSC or Calvet experiments evaluated using the isoconversional approach, can also provide these data with accuracy and an excellent reliability for the temperature increase rate [27], as well as for the pressure increase [28, 29].

10.5.1.2 Design of the Relief Device

First, a choice must be made between safety valve and bursting disk, or even both in series or in parallel. A safety valve has the advantage of closing again when the pressure decreases below its set pressure. Thus, the reactor, that is, the remaining contents, can be made safe by this pressure release. The drawback of safety valves is their sensitivity against plugging or corrosion. With a bursting disk, the full relief section is available immediately after bursting, providing immediate high relief capacity. Bursting disks are more resistant against corrosion than safety valves, if correctly chosen.

Another drawback is that the bursting disk obviously does not close again, consequently the relief only ending when atmospheric pressure is reached. The set pressure of bursting disks may also vary with temperature, or even decrease with time, when submitted to frequent temperature or pressure variations. This may cause inadvertent relief. In certain cases, a combination in series, i.e. bursting disk followed by safety valve, is used to protect the safety valve against corrosion or splashes that may hinder its correct function. In such cases, the space between both devices must be monitored to avoid any pressure build up in this volume that would cause a shift of the set pressure of the whole system towards higher pressures [30] (see Case history "nitroaniline" in Chapter 6). Both devices can also be installed in parallel, the safety valve having a lower set pressure protecting against small deviations and the bursting disk having a higher set pressure and higher capacity offering protection against more important overpressures. In such an arrangement, a small over-pressurization does not lead to emptying the reactor.

Computer programs are available for the dynamic simulation of venting [22, 23]. Leung [31, 32] developped simplified methods for hand calculation for this purpose. The design method for safety valves is detailed in the work of Schmidt and Westphal [33, 34]. The design of emergency pressure relief systems is a complex matter and will not be treated in more detail in this book.

10.5.1.3 Design of Relief Devices for Multipurpose Reactors

In the fine chemicals and pharmaceutical industries, reactors are often used for diverse processes. In such a case, it is difficult to define a scenario for the design of the pressure relief system. Nevertheless, this is required by law in many countries. Thus, a specific approach must be found to solve the problem. One possibility, that is applicable for tempered systems, consists of reversing the approach. Instead of dimensioning the safety valve or bursting disk, one can choose a practicable size and calculate its relief capacity for two-phase flow with commonly-used solvents. This relief capacity will impose a maximum heat release rate for the reaction at the temperature corresponding to the relief pressure.

Then we consider a reaction presenting a given accumulation corresponding to a known adiabatic temperature rise. In the case of cooling failure, the reaction proceeds under adiabatic conditions: it is accelerated by the temperature increase, but at the same time, the reactant depletion decreases the reaction rate. Thus, the reaction rate passes a maximum, as described in Figure 10.7 (see also section 10.6.2.1). For the design of the relief system, the maximum heat release rate at

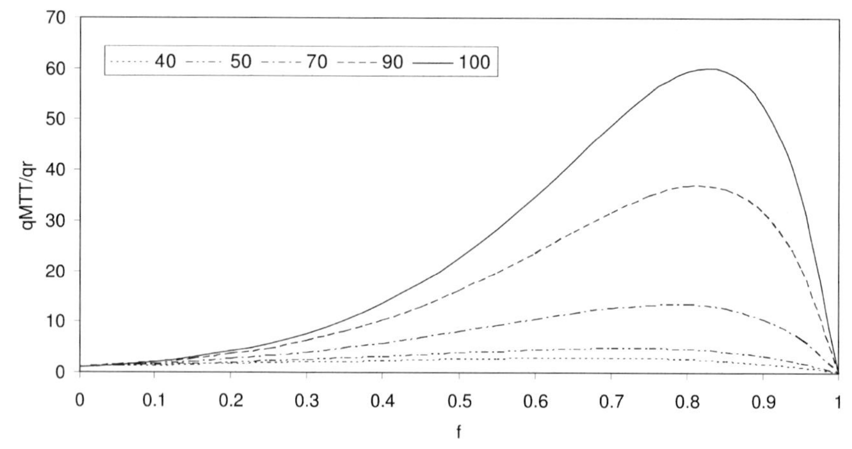

Figure 10.7 Acceleration factor $f_{acc} = q_{(MTT)}/q_{(TP)}$ as a function of the position x of MTT between T_p $(x = 0)$ and $MTSR$ $(x = 1)$, calculated for different values of the accumulation (ΔT_{ad}) based on an activation energy of $100\,kJ\,mol^{-1}$.

relief temperature defines the maximum allowable heat release rate under normal operating conditions using the acceleration factor. Thus knowing the process temperature and cooling capacity of the reactor, it can be verified that the mass flow rate during relief is compatible with the installed capacity.

10.5.1.4 **Design of the Effluent Treatment System**

There are different ways of processing the effluent of a pressure relief [35]. One is atmospheric discharge, which is rare as only harmless effluents can be so used. The second is flaring or incineration, which is only sparsely available in fine chemicals or pharmaceutical industries. Total containment is another approach, but it requires pressure resistant equipment (see Section 10.5.2). Thus, the most used is partial containment.

The aim of the effluent treatment system is to separate the liquid from gasses or vapor, which are allowed to escape either into the atmosphere or to a subsequent treatment unit. This is achieved by various means. The simplest is the gravity separator using an empty tank with enough volume to contain the liquid and to allow droplets to settle. Such a tank may also be used as a passive quench tank, if the vented mixture is sparged through an appropriate liquid. This may have two effects, stopping the reaction by cooling, and condensing the vapor. The separation of gas and liquid can also be achieved by centrifugal force using a cyclone. A review of these methods and design principles is given by McIntosh [36, 37].

10.5.2 Containment

A further method that may mitigate the consequences of a runaway reaction, is containment. Containment can be in the reactor itself, providing it is designed to withstand the maximum pressure after runaway. The contents of the reactor must be treated in an appropriate way afterwards. Where practicable, this passive measure reduces the consequences of a runaway.

There are also other forms of containment to consider. The first is mechanical protection in the sense that it protects the environment from missiles or flying debris when the reactor bursts, for example, by placing the equipment in a bunker. This approach is often used for high-pressure laboratory equipment, but in principle, may also be used for industrial equipment. The bunker is generally open to a "safe place," where nobody can be harmed.

A tight closed room can also be used as a containment area, providing enough volume for the effluent to be contained, giving time for the subsequent treatment of relieved material by, for example, a scrubber system. These types of measures should only be considered as a last resort, since they are linked with heavy damage to the equipment. An alternative is the safe-bag retention system, proposed by Siemens [38].

10.6 Design of Technical Measures

The choice and design of technical protection measures against runaway is in accordance with the risk level. This means that the consequences and controllability of the commencing runaway must be assessed. The criticality classes, based on four characteristic temperatures, are at the root of this assessment and serve in the design of protection measures.

10.6.1 Consequences of Runaway

A runaway reaction may have multiple consequences. The high temperature by itself may be critical, as the higher the final temperature, the worse the effects of the runaway. In case of a large temperature increase, some components of the reaction mixture may be vaporized, or some gaseous or volatile compounds may be produced. This, in turn, may lead to further consequences, a pressure increase in the system and/or release of gases or vapor, which may cause secondary damage due to toxicity or flammability.

10.6.1.1 Temperature

The adiabatic temperature rise, proportional to the reaction energy, represents an easy to use criterion for the evaluation of the severity of an uncontrolled energy release from a runaway reaction (see Section 3.3.2). The adiabatic temperature rise can be easily calculated by dividing the energy of reaction by the specific heat capacity:

$$\Delta T_{ad} = \frac{Q'}{c'_P} \tag{10.2}$$

In criticality classes 1 to 3, the energy to be considered is the reaction energy (Q'_{rx}) only, whereas in classes 4 and 5, the energy to be considered is the total energy, that is, the sum of the reaction and decomposition energies ($Q'_{rx} + Q'_{dc}$). The temperature increase may represent a threat in itself, but in most cases, it will result in a potential pressure increase.

10.6.1.2 Pressure

The pressure increase depends on the nature of the pressure source, that is, gas or vapor pressure. Further, the characteristics of the system, that is, if the reactor is closed or open to the atmosphere will determine the consequences. In an open system, the gas or vapor will be released from the reactor, whereas in a closed system, the result of a runaway will be a pressure increase. The resulting pressure can be compared to the set pressure of the pressure relief system (P_{set}) or to the maximum allowed working pressure (P_{max}), or also to the test pressure (P_{test}) of the equipment.

10.6.1.3 Release

In an open system, since the gas or the vapor will be released from the reactor, the consequences depend on the length of time of the release and on the properties of the gas or vapor (e.g. toxicity or flammability). The extension can be assessed by using the volume of the toxic cloud and calculating its dilution to a critical limit. For toxicity, the limit may be taken as the IDLH or other limits defined by law (e.g. EPRG-2 ...). In case the gas or vapor is flammable, the lower explosion limit (LEL) is the critical limit. Since it is easier to have a good representation of a distance than of a volume, it is proposed to use the radius of a half sphere to describe the extension of the gas or vapor cloud. Such a simple approach has nothing to do with dispersion calculation using complex models and meteorological information, but is useful to assess the risks due to a runaway.

Thus, four different cases must be considered:

1. Closed gassy system: gas release in a closed reactor.
2. Closed tempered system: vapor pressure in a closed system.
3. Open gassy system: gas release in an open system.
4. Open tempered system: vapor release in an open system.

10.6.1.4 Closed Gassy Systems

The volume of gas potentially released by a reaction (including secondary reactions in criticality classes 4 and 5) can be known from the chemistry or measured experimentally by appropriate calorimetric methods, as for example, Calvet calorimetry, mini-autoclave, Radex, or Reaction Calorimetry (as V'_g at T_{mes} and P_{mes}). It must be corrected for the temperature to be considered, MTSR (class 2), MTT (class 3 or 4), or T_f (class 5). Where the gas stems from the main reaction, only the accumulated fraction (X) will be released:

$$V_g = M_r \cdot V'_g \cdot X \cdot \frac{T_{(K)}}{T_{mes(K)}} \tag{10.3}$$

This volume can be converted into a pressure increase by taking the available free volume for the gas in the reactor ($V_{r,g}$):

$$P = P_0 + \frac{V_g}{V_{r,g}} \cdot P_{mes} \tag{10.4}$$

10.6.1.5 Closed Vapor Systems

In this case, the pressure increase is due to the vapor pressure of volatile compounds. Often the solvent can be considered as the volatile compound, so its vapor pressure can be obtained from a Clausius–Clapeyron equation:

$$\frac{P}{P_0} = \exp\left[\frac{-\Delta H_v}{R}\left(\frac{1}{T} - \frac{1}{T_0}\right)\right] \tag{10.5}$$

or from an Antoine equation:

$$\log P = A - \frac{B}{C+T} \tag{10.6}$$

For complex systems it may be obtained from a phase diagram $P = f(x)$.

10.6.1.6 Open Gassy Systems

In an open system with gas production, the volume of gas can be obtained from Equation 10.3, and the volume calculated either for a toxicity limit, as for example the level called "Immediately Dangerous to Life and Health" ($IDLH$):

$$V_{tox} = \frac{V_g}{IDLH} \Rightarrow r = \sqrt[3]{\frac{3 \cdot V_{tox}}{2\pi}} \tag{10.7}$$

or from the lower explosion limit LEL:

$$V_{ex} = \frac{V_g}{LEL} \Rightarrow r = \sqrt[3]{\frac{3 \cdot V_{ex}}{2\pi}} \tag{10.8}$$

In these equations, the extension is calculated as the radius of a half sphere, as it is easier to estimate a distance than a volume. This geometry is used to give an approximate idea of the order of magnitude of the area that may affected by gas or vapor release. This calculation is purely static and has nothing to do with emission and dispersion. Other shapes could also be considered. The assessment is by comparing the extension to characteristic dimensions, for example, of the equipment, plant, and site.

10.6.1.7 Open Vapor Systems

An open tempered system is a system in which the latent heat of evaporation is used to halt the temperature increase, that is, to temper the system. This can be achieved at atmospheric pressure by reaching the boiling point or at higher pressure by applying a controlled pressure relief. The first step is to calculate the mass of vapor that may be relieved from the latent heat of evaporation and the characteristic temperatures:

$$M_v = \frac{(T_{\max} - MTT) \cdot c_p' \cdot M_r}{\Delta H_v'} \quad (10.9)$$

The maximum temperature (T_{max}) can be either the MTSR for class 3, or T_f for classes 4 and 5. This mass is converted into a volume by using the vapor density that may be estimated as an ideal gas:

$$\rho_v = \frac{PM_w}{RT} \quad (10.10)$$

and the extension is calculated in a similar way as for open gassy systems by using either a toxicity limit or the lower explosion limit, as in Equations 10.7 and 10.8. The relevant concentration limits may also be found in material safety data sheets.

10.6.1.8 Extended Assessment Criteria for Severity

The assessment criteria based on energy, pressure, and release extension are summarized in Table 10.1. The energy is assessed using the same criteria as presented in Table 3.1. Additionally, the pressure effect is assessed by using the characteristic pressure limits of the equipment: the set pressure of the pressure relief system (P_{set}), the maximum allowable working pressure (P_{max}), and the testing pressure of the equipment (P_{test}). The extension is assessed using the characteristic dimensions of the situation: equipment (generally several meters), the plant (generally 10–20 m), and the site (generally above 50 m). In case more than one criterion applies, the highest rating is taken (worst case) to assess the severity.

10.6.2 Controllability

For the controllability assessment, no quantitative failure rates will be used, with a semi-quantitative approach based on the probability of keeping a runaway under

Table 10.1 Assessment criteria for the severity, using the energy (ΔT_{ad}), the pressure for closed systems, and the extension for open systems.

Severity	ΔT_{ad}	P	Extension (r)
Serious	>400 K	$>P_{test}$	>Site
Critical	200–400 K	$P_{max} - P_{test}$	Site
Medium	50–200 K	$P_{set} - P_{max}$	Plant
Negligible	<50 K	$<P_{set}$	Equipment

control at the level MTT being used instead. The principle is that the thermal activity at a given temperature as MTT for example, is estimated with the aim of predicting the behavior of the reacting mixture at this temperature level. A low activity means that the temperature course is easy to control, whereas a high activity makes it difficult, so loss of control is probable. It is assumed that the same reaction that releases heat also releases gas and obviously the evaporation of volatiles. This assumption can be verified experimentally by comparing heat release rate and pressure increase rate in closed cell measurements. If the assumption is not valid, the gas release rate must be derived from a kinetic analysis of gas release rate as a function of temperature. The heat release can be compared directly to the cooling capacity of an emergency cooling system. The control of gas or vapor release is assessed using the maximum gas or vapor velocity in the equipment.

10.6.2.1 **Activity of the Main Reaction**

Starting from process temperature (T_p), the reaction is accelerated by the temperature increase to MTT following Arrhenius Law. But, at the same time reactants are converted, which results in a depletion of the reactant concentration and consequently in reduction of the reaction rate. Thus, two antagonistic factors play at the same time: acceleration with temperature and slowing down by the reactant depletion. Both effects may be summarized in an acceleration factor (f_{acc}) that multiplies the heat release rate. Thus:

$$q_{(MTT)} = q_{(T_P)} \cdot \underbrace{\exp\left[\frac{E}{R}\left(\frac{1}{T_P} - \frac{1}{MTT}\right)\right]}_{\text{increases reaction rate}} \cdot \underbrace{\frac{MTSR - MTT}{MTSR - T_p}}_{\text{decreases reaction rate}} = q_{(T_P)} \cdot f_{acc} \tag{10.11}$$

In this equation, the conversion term for a first-order reaction (1–X) is expressed as a function of the characteristic temperature levels of the scenario. First-order is a conservative approximation, since for higher reaction orders the reactant depletion is even higher. Zero-order would even be more conservative, but is generally unrealistic.

The heat release rate of the reaction at process temperature can be estimated from an experiment in a reaction calorimeter. If it is unknown, as in a worst-case assumption, the cooling capacity of the reactor can be used instead, since for an isothermal process the neat heat release rate of the reaction is certainly inferior to the cooling capacity. Some examples of the acceleration factor are represented graphically in Figure 10.7.

10.6.2.2 **Activity of Secondary Reactions**

For cases where the secondary reaction plays a role (class 5), or if the gas release rate must be checked (classes 2 or 4), the heat release rate can be calculated from the thermal stability tests (DSC or Calvet calorimeter). Secondary reactions are often characterized using the concept of Time to Maximum Rate under adiabatic conditions (TMR_{ad}). A long time to maximum rate means that the time available to take risk-reducing measures is sufficient. However, a short time means that the runaway may not be halted at the given temperature. The heat release rate at the temperature at which TMR_{ad} is equal to 24 hours, may be calculated from

$$q'_{D24} = \frac{c'_P \cdot R \cdot T_{D24}^2}{24 \cdot 3600 \cdot E_{dc}} \tag{10.12}$$

Note that in this case, the equation is solved for q, an algebraic solution, whereas solving for T (as required for the determination of T_{D24}) results in a transcendental equation, since the heat release rate is an exponential function of temperature. This would require an iterative procedure. This heat release rate may serve as a reference for the extrapolation:

$$q'_{(T)} = q'_{D24} \cdot \exp\left[\frac{E_{dc}}{R}\left(\frac{1}{T_{D24}} - \frac{1}{T}\right)\right] \tag{10.13}$$

10.6.2.3 Gas Release Rate

If we consider that thermal effects are the driving force of a runaway, we may assume that the gas release is due to the same reaction. Thus, the gas release rate can be calculated from

$$\dot{v}_g = V'_g \cdot M_r \cdot \frac{q'_{(MTT)}}{Q'} \tag{10.14}$$

Here the heat release rate and the energy represents the sum of all active reactions. It may be only the main reaction (class 3) or both main and secondary reactions (class 5). This calculates the gas velocity in the equipment:

$$u_g = \frac{\dot{v}_g}{S} \tag{10.15}$$

The section (S) used in this expression is the narrowest part of the piping system, for example, the gas ventilation. The gas release across the liquid surface in a vessel may lead to swelling. This effect may also be assessed using the section of the vessel. The capacity of a scrubber may also be used as an assessment criterion.

10.6.2.4 Vapor Release Rate

The vapor mass flow rate is proportional to the heat release rate and can be calculated in a similar way, as explained in Section 9.4.2:

$$\dot{m}_v = \frac{q'_{(MTT)} \cdot M_r}{\Delta H_v} \tag{10.16}$$

The vapor mass flow rate can be converted to a volume flow rate by using the vapor density calculated (Equation 10.10), which calculates the vapor velocity in the equipment using the section of the vapor tube:

$$\dot{v}_v = \frac{\dot{m}_v}{\rho_v} \quad u_v = \frac{\dot{v}_v}{S} \tag{10.17}$$

The assessment of the equipment vapor flow capacity should also take the cooling capacity of the condenser into account. This can be directly compared to the heat release rate. Further, the swelling of the reaction mass, due to the presence of bubbles, may also become critical for high degrees of filling (see Section 9.4.4). When both vapor and gas are released, obviously the sum of both velocities must be used in the assessment.

10.6.2.5 Extended Assessment Criteria for the Controllability

The effectiveness of the implemented measures obviously depends on the technical environment in the plant unit where the process is to be implemented. Thus, the technical characteristics must be known for the assessment. The assessment of the controllability is based on the time-scale of the runaway reaction (TMR_{ad}), on the achievable pressure level, and on the gas or vapor velocities. A proposal for these assessment criteria is summarized in Table 10.2:

- The time period for the action of the planned measure is given by the reaction kinetics; TMR_{ad} gives a good indication; the longer the time, the more controllable the process.
- In a closed system, the characteristic pressure limits will serve in the assessment. Unfortunately, this criterion is usually not discriminating, since when the equipment presents a high degree of filling, even with a very small gas release, the pressure may increase critically.
- For an open system, the vapor or gas velocities serve as assessment criteria. These velocities may be adapted and so are different of those encountered in emergency pressure relief, since here the aim is to control a runaway before it becomes critical. Criteria are provided for the velocities in a piping system and for the velocity across the liquid surface in a vessel, by assessing the swelling effect.

These criteria may be adapted to fit the specific plant conditions.

Table 10.2 Assessment criteria for the probability of loss of control during a runaway reaction.

Controllability	TMR_{ad} (h) From MTT	q' (W kg^{-1}) Stirred	q' (W kg^{-1}) Unstirred	u m^{-1}s^{-1}pipe	u cm^{-1}s^{-1}vessel[a]
Unlikely	<1	>100	>10	>20	>50
Difficult	1–8	50–100	5–10	10–20	20–50
Marginal	8–24	10–50	1–5	5–10	15–20
Feasible	24–50	5–10	0.5–1	2–5	5–15
Easy	50–100	1–5	0.1–0.5	1–2	1–5
Unproblematic	>100	<1	<0.1	<1	<1

a) Allowing the assessment of swelling effects.

Table 10.3 Required data set for the different criticality classes.

Class[a)]	1	2	3	4	5
Gas main reaction $V'_{g,rx}$ [b)]	+	+	+	+	+
Gas sec. reaction $V'_{g,dc}$ [b)]	(+)	(+)		+	+
Vapor (P_{vap})[b)]			+	+	+
Power main reaction q'_{rx}			+	+	(+)
Power sec. reaction q'_{dc}				(+)	+

a) The determination of the class requires the knowledge of four temperature levels: T_p, MTSR (that is X_{ac}), MTT, and T_{D24}.
b) Besides the volume or vapor pressure, the toxicity limit or the LEL must be known. The calculation of the velocities also requires information about the diameter of the piping system.

10.6.3 Assessment of Severity and Probability for the Different Criticality Classes

Obviously not all the parameters described above need to be evaluated for each scenario. In this context, the criticality classes are a useful tool in that they help in selecting the required data for the assessment of severity and probability (see Section 3.3.6). The criticality classes also give backbone to the systematic design procedure (Table 10.3). The procedure to follow for this assessment is presented below for each criticality class.

10.6.3.1 Criticality Class 1

In this class, neither the MTT is reached nor are secondary reactions triggered. Only if the reaction mass is maintained over a longer time under heat accumulation conditions at the MTSR, can the secondary reaction lead to a slow temperature increase. It is recommended to check for gas production, which could lead to a pressure increase if the reactor was closed or to a vapor or gas release if the reactor was opened. This can be done by using the procedure represented in Figure 10.8. In general, the gas release rate will be low due to the fact that $MTT < T_{D24}$.

10.6.3.2 Criticality Class 2

The situation is similar to class 1, except that the MTT is above T_{D24}. This means that under heat accumulation conditions, the activity of secondary reactions cannot be neglected, leading to a slow but significant pressure increase, or gas or vapor release. Nevertheless, the situation may become critical only if the reaction mass is left for a longer time at the level MTT. The assessment can be made using the same procedure as for criticality class 1, represented in Figure 10.8. The gas or vapor flow rate is an important parameter for the design of the required protection measures such as condenser, scrubber, or other treatment units.

For semi-batch reactions, it is important to check if the rating as class 2 is due to the control of the accumulation by the feed rate. Often reactions belong to class

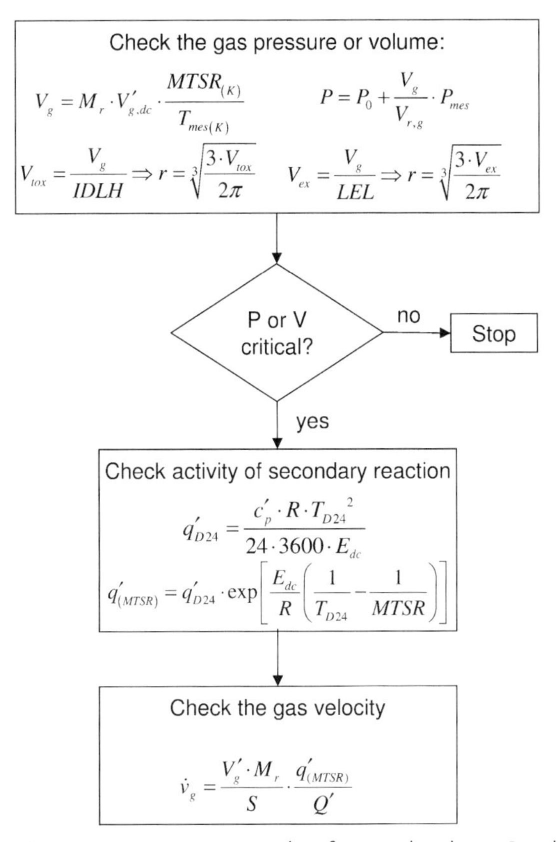

Figure 10.8 Assessment procedure for criticality classes 1 and 2.

5 if the feed is not immediately halted when a failure occurs. It is important to recognize these "shifting class reactions," since they require a reliable interlock to stop the feed, in case of deviation from the desired temperature towards higher and lower temperature, in order to avoid any undesired accumulation of reactant.

10.6.3.3 **Criticality Class 3**

In this class, the level MTT will be reached first, in the case of loss of control of the main reaction. Thus, the potential for pressure rise and gas or vapor release must be assessed using the energy of the main reaction only, since the secondary reactions are not triggered by the loss of control of the main reaction. The thermal activity at MTT can be determined by Equation 10.11, since it is only due to the main reaction. The heat release rate can be converted to gas or vapor release rate, by using Equations 10.14 to 10.17. This assesses the controllability of the runaway by stabilizing the temperature at the level MTT, eventually by using a controlled depressurization or hot cooling (cooling by evaporation). The gas or vapor

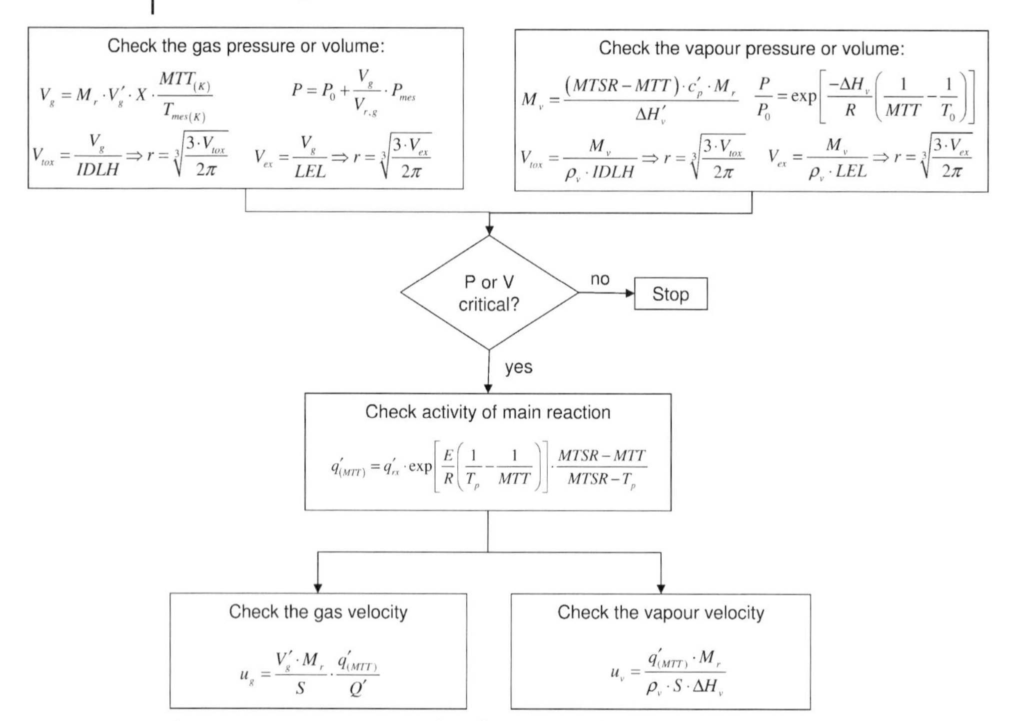

Figure 10.9 Assessment procedure for criticality class 3.

velocities are also useful for the design of protection measures, such as condenser, scrubber, or other gas or vapor treatment equipment. The procedure is depicted in Figure 10.9.

10.6.3.4 **Criticality Class 4**

This situation is similar to class 3, except that the MTSR is above T_{D24}, meaning that if the temperature cannot be stabilized at MTT, a secondary reactions could be triggered. Thus, the potential of the secondary reactions cannot be neglected and must be included in evaluation of the severity. The calculation of the potential produced gas volume must also take into account the secondary reaction. The final temperature is given by

$$T_f = T_p + X_{ac} \cdot \Delta T_{ad,rx} + \Delta T_{ad,d} \tag{10.18}$$

Then the stabilization of the temperature at MTT can be assessed in a similar way as for class 3, by using the determined thermal activity, following the procedure represented in Figure 10.10. Here the thermal activity of the secondary reaction may be ignored, but it should be checked whether the gas production rate by the secondary reaction remains uncritical. This assesses the controllability of the runaway by using controlled depressurization or hot cooling (cooling by evapora-

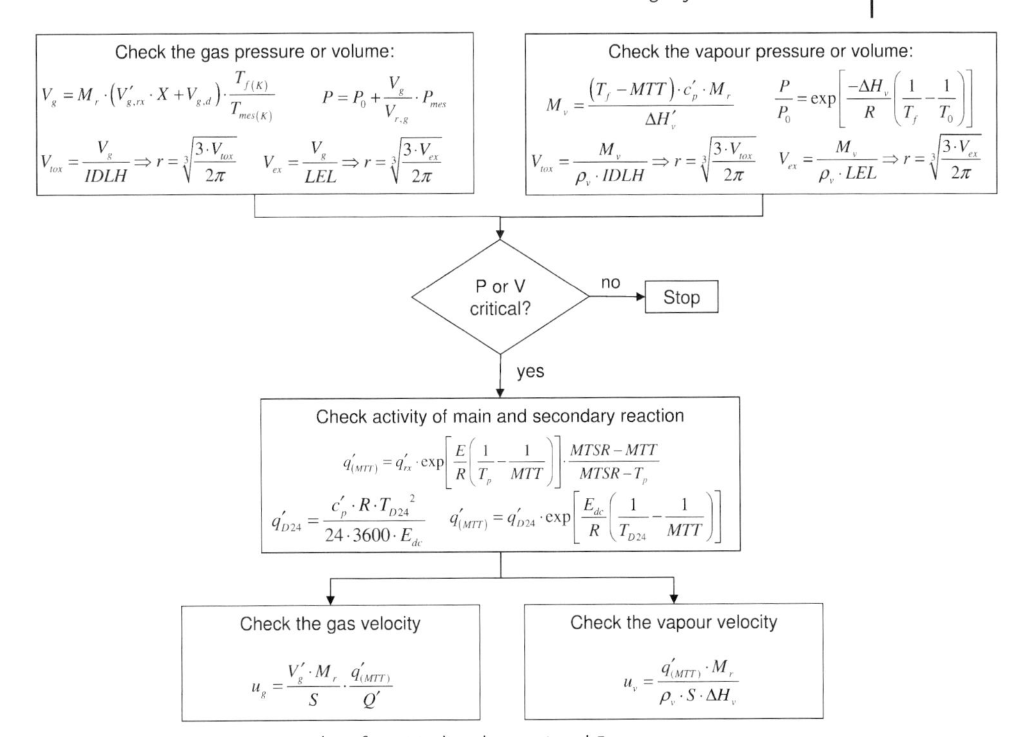

Figure 10.10 Assessment procedure for criticality classes 4 and 5.

tion). The gas or vapor velocities are also useful for the design of protection measures, such as condenser, scrubber, or other gas or vapor treatment equipment.

10.6.3.5 **Criticality Class 5**

In this class, in case of loss of control of the main reaction, the secondary reactions will be triggered. Thus, the severity is assessed by taking the potential of both main and secondary reactions into account, as in class 4. Nevertheless, the difference with Class 4 is that the level MTT is in the range where the secondary reaction is active. Thus, the chance of stabilizing the temperature at MTT is poor. In order to assess the thermal activity at MTT, two cases must be considered. In the first case, the MTT is below MTSR, that is, the main reaction is not completed as MTT is reached, but the secondary reaction is already active. Hence both the main and the secondary reactions must be taken into account. In the second case, MTT is above MTSR, thus the main reaction is completed at MTT and only the secondary reaction is taken into account. Nevertheless, generally the gas or vapor velocities are found to be too high to allow a stabilization at MTT.

Emergency measures, such as pressure relief or containment, must be taken to mitigate the consequences of a runaway that can no longer be avoided. Nevertheless, by far a better measure is to redesign the process to reduce the MTSR to a level below T_{D24}. This may be achieved, for example, by using a semi-batch reactor instead of a batch reactor and ensuring that the feed rate is properly limited and interlocked with the temperature and the stirrer, to maintain the accumulation of reactant at an acceptable level. A lower concentration could achieve the same result, but at a cost to the process economy. Of course, other process changes should be considered, including continuous reactors, and other synthesis route avoiding instable reaction masses (increase T_{D24}). Again, this shows how important it is to perform this assessment during process development, when there is enough time left to change the process.

Worked Example 10.3: Criticality and Control Measures

An exothermal reaction is to be performed in a 2.5 m^3 stirred tank reactor as an isothermal semi-batch process at 80 °C. The specific heat of the reaction is 180 kJ kg^{-1}, the specific heat capacity of the reaction mass is 1.8 kJ kg^{-1} K^{-1}, and the accumulation is 30%. The reaction is to be at atmospheric pressure and boiling point is 101 °C (MTT). There is a secondary reaction (decomposition) that is uncritical below 105 °C, that is, T_{D24} = 105 °C. The decomposition energy is 150 kJ kg^{-1} and this decomposition releases 5 liters of a toxic, but not flammable, gas per kg reaction mass, measured at 25 °C and atmospheric pressure.

Data for the solvent, methyl cyclohexane:

T_b = 101 °C M_w = 98.2 g mol^{-1}
$\Delta H'_v$ = 357 kJ kg^{-1}
IDLH = 1200 ppm
LEL = 1.2% vol.

Data for the decomposition gas

IDLH = 200 ppm, not flammable

Data for the reactor:

Charge 2000 kg (final reaction mass)
Void volume 1 m^3
Vapor tube diameter 150 mm from reactor to condenser
Gas tube diameter following the condenser: 80 mm
Set pressure of bursting disk 1 bar g
Maximum allowed working pressure 3.2 bar g
Test pressure 6 bar g

The cooling capacity of the reactor at 80 °C is estimated to be 60 kW.

Question:
Assess the risk linked to triggering a runaway of the reaction and propose risk-reducing measures.

Solution:
The first step is to assess the energy potential and the severity linked to vapor and gas release.

From the specific heat of reaction of 180 kJ kg^{-1} and the specific heat capacity of the reaction mass of 1.8 kJ kg^{-1} K^{-1}, we find an adiabatic temperature rise of 100 K. Since the accumulated energy is 30%, the $MTSR$ is

$$MTSR = 80 + 0.3 \times 100 = 110°\text{C}$$

The criticality class is 4 ($T_p < MTT < T_{D24} < MTSR$). Thus, the secondary reaction could theoretically be triggered and the total energy release is

$$Q' = X_{ac} Q'_{rx} + Q'_d = 0.3 \times 180 + 150 = 204\,\text{kJ kg}^{-1}$$

Then from Equation 10.18 the final temperature is

$$T_{max} = T_r + X_{ac} \cdot \Delta T_{ad,rx} + \Delta T_{ad,d} = 80 + 0.3 \times 100 + 83 \cong 193°\text{C}$$

This represents a temperature rise of 113 °C, corresponding to a medium severity (Table 10.1).

The potential due to vapor release can be assessed by calculating the volume of flammable vapor that may be released (Equations 10.8 to 10.10).

The mass of vapor is

$$M_v = \frac{(T_{\max} - MTT) \cdot c'_P \cdot M_r}{\Delta H'_v} = \frac{(193 - 101) \times 1.8 \times 2000}{357} \cong 928\,\text{kg}$$

Its density at 101 °C is

$$\rho_v = \frac{PM_w}{RT} = \frac{1.013 \cdot 10^5\,\text{Pa} \times 0.0982\,\text{kg mol}^{-1}}{8.314\,\text{J mol}^{-1}\,\text{K}^{-1} \times 374\,\text{K}} \cong 3.2\,\text{kg m}^{-3}$$

Thus the volume of pure methylcyclohexane vapor is

$$V = \frac{M_v}{\rho_v} = \frac{928}{3.2} = 290\,\text{m}^3$$

and the volume diluted to the LEL is

$$V_{ex} = \frac{V_g}{LEL} = \frac{290}{0.012} = 24167\,\text{m}^3$$

This represents an extension calculated as half sphere of

$$r = \sqrt[3]{\frac{3 \cdot V_{ex}}{2\pi}} = \sqrt[3]{\frac{3 \times 24167}{2\pi}} \cong 23\,\text{m}$$

The same extension is obtained for the toxicity limit of the vapor. This extension is larger than the plant (20 m), but would remain within the site perimeter (50 m). Thus, the severity based on flammable or toxic vapor is rated as "critical," thus, it is important to control the vapor release.

From Equation 10.3, the volume of gas released by the decomposition is

$$V_g = M_r \cdot V_g' \cdot X \cdot \frac{T_{(K)}}{T_{mes(K)}} = 2000 \times 0.005 \times 1 \times \frac{466}{303} \cong 15.4\,\text{m}^3$$

The same calculation at boiling point (MTT) gives 12.3 m^3. Diluted to the IDLH, it becomes

$$V_{tox} = \frac{V_g}{IDLH} = \frac{15.4}{200 \cdot 10^{-6}} \cong 77\,000\,\text{m}^3$$

which gives an extension of about 33 m calculated as a half sphere.

In the system was closed, the gas volume would be compressed to the available volume in the reactor, resulting in a pressure of

$$P = P_0 + \frac{V_g}{V_{rg}} = 1.013 + \frac{15.4}{1} \cong 15.4\,\text{bar g}$$

The vapor pressure can be estimated from a Clausius–Clapeyron equation:

$$P = P_0 \cdot e^{\frac{-H_V' \cdot M_W}{R}\left(\frac{1}{T_f} - \frac{1}{MTT}\right)} = 1013 \times e^{\frac{-357\,000 \times 0.0982}{8.314}\left(\frac{1}{466} - \frac{1}{374}\right)} = 9.4\,\text{bar abs}$$

Since these pressures (gas or vapor) are far above the maximum allowed working pressure, and even the test pressure of the reactor, the severity based on this criterion is rated as "serious."

In summary, the severity is rated:

- "low" in terms of energy release
- "critical" in terms of toxic and flammable vapor
- "critical" in terms of toxic gas release
- "serious" in terms of pressure in a closed system

Thus, the system must be kept open to allow vapor to condense and escape. Since the gas is toxic, a scrubber, which works in the case of cooling failure, must be provided. The condenser must also work after a cooling failure, such as an independent coolant. In order to check the feasibility of these measures, it is important to assess the controllability of the runaway at MTT: The objective is to control the reaction course by providing evaporative cooling.

Since the system is in criticality class 4, the main contribution to the activity at MTT stems from the synthesis reaction. Nevertheless, the contribution of the decomposition reaction should also be checked, since MTT and T_{D24} are close together.

To use Equation 10.11, we must know the heat release of the reaction under normal operating conditions and its activation energy. Since the heat release rate is not given, we assume that the heat release rate is no higher than the cooling capacity. Thus:

$$q'_{rx} = \frac{q_{ex}}{M_r} = \frac{60\,\text{kW}}{2000\,\text{kg}} = 30\,\text{W kg}^{-1}$$

The activation energy is used to extrapolate the heat release rate to higher temperatures, thus a high value should be used to be conservative, 100 kJ mol^{-1}:

$$q_{(MTT)} = q_{(T_P)} \cdot \underbrace{\exp\left[\frac{E}{R}\left(\frac{1}{T_P} - \frac{1}{MTT}\right)\right]}_{\text{increases reaction rate}} \cdot \underbrace{\frac{MTSR - MTT}{MTSR - T_P}}_{\text{decreases reaction rate}}$$

$$q_{(MTT)} = 30 \times \exp\left[\frac{100\,000}{8.314}\left(\frac{1}{353} - \frac{1}{374}\right)\right] \times \frac{110 - 101}{110 - 80} = 30 \times 6.78 \times 0.3 = 61\,\text{W kg}^{-1}$$

The heat release rate of the secondary reaction can be estimated from the T_{D24}, using Equation 10.12. Since the activation energy is unknown, we use a low value of 50 kJ mol^{-1} when extrapolating to lower temperatures:

$$q'_{D24} = \frac{c'_P \cdot R \cdot T^2_{D24}}{24 \cdot 3600 \cdot E_{dc}} = \frac{1800 \times 8.314 \times 378^2}{24 \times 3600 \times 50\,000} \cong 0.5\,\text{W kg}^{-1}$$

Extrapolating to MTT we obtain

$$q'_{(T)} = q'_{D24} \cdot \exp\left[\frac{E_{dc}}{R}\left(\frac{1}{T_{D24}} - \frac{1}{MTT}\right)\right] = 0.5 \times \exp\left[\frac{50\,000}{8.314}\left(\frac{1}{378} - \frac{1}{374}\right)\right] = 0.4\,\text{W kg}^{-1}$$

Thus, the main contribution is by the synthesis reaction, but the heat release rate of the decomposition will be used to calculate the gas release rate. The heat release rate of 64 W kg^{-1} would lead to a fast temperature increase under adiabatic conditions (TMR_{ad} < 1 hour). The question is as to whether or not the reaction may be controlled at the boiling point.

Vapor mass flow rate:

$$\dot{m}_v = \frac{q' \cdot M_r}{\Delta H'_v} = \frac{61.5\,\text{W kg}^{-1} \times 2000\,\text{kg}}{357\,\text{kJ kg}^{-1}} \cong 0.35\,\text{kg s}^{-1} = 1240\,\text{kg h}^{-1}$$

The cross-section of the vapor tube (diameter 150 mm) is 177 cm², which gives a vapor velocity of

$$u = \frac{\dot{m}_v}{\rho_v \cdot S} = \frac{0.35\,\mathrm{kg\,s^{-1}}}{3.2\,\mathrm{kg\,m^{-3}} \times 0.0177\,\mathrm{m^2}} = 6.2\,\mathrm{m\,s^{-1}}$$

This vapor velocity may lead to flooding of the vapor tube and so requires an active condenser.

The gas release rate is calculated from the conversion rate of the decomposition reaction. This is equivalent to assuming that the same reaction, which releases heat, also producing the gas:

$$\dot{v}_g = V_g \cdot \frac{dX}{dt} = V_g \cdot \frac{q'_d}{Q'_d}$$

$$\dot{v}_g = 12.3\,\mathrm{m^3} \times \frac{0.4\,\mathrm{W\,kg^{-1}}}{150000\,\mathrm{J\,kg^{-1}}} = 3.3 \cdot 10^{-5}\,\mathrm{m^3\,s^{-1}} = 118\,\mathrm{l\,h^{-1}}$$

Thus, the gas release is very slow and the gas velocity in the vent line (diameter 80 mm) is negligible.

The assessment of the controllability of the runaway at the boiling point, using the criteria in Table 10.2, gives:

- Thermal runaway: $TMR_{ad} < 1$ hour: the controllability is rated "Unlikely." Runaway will occur, so adiabatic conditions must be avoided.
- Emergency cooling with stirrer requires 65 W kg^{-1}: the controllability is rated "difficult." Emergency cooling will probably not work.
- Evaporation cooling: vapor velocity 6.2 m s^{-1}: the controllability is rated "marginal." Evaporation cooling may work, but the velocity it somewhat beyond flooding.
- Gas release rate: velocity < 1 m s^{-1}: the controllability is rated "unproblematic." The gas release will not cause any pressure build-up. Due to its toxicity, it must be treated before release to atmosphere. The low gas flow rate makes this operation "unproblematic".

Due to the high vapor velocity at boiling point, it is recommended to reduce the accumulation of reactant to a lower value. As an example, an accumulation of 25% instead of 30% reduces the vapor flow rate to 3.3 m s^{-1}, which should not cause any flooding.

10.6.4
Protection System Based on Risk Assessment

Once the severity and the probability corresponding to a scenario are estimated, that is, the risk is assessed, a decision can be made on the nature of the protection system to be implemented. If a safety instrumented system (SIS) is to be used, consisting of one or more independent protection levels (IPL), the required reliability of the protection system, constituting a so-called Safety Integrated Level (SIL) can be determined by using this risk assessment, respective of the required risk reduction.

10.6.4.1 Risk Assessment

The four severity levels and the six probability levels described above, can be arranged in a risk diagram, sometimes called risk matrix or risk profile (see Section 1.3.1.6). The matrix presented in Figure 10.10 is derived from an example given in the IEC 61511 standard [2]. It was adapted to the assessment of runaway reactions with the criteria defined above. In such a risk matrix, the different fields corresponding to accepted (white) and non-accepted risks (dark gray) can be identified. Often an intermediate field (light gray) is used, corresponding to risk that should be reduced as far as the costs are in relation to the risk reduction, following the ALARP principle (as low as reasonably practicable). For the upper right-hand corner of the diagram, the three fields requiring more than four independent protection levels (IPL), protection should not be realized on the basis of automated systems only.

Quantitative failure frequency data are difficult to obtain for multipurpose batch plants in the way that they are often used in the fine chemicals and pharmaceutical industries. Moreover, a quantitative assessment requires detailed knowledge of the control instruments, which may not be available during process development. Therefore, a semi-quantitative approach is proposed, providing the required reliability for future plant equipment.

10.6.4.2 Determination of the Required Reliability for Safety Instrumented Systems

Considered here is the probability that a runaway may not be stopped at the level *MTT*. As explained above, this probability increases with the thermal activity at this temperature. The criticality classes were used to describe the behavior of the reaction mass at this temperature and to determine the appropriate type of measure that should be implemented. Such a measure, for example, quenching a reaction mass will be triggered by an alarm (e.g. temperature) that opens a valve allowing the quenching medium to be flushed into the reactor. Such a device comprising a sensor, a logical unit (alarm), and an actuator (the valve), is called a safety instrumented system (SIS). Such a system provides one independent protection layer (IPL). For a high risk, more than one IPL may be required. Moreover, the reliability of the SIS is defined by the standard IEC 61511 as the safety integrity level (SIL).

Unlikely	1:1	1:3 + 1:1 2:2		
Difficult		1.2 + 1:1 3:1	1:3 + 1:1 2:2	
Marginal		1:2 2:1	1.2 + 1:1 3:1	1:3 + 1:1 2:2
Feasible		1:1	1:2 2:1	1.2 + 1:1 3:1
Easy			1:1	1:2 2:1
Unproblematic				1:1
	Negligible	Medium	Critical	Serious

Figure 10.11 Risk matrix adapted from the IEC 61511 standard, indicating the accepted and non-accepted risks, as well as an intermediate field. The numbers represent the number of required IPLs together with the required SILs.

The design of a protection system against runaway comprises defining the nature of the system as well as its reliability. Let us consider that probability decreases by one order of magnitude, from each level to the level below, for example, "probable" means a ten times higher probability than "occasional," and so forth. Then a risk that should be reduced from "frequent" to "remote" corresponds to a reduction by a factor 10^{-4} and requires, for example, two IPLs with an SIL 2, coded 2:2 or 1 SIL 3, and 1 SIL 1 coded 1:3 + 1:1, in Figure 10.11. When more than one IPL is required, one of them may be a non-instrumented system requiring human intervention as procedural measures do. The assessment scales given in the matrix, as well as the required IPL and SIL, are given as examples to show how thermal data may lead to a systematic definition of protection systems and the corresponding SIL levels to be used. They should be defined according to a company's own safety policy.

10.7 Exercises

► Exercise 10.1

This is a continuation of Exercise 3.1.

A diazotization is to be performed in aqueous phase, by slow addition of sodium nitrite in a 2.5 mol kg^{-1} solution of an aniline derivative. The process temperature is

5 °C and the reaction considered fast at this temperature. Nevertheless, for the safety analysis, an accumulation of 10% is considered realistic. The industrial charge is 4000 kg of final reaction mass in a 4 m^3 glass-lined reactor. This vessel is protected against overpressure by a safety valve with a set pressure of 0.3 bar g. The total empty volume of the vessel is 5.5 m^3. The vent line has an internal diameter of 50 mm and the maximum allowed working pressure of the reactor is 0.3 bar (g).

Thermal data:

Reaction:	$-\Delta H_r = 65\,\text{kJ mol}^{-1}$	$c'_P = 3.5\,\text{kJ kg}^{-1}\,\text{K}^{-1}$
Decomposition:	$-\Delta H_{dc} = 150\,\text{kJ mol}^{-1}$	$T_{D24} = 30\,°\text{C}$

Questions:

1. Estimate the gas release rate at *MTSR* (assess its controllability).
2. Does the process require protection measures against runaway?

▶ Exercise 10.2 Crit 2

This is a continuation of Exercise 3.2.

A condensation reaction is to be performed in a stirred tank reactor in the semi-batch mode. The solvent is acetone, the industrial charge (final reaction mass) is 2500 kg, and the reaction temperature is 40 °C. The second reactant is added in a stoichiometric amount at a constant rate over two hours. Under these conditions, the maximum accumulation is 30%. The reaction does not produce any gas and its heat release rate is 20 W kg^{-1}. The reactor is equipped with a condenser with a cooling power of 250 kW and the vapor tube has a diameter of 250 mm. The reactor can be considered open.

Data:

Reaction:	$Q'_r = 230\,\text{kJ kg}^{-1}$	$c'_P = 1.7\,\text{kJ kg}^{-1}\,\text{K}^{-1}$
Decomposition:	$Q'_{dc} = 150\,\text{kJ kg}^{-1}$	$T_{D24} = 130\,°\text{C}$
Physical data:	Acetone	$T_b = 56\,°\text{C}$ $M_w = 58\,\text{g mol}^{-1}$
	$\Delta H'_v = 523\,\text{kJ kg}^{-1}$	LEL = 1.6% vol

Questions:

1. Estimate the volume and extension of the flammable vapor cloud, approximated as an isotropic half-sphere, which would be released. Assess the consequences.
2. Estimate the vapor flow rate at MTT (assess its controllability).
3. Suggest appropriate risk reducing measures.

▶ Exercise 10.3

This is a continuation of Exercise 3.3.

A sulfonation reaction is performed as a semi-batch reaction in 96% sulfuric acid as a solvent. The total charge is 6000 kg with a final concentration of 3 mol l^{-1}. The reaction temperature is 110 °C and Oleum 20% is added in stoichiometric excess of 30% at a constant rate over 4 hours. Under these conditions, the maximum accumulation of 50% is reached after approximately 3 hours addition. At this time, the heat release rate is $10\,\mathrm{W\,kg^{-1}}$. The stainless steel reactor is equipped with a 50 mm diameter vent line. The maximum allowed working pressure is 6 bar g and the test pressure is 8 bar g.

Data:

Reaction:	$Q'_r = 150\,\mathrm{kJ\,kg^{-1}}$	$c'_P = 1.5\,\mathrm{kJ\,kg^{-1}\,K^{-1}}$
Decomposition:	$Q'_{dc} = 350\,\mathrm{kJ\,kg^{-1}}$	$T_{D24} = 140\,°\mathrm{C}$

The decomposition of the sulfonic acid produces SO_2 (IDLH = 100 ppm).

Questions:

1. Estimate the resulting toxic cloud volume (diluted to IDLH) and the extension.
2. Estimate the gas release rate at MTT. What will the consequences be?

References

1 Kletz, T.A. (1996) Inherently safer design: the growth of an idea. *Process Safety Progress*, **15** (1), 5–8.

2 IEC (2004) *Funktionale Sicherheit – Sicherheitstechnische Systeme für die Prozessindustrie IEC 61511*. DIN VDE.

3 Bou-Diab, L. (2003) *Mögliche Schutzkonzepte für Batch-Reaktoren*, Diploma work, ETH-Zurich, Basle.

4 Hendershot, D.C. (1997) Inherently safer chemical process design. *Journal of Loss Prevention in the Process Industries*, **10** (3), 151–7.

5 Lutz, W.K. (1997) Advancing inherent safety into methodology. *Process Safety Progress*, **16** (2), 86–7.

6 Crowl, D.A., (ed.) (1996) *Inherently Safer Chemical Processes. A Life Cycle Approach*, CCPS Concept book, Center for Chemical Process Safety, New York, p. 154.

7 Hungerbühler, K., Ranke, J. and Mettier, T. (1998) *Chemische Produkte und Prozesse; Grundkonzepte zum umweltorientierten Design*, Springer, Berlin.

8 Koller, G., Fischer, U. and Hungerbühler, K. (2000) Comparison of methods for assessing human health and the environmental impact in early phases of chemical process development, in *European Symposium on Computer Aided Process Engineering-10*, S. Pierucci.

9 Koller, G., Fischer, U. and Hungerbühler, K. (2001) Comparison of methods suitable for assessing the hazard potential of chemical processes during early design phases. *Transactions of the Institution of Chemical Engineers*, **79** (Part B), 157–66.

10 Koller, G., Fischer, U. and Hungerbühler, K. (2000) Assessing safety, health, and environmental impact early during process development. *Industrial and Engineering Chemistry Research*, **39**, 960–72.

11 Benaissa, W. (2006) *Développement d'une méthodologie pour la conduite en sécurité d'un réacteur continu intensifié*, Institut National Polytechnique de Toulouse, Toulouse, France.

12 Schneider, M.-A., Maeder, T., Ryser, P. and Stoessel, F. (2004) A microreactor-based system for the study of fast exothermic reactions in liquid phase: characterization of the system. *Chemical Engineering Journal*, **101** (1–3), 241–50.

13 Schneider, M.A. and Stoessel, F. (2005) Determination of the kinetic parameters of fast exothermal reactions using a novel microreactor-based calorimeter. *Chemical Engineering Journal*, **115**, 73–83.

14 Bourne, J.R., Brogli, F., Hoch, F. and Regenass, W. (1987) Heat transfer from exothermally reacting fluid in vertical unstirred vessels-2 free convection heat transfer correlations and reactor safety. *Chemical Engineering Science*, **42** (9), 2193–6.

15 Hub, L. (1983) On-Line Überwachung exothermer Prozesse. *Swiss Chemistry*, **5** (9a), 53–7.

16 Hub, L. and Jones, J.D. (1986) Early on-line detection of exothermic reactions. *Plant Operation Progress*, **5** (4), 221–3.

17 Zaldivar, J.M. (1991) Fundamentals on runaways reactions: prevention and protection measures, in *Safety of Chemical Batch Reactors and Storage Tanks*, Benuzzi, A. and Zaldivar, J.M. (eds.) ESCS, EEC, EAEC, Brussels, 19–47.

18 Bosch, J., Strozzi, F., Zbilut, J.P. and Zaldivar, J.M. (2004) On-line runaway detection in isoperibolic batch and semibatch reactors using the divergence criterion. *Computers and Chemical Engineering*, **28** (4), 527–44.

19 Zaldivar, J.M., Cano, J., Alos, M.A., Sempere, J., Nomen, R., Lister, D., Maschio, G., Obertopp, T., Gilles, E.D., Bosch, J. and Strozzi, F. (2003) A general criterion to define runaway limits in chemical reactors. *Journal of Loss Prevention in the Process Industries*, **16** (3), 187–200.

20 Zaldivar, J.-M., Bosch, J., Strozzi, F. and Zbilut, J.P. (2005) Early warning detection of runaway initiation using non-linear approaches. *Communications in Nonlinear Science and Numerical Simulation*, **10** (3), 299.

21 Westerterp, K.R. (2006) Safety and runaway prevention in batch and semibatch reactors – a review. *Chemical Engineering Research and Design*, **84** (A7), 543–52.

22 CCPS. (1998) *Guidelines for Pressure Relief and Effluent Handling Systems*, CCPS, AICHE.

23 Fisher, H.G., Forrest, H.S., Grossel, S.S., Huff, J.E., Muller, A.R., Noronha, J.A., Shaw, D.A. and Tilley, B.J. (1992) *Emergency Relief System Design Using DIERS Technology, The Design Institute for Emergency Relief Systems (DIERS) Project Manual*. AICHE, New York.

24 Etchells, J. and Wilday, J. (1998) *Workbook for chemical reactor relief system sizing*, HSE, Norwich.

25 Gustin, J.L. (1991) Calorimetry for emergency relief systems design, in *Safety of Chemical Batch Reactors and Storage Tanks*, Benuzzi, A. and Zaldivar, J.M. (eds.) ECSC, EEC, EAEC, Brussels, 311–54.

26 Schmidt, J. and Westphal, F. (1997) Praxisbezogenes Vorgehen bei der Auslegung von Sicherheitsventilen und deren Ablaseleitungen für die Durchströmung mit Gas/Dampf-Flüssigkeitsgemischen, Teil 1. *Chemie Ingenieur Technik*, **69** (6), 776–92.

27 Roduit, B., Borgeta, C., Berger, B., Folly, P., Alonso, B., Aebischer, J.N. and Stoessel, F. (2004) Advanced kinetic tools for the evaluation of decomposition reactions, *Journal of Thermal Analysis and Calorimetry*, **80**, 229–36.

28 Stoessel, F. (2006) Novel approach to emergency pressure relief design using calorimetric methods, in *STK-Meeting (Swiss Society of Thermal Analysis and Calorimetry)*, STK meeting, Basle.

29 Reuse, P., Wehrli, V., Fierz, H. and Roduit, B. (2004) Characterisation of pressure and temperature rise of runaway reactions using temperature-programmed measurements, in *CHISA*, Prag.

30 Kletz, T. (1988) *Learning from Accidents in Industry*, Butterworths, London.

31 Leung, J.C. (1995) The omega method for discharge rate evaluation, in *International Symposium on Runaway Reactions and Pressure Relief Design*, AIChE.

32 Leung, J.C. (1996) Easily size relief devices and piping for two-phase flow. *Chemical Engineering Progress*, **12**, 28–50.

33 Schmidt, J. and Giesbrecht, H. (2001) Evaluation of uncertainties for safety valve vent system compared to a model-based on-line vessel protection in a (certified) process controller, in *ECCE-3*, Nuremberg Germany.

34 Schmidt, J., Friedel, L., Westphal, F., Würsig, G., Wilday, J., Gruden, M. and Geld, C.v.d. (2001) Sizing of safety valves and connected inlet and outlet lines for gas/liquid two-phase flow, in *10th Int. Symposium on Loss Prevention and Safety Promotion in the Process Industries*, EFCE, Stockholm.

35 McIntosh, R.D., Nolan, P.F., Rogers, R.L. and Lindsay, D. (1995) The design of disposal systems for runaway chemical reactor relief. *Journal of Loss Prevention in the Process Industries*, **8** (3), 169–83.

36 McIntosh, R.D. and Nolan, P.F. (2001) Review and experimental evaluation of runaway chemical reactor disposal methods. *Journal of Loss Prevention in the Process Industries*, **14**, 17–26.

37 McIntosh, R.D. and Nolan, P.F. (2001) Review of the selection and design of mitigation systems for runaway chemical reactions. *Journal of Loss Prevention in the Process Industries*, **14**, 27–42.

38 Siemens A. (2006) *Safe-Bag Retention System*. http://www.automation.siemens.com (accessed on 21.11.2007).

Part III
Avoiding Secondary Reactions

11
Thermal Stability

Case History "Storage During Repair"

The condensation of an aromatic nitro compound with a second reactant should have been performed in an aqueous solution with DMSO in the semi-batch mode. The nitro-compound is initially charged into the reactor with water and DMSO as solvent. Before the progressive addition of the second reactant had been started, the initial mixture was heated to the process temperatures of 60–70 °C. Then a failure of the cooling water system of the plant occurred. It was decided to interrupt the process at this stage and to maintain the mixture under stirring until the failure had been repaired. The feed of the second reactant was postponed and the jacket of the reactor had been emptied.

After 5 days, a thick plume was observed escaping from the ventilation system of the reactor. The temperature of the reactor was then checked and found to be 118 °C. Later, a thick tar with a temperature of 160 °C flowed from the open manhole of the reactor. Immediate application of emergency cooling had no effect. All personnel were evacuated and the reactor then exploded, rupturing into four fragments. The building was seriously damaged over three floors and the control room was totally destroyed. The damage was over one million US $.

The inquiry revealed that the process had been interrupted several times before starting the feed of the second reactant, without any apparent problems. The steam valve of the jacket was found to be leaking. Thus, the reactor had been slowly heating. Since the reaction mass boils at 118 °C the solvent mixture could be progressively evaporated. In this temperature range, a secondary exothermal reaction is active, contributing to the heat input, enabling evaporation and then temperature increase of the remaining reaction mixture. The energy dissipated by the stirrer was found to be insufficient for the observed temperature increase. The initial reaction mixture was shown to decompose following an autocatalytic mechanism.

Thermal Safety of Chemical Processes: Risk Assessment and Process Design. Francis Stoessel

ISBN: 978-3-527-31712-7

Lessons drawn

Secondary decomposition reactions may have serious consequences when they get out of control. In this case history, the thermal stability of the reaction mass was not known before the incident. If only the energy released by this decomposition had been known, the production staff would not have decided to maintain this reaction mass without active temperature control and monitoring. Thus, assessing the consequences and the triggering conditions of secondary decomposition reactions and predicting their behavior requires a specific knowledge and a systematic approach.

11.1 Introduction

This chapter describes a runaway scenario. The first section presents a general review of the decomposition reaction characteristics. The second section is devoted to the energy release that defines the consequences of a runaway. The third section deals with triggering conditions of undesired reactions, based on the concept of TMR_{ad}. The next section reviews some important aspects for the experimental characterization of decomposition reactions. Finally, the last section gives some examples stemming from industrial practice.

11.2 Thermal Stability and Secondary Decomposition Reactions

Often safety data or material safety data sheets mention the thermal stability as an intrinsic property of a substance or mixture. In fact, this is an oversimplification of a concept that must be defined in a more comprehensive way. Basically, a substance or a mixture is thermally stable in a situation where the heat released can be removed in such a way that no temperature increase occurs. This definition

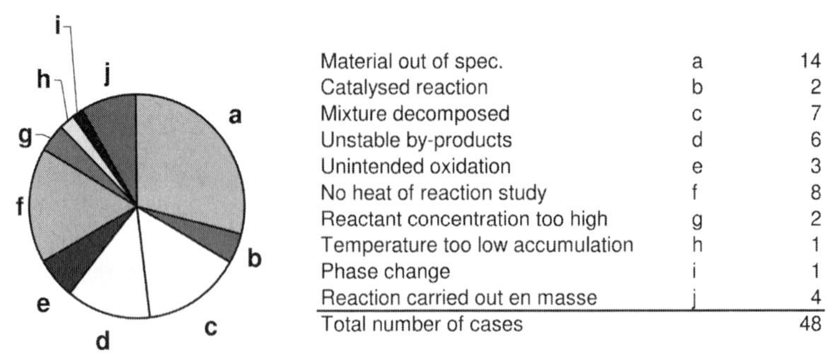

Figure 11.1 Causes of runaway reactions from a statistical survey, showing the importance of secondary or decomposition reactions as a cause of runaway.

implies a reference to a heat balance, since we compare the heat release rate with the heat removal.

Decomposition reactions are often involved in thermal explosions or runaway reactions, in certain cases as a direct cause, in others indirectly as they are triggered by a desired synthesis reaction that goes out of control. A statistical survey from Great Britain [1, 2] revealed that out of 48 runaway reactions, 32 were directly caused by secondary reactions, whereas in the other cases, secondary reactions were probably involved too, but are not explicitly mentioned (Figure 11.1). Therefore, characterizing secondary decomposition reactions is of primary importance when assessing the thermal hazards of a process.

Characterization of decomposition reactions or the evaluation of risks linked with triggering such reactions means that, as for any thermal risk, both the severity and the probability of triggering them must be evaluated (Figure 11.2):

- consequences of secondary decomposition reactions: the damage caused by an uncontrolled decomposition reaction is proportional to the energy released. Therefore, the adiabatic temperature rise may serve as a criterion, as explained in Section 3.3.2.
- probability of triggering: reasons for triggering a secondary reaction may be diverse. It may be due to thermal triggering, that is, too high a temperature. Moreover, catalytic effects or the presence of impurities may also trigger such reactions. The probability of triggering or loss of control may be assessed using the time to explosion, that is, time to maximum rate (TMR_{ad}) by considering that the shorter this time the more likely the loss of control, as explained in Section 3.3.3.

These points will be examined in detail in the following sections.

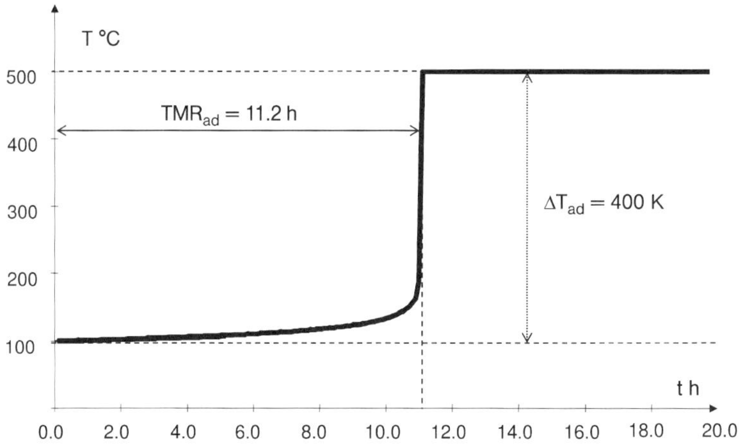

Figure 11.2 Characterization of thermal risks linked with decomposition reactions: the temperature increase is a measure of the severity and the time-scale gives a measure of the probability of triggering a runaway reaction.

11.3
Consequences of Secondary Reactions

11.3.1
Stoichiometry of Decomposition Reactions

A major difference with desired reactions is that the stoichiometry is often unknown, that is, the decomposition products are unknown. The reason is that decomposition reactions are often affected by the triggering conditions and thus often run along different reaction paths. This is a major difference compared to a total combustion, for example. The consequence is that the decomposition enthalpy cannot be predicted using standard enthalpies of formation (ΔH_f) taken from, for example, tables or estimated by group increment methods, such as Benson groups [3, 4]:

$$\Delta H_r = \sum_{\text{Products}} \Delta H_f - \sum_{\text{Reactants}} \Delta H_f \tag{11.1}$$

Nevertheless, computing methods hare been developed for the estimation of decomposition energies.

11.3.2
Estimation of Decomposition Energies

A well-known tool for the estimation of reactivity hazards of organic material is called CHETAH [5]. The method is based on pattern recognition techniques, based on experimental data, in order to infer the decomposition products that maximize the decomposition energy, and then performs thermochemical calculations based on the Benson group increments mentioned above. Thus, the calculations are valid for the gas phase, but this may be a drawback, since in fine chemistry most reactions are performed in the condensed phase. Corrections must be made, but in general they remain small and do not significantly affect the results.

The CHETAH method, as well as other estimation techniques, is not intended to replace experimental testing of material and never should be used for this purpose. Responsible use of such software tools means using them only as screening tools in an overall scheme involving both physical testing and other predictive tools [6]. Therefore, it is strongly recommended to determine decomposition enthalpies experimentally using dynamic DSC, for example. This technique allows simulating most severe confinement conditions, while heating the sample from room temperature to approximately 500 °C (see Section 11.5).

11.3.3
Decomposition Energy

Decomposition energies are often high and since decomposition products comprise small fragments they are volatile or even gaseous. Consequently, decomposi-

Table 11.1 Typical values of decomposition enthalpies for different functional groups.

Functional group		ΔH_d **kJ mol^{-1}**
Diazonium salt	$-N{=}N^+$	−160 to −180
Diazo	–N=N–	−100 to −180
Isocyanate	–N=C=O	−50 to −75
N-Hydroxide	>N–OH	−180 to −240
Peroxide	>C–O–O–C<	−350
Nitro	$Ar{-}NO_2$ *or* $R{-}NO_2$	−310 to −360
Nitrate	$-O{-}NO_2$	−400 to −480
Epoxide	$>C\overset{O}{-}C<$	−70 to −100

tion reactions are accompanied by large energy release and perhaps by significant pressure increase. This explains the high severity of incidents where decomposition reactions are involved. Some typical values of decomposition energies for representative functional groups were experimentally determined and compiled by Grewer [7] and are summarized in Table 11.1. Estimations based on isolated functional groups may lead to erroneous conclusions, since different functional groups may react together in an unwanted way.

A special mention must be made for the catalytic effect of impurities, which may have a great influence on the energy of decomposition, since often the reaction path is affected by such impurities [8]. Further, the presence of different functional groups in one molecule may destabilize the molecule, or allow polycondensation reactions, such as in the case of aromatic chloro-anilines. A polymerization reaction may also be difficult to predict. Here again, it is recommended to measure the energies of decomposition of the compounds in the reaction mixture or in the same state as they are handled in the process to be studied. This especially means that the thermal analysis must be performed together with the solvent. Since impurities may catalyze decomposition reactions, it is also important to use a "technical grade" as will be used in the plant. In other words, performing measurements on previously purified samples should be avoided as they should be used as they are.

Therefore, priority must be given to experimental determination of decomposition energies. It is also essential to perform the experiment under conditions that are as close as possible to plant conditions.

Often chemical compounds are sensitive to oxygen and undergo oxidation reactions. In such a case, the energy release is by far higher and may reach values of the heat of combustion (Table 11.2). These values can easily be found in tables, for example, [9, 10]. Oxidative decompositions may become a major problem in physical unit operations, where intensive air contact takes place during drying, milling, or blending of solids. In these operations, the product is submitted to energy input (thermal or mechanical) and simultaneous air contact. To assess such situations, specific testing procedures must be used [6, 11–13].

Table 11.2 Typical values of combustion energies.

Compound	ΔH_{Comb} kJ mol^{-1}	Compound	ΔH_{Comb} kJ mol^{-1}
Methane	−800	*n*-Heptane	−4470
Ethane	−1430	Toluene	−3630
n-Propane	−2040	Naphtalen	−4980
n-Butane	−2660	Hydrocarbon (C20)	−12400

The high energy release accompanying decomposition reactions leads to a high temperature increase if the system is not, or only poorly, cooled. Therefore, a runaway reaction is likely to occur. The consequences can be assessed, using the criteria described in Section 3.3.2.

Besides the temperature, other possible consequences of decomposition reactions are flammable or toxic gas release, solidification, swelling, foaming, carbonation, that may cause the loss of a batch, but also damage leading to the loss of a plant unit and impinging on the production of the desired product. These consequences should also be considered in the assessment. Therefore, the determination of the decomposition energies is a preliminary to any assessment of thermal risks.

11.4 Triggering Conditions

11.4.1 Onset: A Concept without Scientific Base

There may be great temptation to derive safe process conditions directly from the temperature at which a peak is detected in a dynamic DSC experiment. As an example, a so-called "50 K rule" can be found in industrial practice. In fact, such a rule is equivalent to considering that at 50 K below the onset in DSC, no reaction occurs. This is scientifically wrong and may lead to catastrophically erroneous conclusions for two reasons:

1. The temperature determined in a dynamic DSC experiment strongly depends on the experimental conditions, especially on the scan rate (Figure 11.3), on the sensitivity of the experimental set up, and on the sample mass used.
2. There is no defined onset temperature or starting temperature for a reaction. The reaction rate simply increases exponentially with temperature or decreases by lower temperature according to Arrhenius law.

For example, a first-order reaction with an activation energy of 75 kJ mol^{-1} is detected at 209 °C with an instrument having a detection limit of 10 W kg^{-1}, at

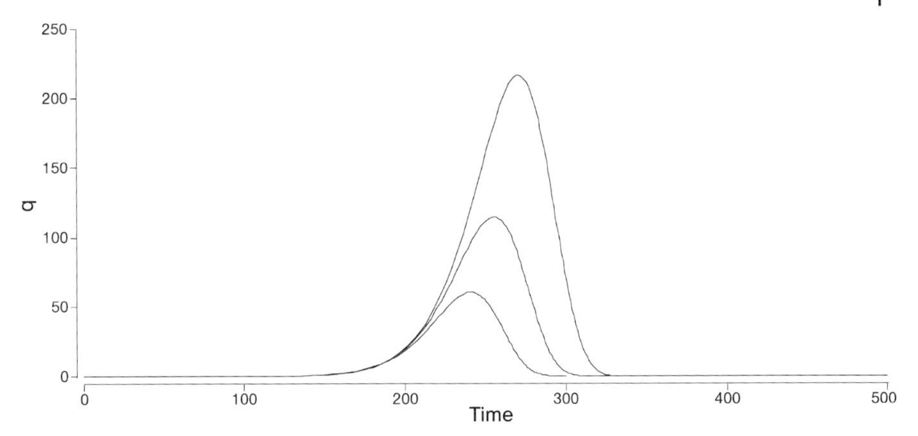

Figure 11.3 Dynamic DSC experiments for the same decomposition reactions with different scan rates. The peak position is a function of the scan rate. The apparent change of the peak surface is due to the fact that the temperature is a function of the scan rate that was changed.

150 °C with an instrument having a detection limit of 1 W kg^{-1}, and at 109 °C with an instrument having a detection limit of 0.1 W kg^{-1}. Thus, it becomes obvious that the "distance rule" must be replaced by a more scientifically sound concept, as with the time to maximum rate based on reaction kinetics.

11.4.2
Decomposition Kinetics, the TMR$_{ad}$ Concept

The probability of triggering a secondary decomposition reaction may be assessed using the time-scale as defined in Section 3.3.3. The principle is that the longer the time available for taking protective measures, the lower the probability of triggering a runaway reaction. The concept of Time to Maximum Rate (TMR$_{ad}$) was developed for this purpose and is presented in Section 2.5.5. The TMR$_{ad}$ under adiabatic conditions is given by

$$TMR_{ad} = \frac{c'_p R T_0^2}{q'_0 E} \tag{11.2}$$

Even if this equation implicitly assumes a zero-order reaction, it was initially developed for zero-order kinetics, and may also be used for other reaction kinetics, giving a conservative approximation since the concentration depletion that would slow down the reaction is ignored (see Section 2.4.3). It gives realistic values for strongly exothermal reactions, as decomposition reactions often are. The calculation of TMR$_{ad}$, according to Equation 11.2, requires the specific heat capacity, the specific heat release rate (q'_0) at the runaway initial temperature T_0, and the activation energy.

These parameters, that is, the heat release rate as a function of temperature and activation energy, can easily be determined by calorimetric experiments. In the following sections, different methods, based on a series of isothermal experiments and estimation techniques using only one dynamic experiment, are presented.

11.4.2.1 Determination of $q' = f(T)$ from Isothermal Experiments

A series of isothermal experiments are performed at different temperatures in a calorimeter, for example, in a DSC. On each curve, the maximum heat release rate, which represents the worst case, is measured (Figure 11.4). The experimental procedure must reach the desired temperature as fast as possible. For this purpose, in a DSC, the oven is preheated to the desired temperature with the reference crucible in place. The sample crucible is then placed into the oven and the measurement begins once thermal equilibrium is achieved, which usually takes approximately 2 minutes. During this time no measurement is possible, but 2 minutes is a very short time relative to the total experimental time of several hours. Thus, the achieved conversion before the measure really starts is negligible. It is left as an exercise for the reader to verify this point. Moreover, the difference may be corrected graphically.

A further point, which must be verified in every case, is that total energy determined by integration of the isothermal thermogram corresponds to total energy determined in a previous dynamic experiment. If this is not the case, it may be due to an experimental error or to a complex reaction presenting an autocatalytic step that is not triggered during the experiment under the conditions used. Hence it is good practice to cool down the instrument after the isothermal period and run a subsequent dynamic experiment in order to determine the residual energy.

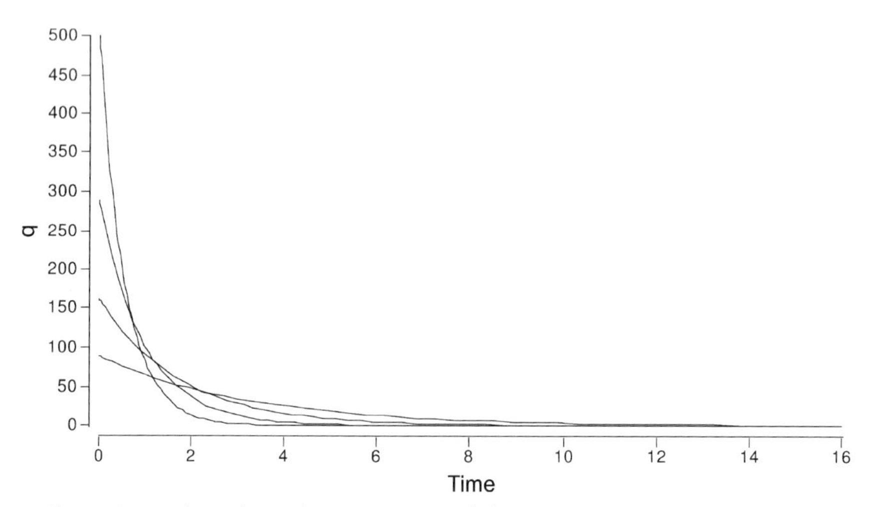

Figure 11.4 Isothermal DSC thermograms recorded at 170, 180, 190, and 200 °C. The measured maximum heat release rate are 90, 160, 290, and 500 $W\,kg^{-1}$.

Table 11.3 Heat release rate as a function of temperature in Arrhenius coordinates.

Temp. °C	T(K)	$1/T$	q W kg^{-1}	ln$q$
170	443.15	0.002 257	90	4.50
180	453.15	0.002 207	160	5.08
190	463.15	0.002 159	290	5.67
200	473.15	0.002 113	500	6.21

This should either deliver a flat thermogram, or show the residual energy due to an autocatalytic reaction step. This practice ensures that the maximum heat release rate measured is correct.

These measured heat release rates calculate the activation energy by performing a linear regression between the natural logarithms of the heat release rates and the reciprocal absolute temperatures (Arrhenius coordinates), as shown in Table 11.3. The slope obtained from the linear regression is an activation temperature of 12 000 K corresponding to an activation energy of about 100 kJ mol^{-1}.

An alternative method graphically determines the slope in an Arrhenius plot: The natural logarithm of the heat release rate is plotted as a function of the reciprocal temperature (in K). Hence, it can be verified that the points obtained are on a straight line, meaning that they follow Arrhenius law and the slope corresponds to the ratio E/R that delivers the activation energy. In Figure 11.5, the abscissa is

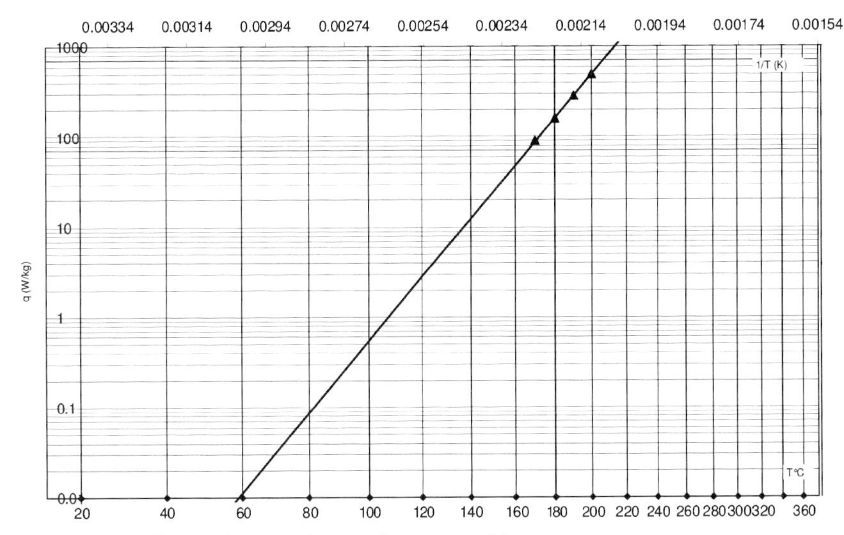

Figure 11.5 Arrhenius diagram showing the measured heat release rates and a linear fit allowing the extrapolation of the data. The abscissa is scaled as $1/T$ (K) and marked in °C.

scaled as $1/T$(K), but marked as the temperature in °C (hyperbolic scale), giving an easy to use representation.

Common to these methods, is that only one point, that is, the maximum heat release rate of the thermogram, is used. Even if the method by itself is efficient in terms of simplicity, experimental work, and evaluation time, it represents a waste of information, in the sense that all the available information is not used in the procedure. Nevertheless, from the point of view of safety it is conservative, since it is based on the zero-order approximation. More complex methods are based on a kinetic analysis of the thermograms presented in Section 11.4.3 below.

As a first approximation, the heat release rate can be extrapolated from one isothermal thermogram using van't Hoff rule (see Section 2.3.1). Nevertheless, this practice assumes a given activation energy and may lead to inaccurate results for an extrapolation over a broad temperature range. Such estimations should be reserved as a guide for decision making concerning the need for more extensive experimental work. Therefore, it should only be used to indicate clearly non-critical cases.

11.4.2.2 Determination of T_{D24}

From one reference point and knowing the activation energy, the heat release rate can be calculated for different temperatures:

$$q_{(T)} = q_{ref} \cdot \exp\left[\frac{E}{R}\left(\frac{1}{T} - \frac{1}{T_{ref}}\right)\right] \tag{11.3}$$

The heat release rate $q_{(T)}$ can, in turn, be used to calculate the time to maximum rate (TMR_{ad}) at the temperature T:

$$TMR_{ad} = \frac{c'_p \cdot R \cdot T^2}{q_{ref} \cdot e^{\frac{E}{R}\left(\frac{1}{T} - \frac{1}{T_{ref}}\right)} \cdot E} \tag{11.4}$$

This equation calculates the TMR_{ad} as a function of temperature.

11.4.2.3 Estimation of T_{D24} from One Dynamic DSC Experiment

A simplified procedure may also be used as a rule of thumb. Its principle is as follows: If the detection limit of an instrument working in the dynamic mode under defined conditions is known, then at the beginning of the peak, the conversion is close to zero and the heat release rate is equal to the detection limit, that is, the temperature at which the thermal signal differs from the signal noise. Thus, the detection limit can serve as a reference point in the Arrhenius diagram. By assuming activation energy and zero-order kinetics, the heat release rate may be calculated for other temperatures.

Worked Example 11.1: TMR_{ad} and T_{D24} from Isothermal DSC Experiments

The estimation of TMR_{ad} from isothermal DSC experiments is illustrated by an example taken from Figures 11.4 and 11.5. For calculation of the activation energy, two reference points, 500 W kg^{-1} at 200 °C and 90 W kg^{-1} at 170 °C, are taken:

$$E = \frac{R \cdot \ln(q_2/q_1)}{\dfrac{1}{T_1} - \dfrac{1}{T_2}} = \frac{8.314 \cdot \ln(500/90)}{\dfrac{1}{443} - \dfrac{1}{473}} \approx 99650\,\text{J} \cdot \text{mol}^{-1}$$

If we want to estimate the TMR_{ad} of the decomposition characterized above, at a lower temperature, for example, 120 °C, we can extrapolate the heat release rate as a function of temperature from 170 °C to 120 °C by using the calculated activation energy in Equation 11.3:

$$q_{(T)} = q_{ref} \cdot \exp\left[\frac{E}{R}\left(\frac{1}{T} - \frac{1}{T_{ref}}\right)\right] = 90 \times \exp\left[\frac{99650}{8.314}\left(\frac{1}{120+273} - \frac{1}{170+273}\right)\right]$$
$$= 2.9\,\text{W} \cdot \text{kg}^{-1}$$

From the specific heat release rate of 2.9 W kg^{-1} at 120 °C, and with a specific heat capacity of 1.8 kJ kg^{-1} K^{-1}, the TMR_{ad} can be calculated for this temperature using Equation 11.4:

$$TMR_{ad} = \frac{C_P' \cdot R \cdot T_0^2}{q_0' \cdot E} = \frac{1800 \times 8.314 \times 393^2}{2.9 \times 99650} = 7998\,\text{s} \approx 2.2\,\text{hrs}$$

Equation 11.4 can also be solved for the temperature corresponding to TMR_{ad} = 24 hours. Since it is a transcendental equation, a graphical or iterative procedure must be used. The resolution of the equation for 8 and 24 hours gives 103 and 89 °C, respectively. Table 11.4 presents the extrapolation of the heat release rate together with the corresponding TMR_{ad} of the example decomposition used above. Such a table is easy to compute on a spreadsheet and immediately identifies the relevant temperature limits. As an alternative, van't Hoff rule: "the reaction rate is multiplied by two for a temperature increase of 10 K" would give the values summarized in Table 11.5. The result is a fairly good approximation: in fact, an activation energy of 100 kJ mol^{-1} corresponds to a factor of 2 for a temperature increase of 10 K in the temperature range around 150 °C. These procedures determine the maximum temperature allowed for a given substance or mixture.

Table 11.4 Calculation of the heat release rate and TMR_{ad} as a function of temperature.

Temp. °C	Power W kg^{-1}	TMR_{ad} h
50	0.004	1113.158
60	0.012	388.637
70	0.034	144.520
80	0.091	56.932
90	0.233	23.646
100	0.563	10.309
110	1.303	4.701
120	2.887	2.233
130	6.150	1.103
140	12.629	0.564
150	25.066	0.298
160	48.203	0.162
170	90.000	0.091
180	163.471	0.052
190	289.366	0.031
200	500.000	0.019

Table 11.5 Approximation using van't Hoff rule.

T °C	170	160	150	140	130	120
q' W kg^{-1}	90	45	22.5	11.25	5.675	2.8

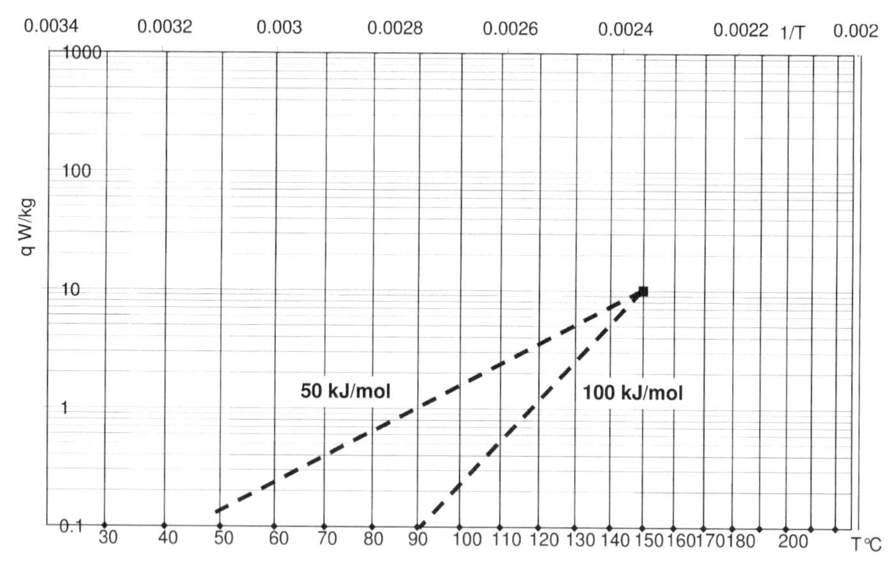

Figure 11.6 Extrapolation of the heat release rate from 10 W kg^{-1} at 150 °C towards lower temperatures. The lower activation energy delivers higher values of the heat release rate.

As an example, one may consider that a DSC operated at 4 or $5\,K\,min^{-1}$ with high pressure crucibles presents a detection limit of $10\,W\,kg^{-1}$. For extrapolation towards lower temperatures, a low activation energy must be chosen, for example, $50\,kJ\,mol^{-1}$ (Figure 11.6). This delivers conservative, that is, high values for the extrapolated heat release rate.

Knowing a reference heat release rate at a reference temperature and an activation energy, extrapolates the heat release rate as a function of temperature and therefore also computing the TMR_{ad} as a function of temperature, as explained in Section 11.4.2.2 above. Thus, it is possible to calculate at which temperature the TMR_{ad} is equal to 24 hours (T_{D24}), by solving Equation 11.4 by iteration. The results for a detection limit of $10\,W\,kg^{-1}$ and a specific heat capacity of $1.0\,kJ\,kg^{-1}\,K^{-1}$ are represented graphically for different activation energies in Figure 11.7. It becomes clear that the solid line corresponding to an activation energy of $50\,kJ\,mol^{-1}$ is more conservative. The solutions of Equation 11.4 can be fitted by linear regression and lead to a simple relationship between the onset observed in a dynamic DSC experiment and the temperature at which the TMR_{ad} becomes longer than 24 hours (T_{D24}):

$$T_{D24} = 0.7 \cdot T_{(q'=10\,W\cdot kg^{-1})} - 46 \tag{11.5}$$

where the temperatures are expressed in °C. This equation can be considered to deliver a conservative prediction of T_{D24} with a reasonable safety margin. Nevertheless, as a semi-empirical rule, it must be validated.

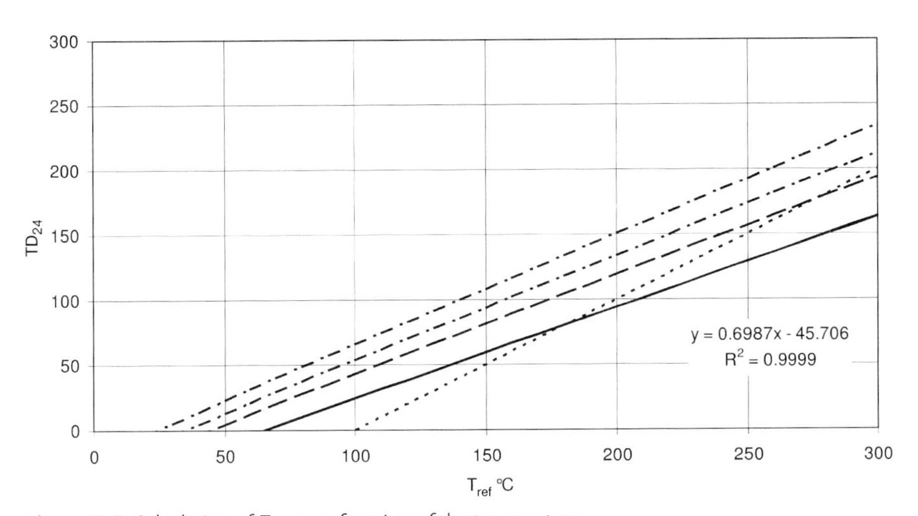

Figure 11.7 Calculation of T_{D24} as a function of the temperature at which the detection limit of $10\,W\,kg^{-1}$ is reached with different values of the activation energy of $50\,kJ\,mol^{-1}$ (solid line) 75, 100, and $150\,kJ\,mol^{-1}$ (dashed lines) and a specific heat capacity of $1\,kJ\,kg^{-1}\,K^{-1}$. The dotted line represents the "100 K-rule."

11.4.2.4 **Empirical Rules for the Determination of a "Safe" Temperature**

The procedure defining a safe temperature from dynamic DSC experiments, by subtracting a given "distance" in temperature from the onset temperature, is called a "distance rule." Such a rule means that the reaction is no longer active below this safe temperature. In the early 1970s, a 50 K rule was common, but experience showed that the prediction were unsafe. The safe distance was then increased to 60 K and finally 100 K, as shown in the literature [14]. This rule consists of subtracting 100 K from the "onset temperature" in a dynamic DSC experiment. Such a rule can be validated by comparing its predictions to experimental values. A comparison was made by Keller [15] who used a great number of experiments, stemming from Grewer and Klais, with pure substances [7] and also with non-published data of reaction mixtures and distillation residues [16] measured between 1994 and 1997 by the same authors at Hoechst AG. These data comprised dynamic DSC experiments and determination of the adiabatic decomposition temperature in 24 hours (ADT_{24}) equivalent to the T_{D24} determined by adiabatic experiments realized under pressure. Thus, it is an ideal source of data for validation of the rule, which is considered valid only if all the experimental values present the same or higher T_{D24} than predicted by the rule.

The comparison represented in Figure 11.8 clearly shows that the 50 K rule (dashes and dots) is not valid and that a significant number of experimental points lie below the dashed line representing the 100 K rule. Thus, the predictions of the simple distance rules are not safe. Increasing the distance further would lead to too a great safety margin in a number of cases, which would impinge on the economy of many processes.

A systematic study of the validity of such a procedure was performed in collaboration with ETH-Zürich [15]. The validation of the procedure was based on numerical simulations of dynamic experiments and adiabatic runaway curves. These simulations were carried out using different rate equations: nth-order, consecutive, branched, and autocatalytic reactions. Moreover, the results were compared to experimental results obtained with over 180 samples of single technical chemical compounds, reactions masses, and distillation residues [17] (Figure 11.8). Thus, they are representative for industrial applications. The line corresponding to this rule (Equation 11.5) is also represented (full line) in Figure 11.8. All experimental points lie above the line and the safety margin remains reasonable. Thus, the method is conservative, but delivers a reasonable safety margin.

In the original work, Keller [15] used a detection limit of $20\,W\,kg^{-1}$ and a specific heat capacity of $1.8\,kJ\,kg^{-1}\,K^{-1}$, obtaining the same result. The explanation is left to the reader as an exercise.

This procedure for estimating T_{D24} from the temperature at which the peak onset is detected in a dynamic experiment is justified, since the TMR_{ad} is based on a zero-order approximation and at the beginning of the DSC peak, the conversion is close to zero. Thus, the heat release rate determined by the procedure is not affected by the rate equation, at least for non-autocatalytic reactions, and may be used for the estimation.

These estimations techniques should only be used as a first approximation, since they are extremely sensitive to the definition of the baseline and hence must

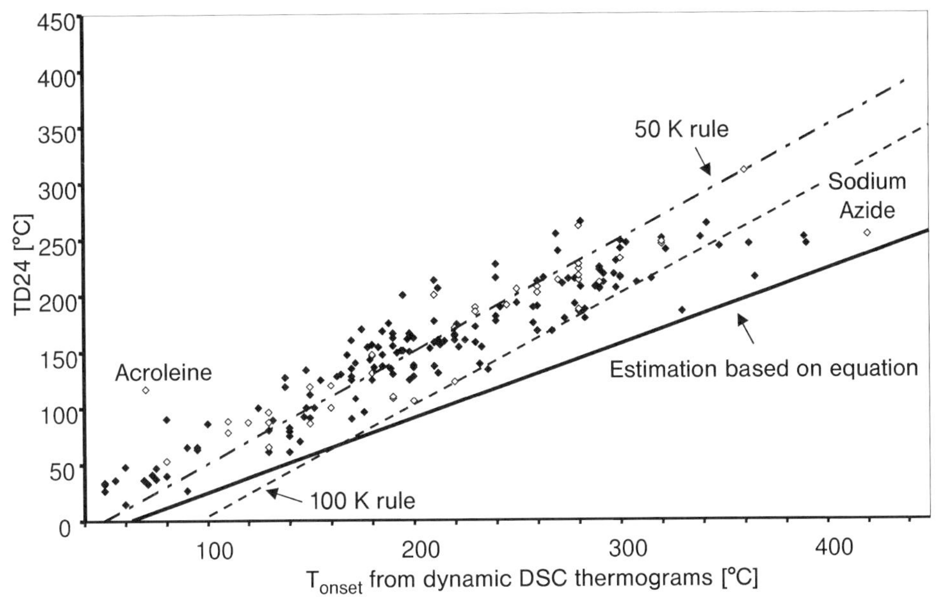

Figure 11.8 Comparison of experimental T_{D24} with the onset in DSC experiments. A prediction rule (lines) can be considered safe if all experimental values are located above the line representing the rule. White dots are pure compounds, black dots are reaction mixtures and distillation residues.

be employed with care. They cannot be used in cases where the exothermal peak starts from an endothermal signal, since the latter may be due to melting. In such a case, the decomposition is not thermally initiated, but initiated by the melting, that is, by a physical phenomenon that does not behave following Arrhenius law. In cases where the estimated TMR_{ad} is found close to, or even shorter than 24 hours, it is strongly recommended to perform a more thorough determination, using the methods presented in Section 11.4.2.2 above.

11.4.3 Complex Secondary Reactions

The procedure presented above assumes a single reaction that is approximated by a zero-order rate equation. Nevertheless, thermograms recorded in practice often show complex behavior with multiple peaks, which even may be lumped. This reveals a more complex kinetic scheme that can no more be described solely by one rate equation. Moreover, the activation energy may be different for the different elementary steps involved, which makes the extrapolation hazardous (Figure 11.9). Depending on the temperature range of interest, one or the other step may dominate the heat release rate. This must be considered when assessing the probability of triggering a secondary reaction.

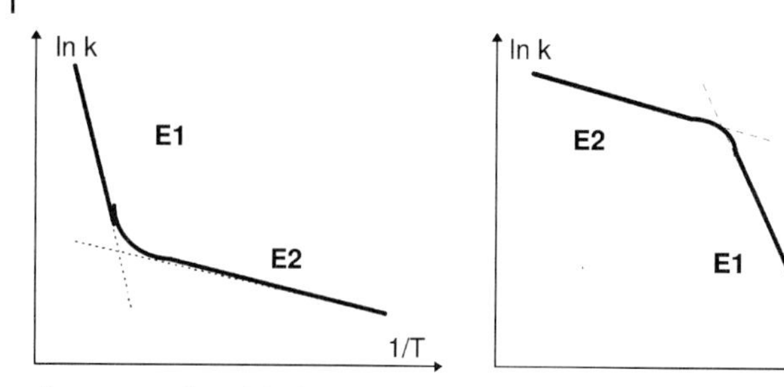

Figure 11.9 Plot of the logarithm of the heat release rate as a function of the reciprocal temperature, showing a change in activation energy as a function of temperature.

11.4.3.1 Determination of TMR_{ad} from Isothermal Experiments

With the approach using isothermal thermograms, the different thermograms must be checked for consistency. In certain cases when the peaks are well separated, as for consecutive reactions, they may be treated individually and the heat release rates can be extrapolated separately, and used for the TMR_{ad} calculation. The reaction that is active at lower temperature will raise the temperature to a certain level where the second becomes active, and so on. So under adiabatic conditions, one reaction triggers the next as in a chain reaction. In certain cases, in particular for the assessment of stability at storage, it is recommended to use a more sensitive calorimetric method as, for example, Calvet calorimetry or the Thermal Activity Monitor (see Section 4.3), to determine heat release rates at lower temperatures and thus to allow a reliable extrapolation over a large temperature range. Complex reactions can also easily be handled with the iso-conversional method, as mentioned below.

11.4.3.2 Determination of $q' = f(T)$ from Dynamic Experiments

In a dynamic experiment, the temperature and the conversion vary with time. Since the temperature is forced to follow the imposed scan rate, by varying the scan rate, the peak appears at different times, that is, at different temperatures (Figure 11.10). This allows for kinetic analysis of the thermograms. The principles of such evaluations can be demonstrated on a single first-order reaction, as an example. The temperature varies linearly with time:

$$T = T_0 + \beta t \tag{11.6}$$

The heat release rate is a function of the rate constant, that is, temperature and conversion:

$$q' = k_0 \cdot e^{-E/RT}(1 - X) \cdot Q' \tag{11.7}$$

The thermal conversion is obtained through integration of the area comprised between the signal and the base line:

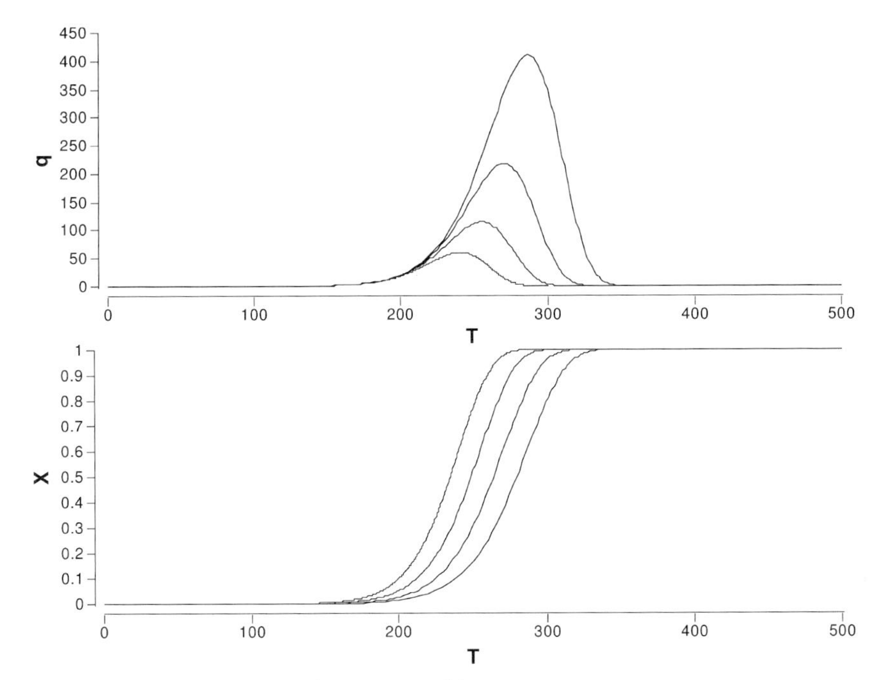

Figure 11.10 Dynamic DSC thermograms of the same reaction recorded with different scan rates (1, 2, 4, and 8 K min^{-1}). The heat release rate q and the conversion are plotted as a function of temperature.

$$X = \frac{\int_0^t q' \cdot d\tau}{\int_0^\infty q' \cdot d\tau} = \frac{\int_0^t q' \cdot d\tau}{Q'} \tag{11.8}$$

Thus, for a point on a thermogram, the heat release rate, the temperature, and the conversion are known. If the same conversion is considered in different thermograms, it is reached at different temperatures and with different heat release rates, solving the system of equations for the pre-exponential factor k_0 and the activation energy E. Many such evaluation methods, such as Borchardt and Daniels [18], Kissinger [19, 20], and Flynn and Wall [21, 22] or Osawa [23, 24], also referred to as ASTM, were developed in the past. These methods allow a quantitative determination of the kinetic parameters for single reactions. Presently, the powerful computing capacity available to every professional uses so called isoconversional methods [42] for multiple or complex reactions and thus gives access to the heat release rate as a function of temperature and conversion. Here, a specific method developed by Roduit [25] allows kinetic analysis for complex DSC signals based on the iso-conversional method. It also provides a means for simulating the corresponding reactive system under adiabatic conditions, which gives direct access to the time to maximum rate in a very efficient way [26].

11.5
Experimental Characterization of Decomposition Reactions

11.5.1
Experimental Techniques

For the assessment of thermal risks linked with the performance of a process at industrial scale, the decomposition reactions that could potentially be triggered should be known and characterized in terms of energy and triggering conditions. Diverse calorimetric or better micro-calorimetric methods are available for this task. As often used-techniques, we find Differential Scanning Calorimetry (DSC) [27–30], Calvet calorimetry, Differential Thermal Analysis (DTA) [31], and adiabatic methods, such as Accelerating Rate Calorinetry (ARC) [32, 33] or Vent Sizing Package (VSP) [34–36], and similar instruments. Semi-quantitative techniques, such as Lütolf-Test [12], Radex [37], or Sedex are also often used. These techniques are described in Section 4.3. The most commonly-used methods belong to the micro-calorimetric techniques (DSC, DTA, and Calvet). This is essentially due to the fact that they use small samples that allow a quantitative measurement even on strongly exothermal processes, without putting the equipment at risk.

One essential experimental feature is that pressure resistant closed crucibles are used. This is because, at heating, volatile compounds may evaporate that may mask an exothermal phenomenon occurring in the same temperature range, and so the sample mass is no longer defined (see Section 4.3.2.1).

The scanning or dynamic mode of operation ensures that the whole temperature range of interest is explored. This must be ensured also in adiabatic experiments, where it is essential to “force” the calorimeter to higher temperatures, in order to avoid missing an important exothermal reaction (see Exercise 2 in Chapter 4).

Nevertheless, two key questions remain: “Which sample should be analysed and which process conditions should be explored?” In order to give some hints on these important points, several typical examples from industrial practice are reviewed in the following subsections. In a first subsection, a choice of typical samples to be analysed is shown. In the next sub-sections, credible process deviations will be reviewed and experimental techniques are presented in further sections.

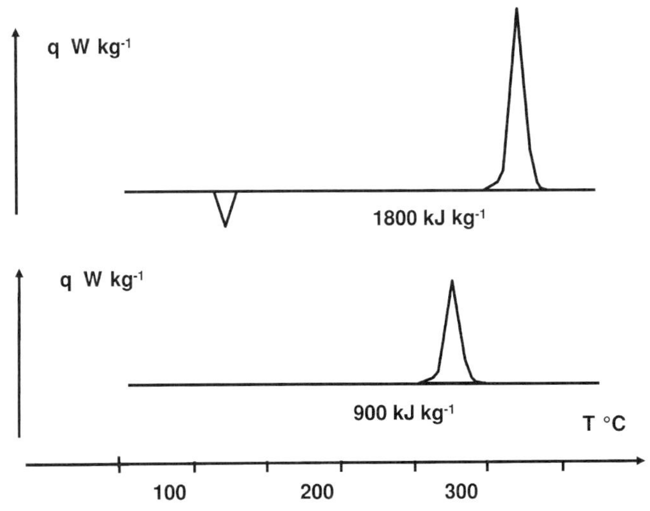

Figure 11.11 DSC thermograms recorded at $4\,K\cdot min^{-1}$ of a pure solid nitro aromatic compound (upper thermogram) and of the same solid in a 50% (w/w) solution in a solvent (lower thermogram).

11.5.2 Choosing the Sample to be Analysed

11.5.2.1 Sample Purity

A general but essential rule in thermal analysis for process safety is to use samples that are as representative as possible for the industrial problem to be solved. As an example, the use of purified samples should be avoided (see Exercise 11.2). If a solid is to be used, either in a solution or as a suspension, the thermal analysis should also be performed on a solution or suspension. It is often observed that a solid compound is destabilized when dissolved in a solvent (Figure 11.11).

The dissolution in a solvent obviously reduces the decomposition energy. Nevertheless, this reduction is not always proportional to the concentration, since a solvent may interfere in the decomposition mechanism. Moreover, the position of the peak in the thermogram, being the temperature range where it is detected, is often shifted towards a lower temperatures, which means loss of stability.

A particularly critical case is the exothermal decomposition immediately following the endothermal melting peak. In such cases, the decomposition is faster in the liquid than in the solid state. In an industrial environment, this could mean that a hot spot may melt a small part of the solid, which begins to decompose and the decomposition may propagate through the entire volume of the product. In such cases, the definition of a safe operation temperature becomes critical.

11.5.2.2 Batch or Semi-batch Process

For a first assessment of the thermal risks linked with the performance of a batch or semi-batch process, the thermal stability of a reacting mixture is of primary

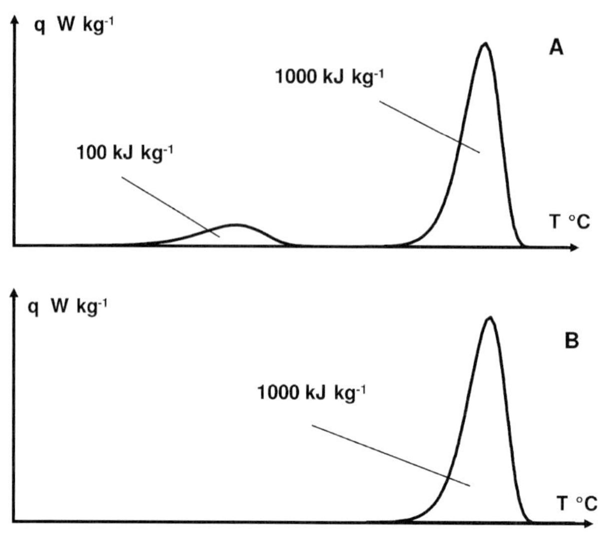

Figure 11.12 Typical DSC thermograms obtained from samples of a reaction mass. The reactant's mixture (A) shows two peaks corresponding to the main and secondary reactions. The thermogram from the final reaction mass (B) only shows the peak corresponding to the secondary reaction.

interest. As a kind of "minimum program", a sample of the initial reactant mixture, that is, before the reaction takes place, and a sample of the final reaction mass, that is, after the reaction is completed, can be analysed by micro-calorimetric methods. This type of experiment delivers a "finger print" of the energies to be released in the reaction mass, thus it is a very efficient technique to identify potentially critical samples as a screening.

For the analysis of an initial reaction mass by DSC, the different reactants must be prepared in the crucible at a temperature where no reaction takes place. The reaction is then initiated during the temperature ramp, giving a first peak (Figure 11.12A), and if there is one, the decomposition of the final reaction mass will be triggered at higher temperatures and deliver a second peak on the thermogram. A sample of the final reaction mass obtained from a laboratory experiment can be analysed in the same way. In this second thermogram, only the decomposition of the final reaction mass will appear (Figure 11.12B).

For semi-batch reactions, it is wise to analyze a sample of the mixture present in the reactor before feed of the reactant is started. In fact, this mixture is often preheated to the process temperature before feeding, hence it is exposed to the process temperature and it could be useful to interrupt the process at this stage, in case of necessity. Such a thermogram assesses the thermal risks linked with such a process interruption (see case history at the beginning of this chapter).

Another powerful technique to provide thermal information on main and secondary reactions is to use Calvet calorimetry. The calorimeter C80, commercial-

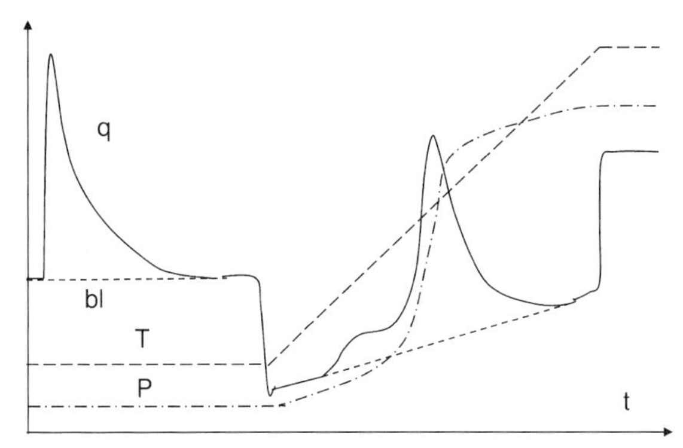

Figure 11.13 Thermogram in a Calvet calorimeter showing the main reaction performed under isothermal conditions and the secondary reaction triggered during the temperature ramp (solid line and baseline). The dashed line is the temperature and the dash-point line the pressure.

ized by SETARAM (see Section 4.3.2.2), provides a mixing-cell that allows two reactants to be brought to reaction temperature in two compartments separated by a membrane. Once the instrument is thermally equilibrated, the membrane is broken and the reactants are allowed to mix, starting the reaction. After the reaction has been completed under isothermal conditions, the instrument is heated in the scanning mode to trigger secondary reactions (Figure 11.13). The advantage of this technique is to provide thermal and pressure information in one experiment in a very efficient way.

The thermal risks identified by these techniques can be assessed using the criteria presented in Chapter 3. Samples showing low energies corresponding to an adiabatic temperature rise below 50 K and no pressure can be considered thermally safe and must not be analysed further. If the energy is significant, the triggering conditions must be assessed by the techniques presented in Section 11.4 above. In this sense, the dynamic (scanned) experiments present an efficient way of screening thermal risks.

11.5.2.3 **Intermediates**

Many reactions do not lead directly from the reactants to the product by a single step. Often intermediates may appear, even if they are not isolated. Such intermediates may be unstable, presenting a specific thermal risk. As an example, we consider the hydrogenation of an aromatic nitro-group to the corresponding amine. Such a reduction runs over different steps and one of the reaction paths leads to the corresponding nitroso compound that is, in turn, reduces to the phenyl hydroxyl amine and finally to the amine [38–40]. Phenyl hydroxyl amines are known to be unstable. In fact, they may lead to a disproportionate reaction (redox reaction) without hydrogen uptake. Since this reaction is exothermal, it presents a thermal risk that must be assessed (Figure 11.14). Such reactions have led to incidents in the past [41].

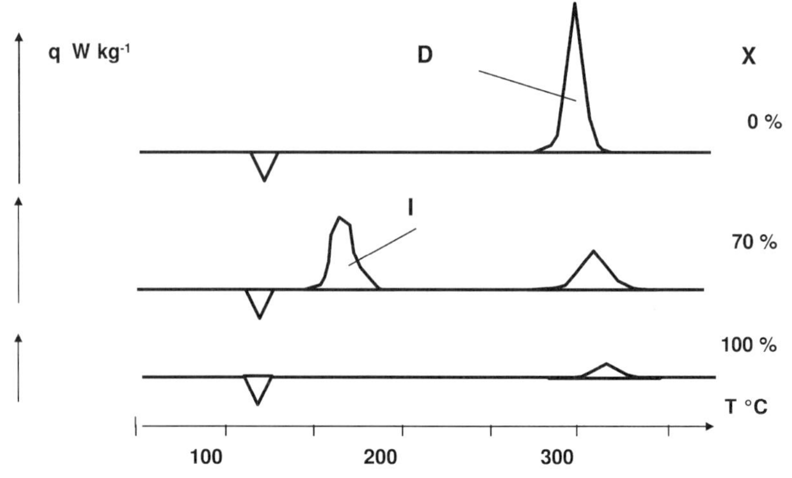

Figure 11.14 Formation of instable intermediates during the catalytic hydrogenation of nitro-aromatic compounds. The decomposition of phenyl hydroxyl amine (I) is shown in the thermogram at 70% of hydrogen uptake. The decomposition of the nitro aromatic compound (D) decreases as the hydrogenation progresses.

11.5.3 Process Deviations

During the risk analysis of chemical processes, diverse deviations from normal operating conditions must be assessed, among them charging errors, especially in batch or semi-batch processes representing a particularly important deviation category. Since the DSC operates with small sample sizes, and requires only a short time to obtain experimental results, it is a method of choice for investigation of the effect of charging error on the thermal stability of reaction masses.

Moreover, the DSC technique investigates other types of deviations, such as the effect of the solvent or catalytic effects due to impurities.

11.5.3.1 Effect of Charging Errors

There are different types of charging errors, which may have an important impact on the thermal stability of the reaction mixture. The nature of the error results from the risk analysis that takes into account the nature of the process, but also the industrial environment. The most probable error in industrial operation is charging the wrong amount of the required reactant. To give an example, when charging a solid reactant from bags, the number of bags charged could deviate if no appropriate measures are taken. The identification of the reactants may also be critical. In this context, the type of packaging used plays a key role. Such deviations can easily be analysed by DSC, since the technique allows the preparation

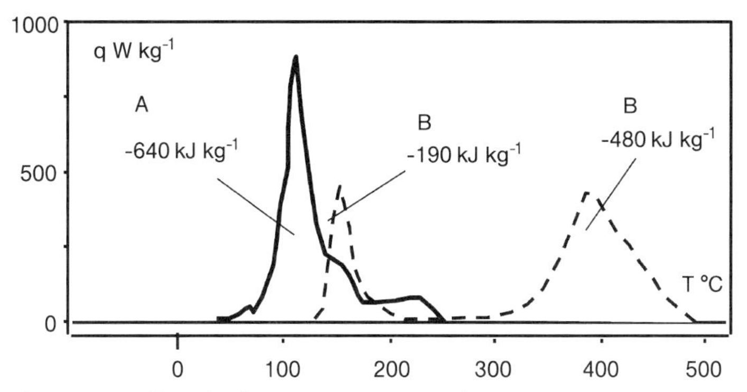

Figure 11.15 Effect of a charging error in a complex recipe. In the first thermogram (solid line), reactant A was omitted, in the second (dashed line), reactant B was omitted.

of different deviating charges. The result can be interpreted on the dynamic thermograms obtained under these deviating conditions.

As an example, in Figure 11.15, three different reactants are to be charged in a solvent. The reaction should be performed as a batch reaction at 30 °C. The first thermogram shows the thermal stability of the reaction mixture with compound (A) omitted. The second thermogram shows the thermal stability of the reaction mass with compound (B) omitted. Both thermograms give approximately the same overall energy, but in thermogram (A) there is only one peak detected from about 30 °C. This would mean immediate triggering of the decomposition when starting the process. In the second case (B omitted), there are two peaks, the first of them detected above 120 °C. Thus, the probability of triggering this decomposition is significantly lower than in the first case.

As a conclusion, it appears clear that measures must be taken to avoid omitting the charging of compound (A); the omission of (B) is not as critical, and requires no special measure. Thus, based on two dynamic DSC thermograms, important conclusions for the safety of the studied process can be drawn.

11.5.3.2 Effect of Solvents on Thermal Stability

It is known that solvents may affect the reaction mechanism or reaction rate. This is often the case for synthesis reactions, but may also occur for secondary reactions. Therefore, the nature of the solvent may affect the thermal stability of reaction masses. Thus, the effect of solvent on thermal stability should be checked in the early stages of process development. Here again, thermal analysis by DSC is a powerful technique, since it allows a rapid screening with some milligrams of reaction mass. An example is given in Figure 11.16.

11.5.3.3 Catalytic Effects of Impurities

Traces of impurities, such as peroxides, rust or, metal ions originating from material corrosion, may catalyze decomposition reactions. Such effects can easily be

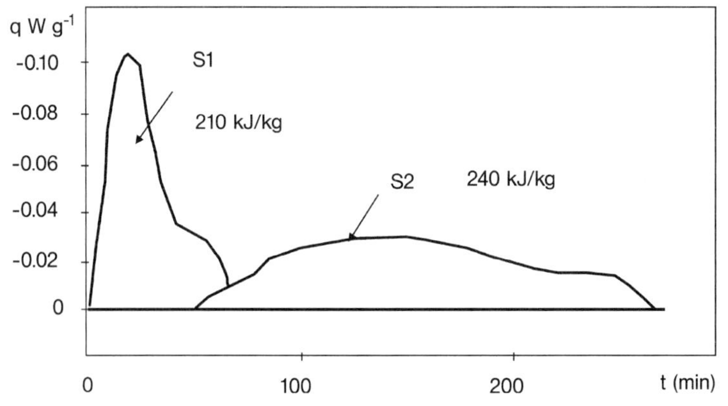

Figure 11.16 Decomposition of 1,3-Dichloro-5,5-dimethyl-hydantoine in heptane (S1) und tetrachloro carbon (S2). Isothermal DSC-thermograms recorded at 140 °C.

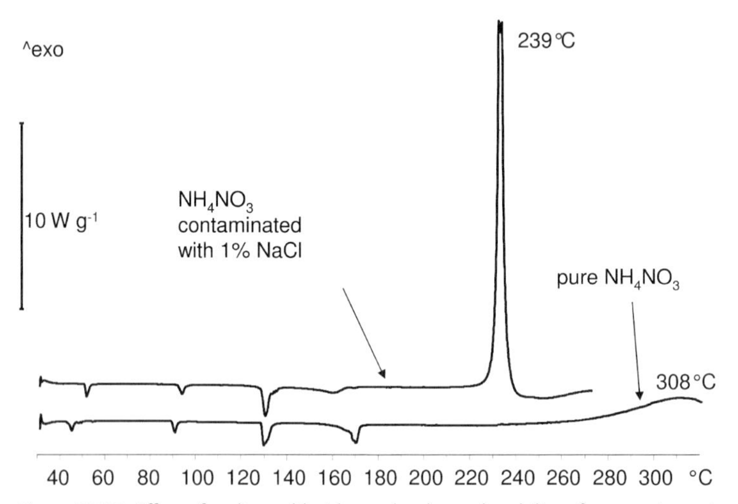

Figure 11.17 Effect of sodium chloride on the thermal stability of ammonium nitrate.

checked by dynamic DSC analysis. The samples can be contaminated by relevant impurities, as identified during risk analysis or from chemical knowledge in general. As an example, Figure 11.17 shows the effect of 1.4 % of sodium chloride on the thermal stability of ammonium nitrate. Numerous examples are shown by Grewer [7].

Inversely, the addition of an inhibitor to stabilize a compound may be an adequate measure to reduce the thermal risks due to decomposition. As an example, in Figure 11.18, the effect of adding zinc oxide to DMSO is shown [42]. This kind of study determines the required inhibitor concentration after, for example, a distillation for solvent recovery.

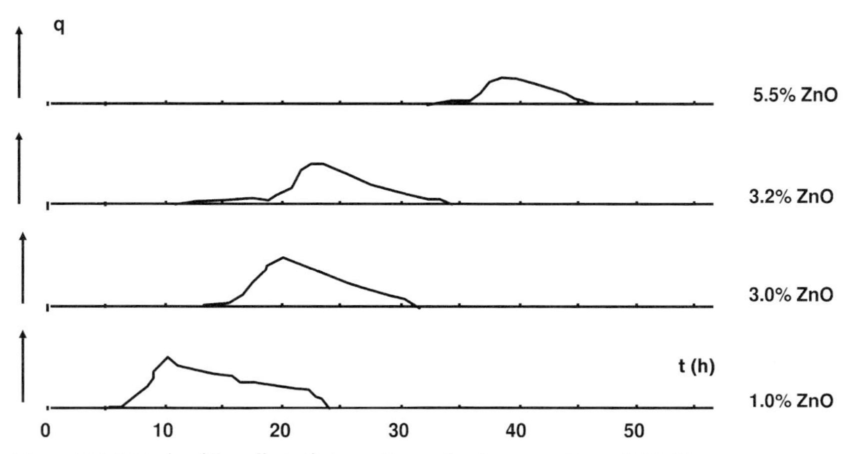

Figure 11.18 Study of the effect of zinc oxide on the decomposition of DMSO.

11.6 Exercises

Exercise 11.1

A diazotization reaction gave rise to a severe explosion, as the concentration of the reactants was increased (see case history in Chapter 4). The corresponding thermograms are represented in Figure 11.19. These thermograms were recorded at a scan rate of $4\,K\,min^{-1}$ with the reactant's mixtures. The process temperature should be 45 °C. The chemist in charge of the process performed a laboratory

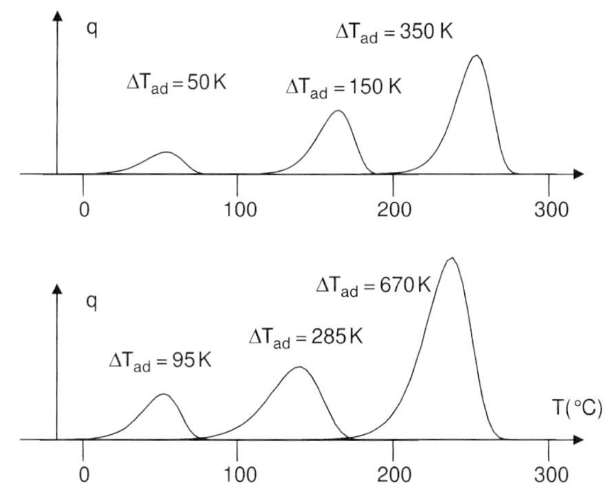

Figure 11.19 Dynamic DSC thermograms of the diazotization reaction mass. Initial concentration in the upper thermogram and increased concentration in the lower.

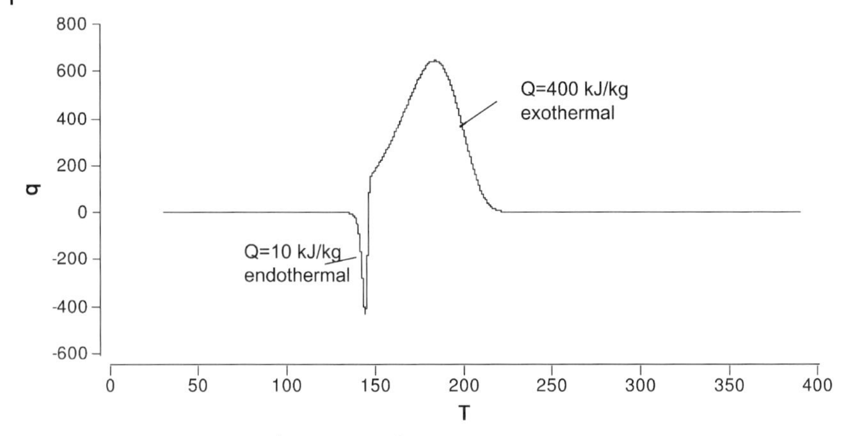

Figure 11.20 Dynamic DSC thermogram of pure (A). Recorded at a scan rate of $4\,K\,min^{-1}$ in a closed pressure resistant crucible.

experiment in a 200 milliliter stirred 3-necked glass flask. He recorded the temperature of the reaction and of the bath during diazotization, but did not notice any temperature difference. Both the bath and the reaction mass remained at 45 °C.

Questions:
1. Give an interpretation of the thermograms.
2. Why did the initial process work properly?
3. Why did the explosion occur, despite the fact that there was no temperature difference in the experiment described above?

► Exercise 11.2

A pure solid reactant (A) is to be used in solution in a solvent for a reaction. The intended process temperature is 80 °C and the reactor is a stirred tank with a nominal volume of $10\,m^3$. The dynamic DSC thermogram of pure (A) is depicted in Figure 11.20.

Question:
By using this thermogram, do you think that the intended operation is possible without any thermal risks? Explain your arguments.

► Exercise 11.3

A reaction has to be performed at 80 °C. Two thermograms were recorded (Figure 11.21). The upper thermogram was obtained with the reactants mixed at room temperature. The lower thermogram was obtained with a sample of the final

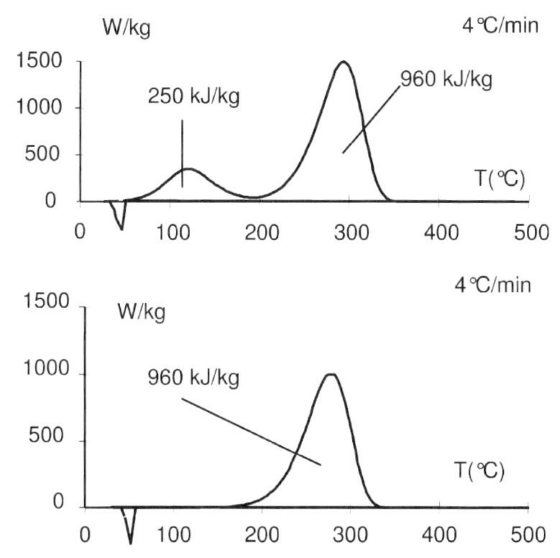

Figure 11.21 DSC thermograms of the reactants mixed at room temperature (upper) and of the final reaction mass (lower).

reaction mass, that is, after the reaction was completed. The specific heat capacity of the reaction mass is $c'_p = 1.7\,\text{kJ}\,\text{kg}^{-1}\,\text{K}^{-1}$.

Questions:

1. Assess the thermal risks linked to the industrial performance of the process.
2. Do you think that a batch process is possible?

References

1 Nolan, P.F. and Barton, J.A. (1987) Some lessons from thermal runaway incidents. *Journal of Hazardous Materials*, **14**, 233–9.

2 Maddison, N. and Rogers, R.L. (1994) Chemical runaways, incidents and their causes. *Chemical Technology Europe*, (11–12) 28–31.

3 Benson, S.W. (1976) *Thermochemical Kinetics. Methods for the Estimation of Thermochemical Data and Rate Parameters*, 2nd edn, John Wiley, New York.

4 Poling, B.E., Prausnitz, J.M. and O'Connell J.P. (2001) *The Properties of Gases and Liquids*, 5th edn, McGraw-Hill, New York.

5 CHETAH. (1975) *Chemical Thermo-dynamic and Energy Release Evaluation Program*, ASTM, Philadelphia.

6 Barton, A. and Rogers, R. (1997) *Chemical Reaction Hazards*, Institution of Chemical Engineers, Rugby.

7 Grewer, T. (1994) *Thermal hazards of chemical reactions*. Industrial safety series, Vol. 4, Elsevier, Amsterdam.

8 Grewer, T. and Hessemer, W. (1987) Die exotherme Zersetzung von Nitro-verbindungen unter dem Einfluss von Zusätzen. *Chemie Ingenieur Technik*, **59** (10), 796–8.

9 Perry, R. and Green, D. (eds) (1998) *Perry's Chemical Engineer's Handbook*, 7th edn, McGraw-Hill, New York.

10 Weast, R.C. (ed.) (1974) *Handbook of Chemistry and Physics*, 55th edn, CRC Press, Cleveland.

11 ESCIS (ed.) (1989) *Sicherheitstest für Chemikalien*, Schriftenreihe Sicherheit. Vol. 1, SUVA, Luzern.

12 Bartknecht, W. and Zwahlen, G. (1987) *Staubexplosionen*, Springer-Verlag, Heidelberg.

13 CCPS (1995) *Guidelines for Chemical Reactivity Evaluation and Application to Process Design*, American Institute of Chemical Engineers, CCPS.

14 Collective work (2000) *Erkennen und Beherrschen exothermer chemischer Reaktionen*, Technischer Ausschuss für Anlagensicherheit, TRNS Nr. 410.

15 Keller, A., Stark, D., Fierz, H., Heinzle, E. and Hungerbuehler, K. (1997) Estimation of the time to maximum rate using dynamic DSC experiments, *Journal of Loss Prevention in the Process Industries*, **10** (1), 31–41.

16 Pastré, J. (2000) *Beitrag zum erweiterten Einsatz der Kalorimetrie in frühen Phasen der chemischen Prozessentwicklung*, ETH-Zürich, Zürich.

17 Pastré, J., Wörsdörfer, U., Keller, A. and Hungerbühler, K. (2000) Comparison of different methods for estimating TMRad from dynamic DSC measurements with ADT 24 values obtained from adiabatic Dewar experiments, *Journal of Loss Prevention in the Process Industries*, **13** (1), 7.

18 Borchardt, H.J. and Daniels, F. (1975) The application of differential thermal analysis to the study of reaction kinetics, *Journal of American Chemical Society*, **79**, 41–6.

19 Kissinger, H.E. (1956) Variation of peak temperature with heating rate in differential thermal analysis, *Journal of Research of the National Bureau of Standards*, **57**, 217–21.

20 Kissinger, H.E. (1959) Reaction kinetics in differential thermal analysis, *Analytical Chemistry*, **29**, 1702.

21 Flynn, J.A. and Wall, L.A. (1967) Initial kinetic parameters from thermogravimetric rate and conversion data. *Polymer Letters*, **5**, 191–6.

22 Flynn, J.A. and Wall, L.A. (1966) A quick and direct method for the determination of activation energy from thermogravimetric data. *Polymer Letters*, **4**, 323–8.

23 Ozawa, T. (1965) A new method of analyzing thermogravimetric data. *Bulletin of the Chemical Society of Japan*, **38**, 1881.

24 Ozawa, T. (1970) Kinetic analysis of derivative curves in thermal analysis, *Journal of Thermal Analysis*, **2**, 301.

25 Roduit, B. (2000) Computational aspects of kinetic analysis. Part E: the ICTAC kinetics project–numerical techniques and kinetics of solid satate processes. *Thermochimica Acta*, **355**, 171–80.

26 Roduit, B., Fierz, H. and Stoessel, F. (2003) The prediction of reaction progress of self-reactive chemicals, in *STK-AFCAT – Meeting*, STK, Basel, Mulhouse.

27 Eigenmann, K. (1976) Sicherheitsuntersuchungen mit thermoanalytischen Mikromethoden, in *Int. Symp. on the Prevention of Occupational Risks in the Chemical Industry*, IVSS, Frankfurt a.M.

28 Brogli, F., Gygax, R. and Meyer, M.W. (1980) DSC a powerful screening method for the estimation of the hazards inherent in industrial chemical reaction, in *6th International Conference on Thermal Analysis*, Bayreuth, Birkhäuser Verlag, Basel.

29 Gygax, R. (1993) *Thermal Process Safety, Data Assessment, Criteria, Measures* (ed. ESCIS), Vol. 8, ESCIS, Lucerne.

30 Frurip, D.J. and Elwell, T. (2005) Effective use of differential scanning calorimetry in reactive chemicals hazard assessment, in *NATAS*, Conference Proceedings.

31 Raemy, A. and Ottaway, M. (1991) The use of high pressure DTA, heat flow and adiabatic calorimetry to study exothermic reactions. *Journal of Thermal Analysis*, **37**, 1965–71.

32 Townsend, I. and Valder, C.E. (1993) Modification of an accelerating calorimeter for operation from sub-ambient temperatures. *Journal of Loss Prevention in the Process Industries*, **6** (2), 75.

33 Townsend, D.I. (ed.) *Accelerating Rate Calorimetry*, Industrial chemical Engineering Series, (ed. IchemE), Vol. 68.

34 Gustin, J.L. (1991) Calorimetry for emergency relief systems design, in *Safety of Chemical Batch Reactors and Storage Tanks*, Benuzzi, A. and Zaldivar, J.M. (eds.) ECSC, EEC, EAEC, Brussels, 311–54.

35 Gustin, J.L. (1993) Thermal stability screening and reaction calorimetry. Application to runaway reaction hazard assessment and process safety management. *Journal of Loss Prevention in the Process Industries*, **6** (5), 275–91.

36 Fisher, H.G., Forrest, H.S., Grossel, S.S., Huff, J.E., Muller, A.R., Noronha, J.A., Shaw, D.A. and Tilley, B.J. (1992) *Emergency Relief System Design Using DIERS Technology, The Design Institute for Emergency Relief Systems (DIERS) Project Manual*. AICHE, New York.

37 Neuenfeld, S. (1993) Thermische Sicherheit chemischer Verfahren. *Chemie Anlagen und Verfahren*, **1993** (9), 34–8.

38 Stoessel, F. (1993) Experimental study of thermal hazards during the hydrogenation of aromatic nitro compounds. *Journal of Loss Prevention in the Process Industries*, **6** (2), 79–85.

39 Gut, G., Kut, O.M. and Bühlmann, T. (1982) Modelling of consecutive hydrogenation reaction affected by mass transfer phenomena. *Chimia*, **36** (2), 96–8.

40 Kut, O.M. and Gut, G. (1980) Einfluss der Absorptionsgeschwindigkeit des Wasser-stoffs auf die Globalkinetik der Flüssigphasenhydrierung von O-Kresol an einem Nickel-Katalysator, *Chimica*, **36** (12), 469–71.

41 McNab, J.I. (1981) The role of thermochemistry in chemical process hazards: catalytic nitro reduction processes, in *Runaway reactions, unstable products and combustible powders*, Institution of Chemical Engineers, Symp. Series, **68**, 3S1.

42 Brogli, F., Grimm, P., Meyer, M. and Zubler, H. (1980) Hazards of self-accelerating reactions, in *3rd International Symposium Loss Prevention and Safety Promotion in the Process Industries*, Swiss Society of Chemical Industry, Basel, 665–83.

12
Autocatalytic Reactions

Case History "DMSO Recovery"

Dimethyl-sulfoxide (DMSO) is an aprotic polar solvent often used in organic chemical synthesis. It is known for its limited thermal stability so usually precautions are taken to avoid its exothermal decomposition. The decomposition energy is approximately 500 J g^{-1}, which corresponds to an adiabatic temperature rise of over 250 K.

This solvent was used for synthesis during a campaign in a pilot plant. It was known to be contaminated with an alkyl bromide. Thus, it was submitted to chemical and thermal analysis, which defined safe conditions for its recovery, that is, a maximum heating medium temperature of 130 °C for batch distillation under vacuum. These conditions were established to ensure the required quality and safe operation. A second campaign, which was initially planned, was delayed and in the mean time the solvent was stored in drums.

One year later, the DMSO was needed again in the pilot plant and it was decided to proceed with distillation to recover a pure solvent. As vacuum was applied to the stirred vessel (4 m^3), difficulties occurred in reaching the desired vacuum. Thus, the operators looked for a leak in the system until someone noticed a sulfide-like smell at the vacuum pump exhaust. It was decided to change the vacuum pump oil that was assumed to be contaminated. To do so, the distillation flap was closed, insulating the vessel from the distillation system, which was brought to atmospheric pressure to allow the oil to be changed. After 30 minutes, the vessel exploded causing extensive material damage, with one operator being injured by flying debris.

What the incident analysis revealed:

- The thermal analysis was repeated on the stored raw DMSO as it was before distillation. The thermal stability was shown to have strongly decreased, when compared to the analysis performed before storage.
- DMSO decomposes following autocatalytic behavior. During storage, decomposition products that catalyze the decomposition were slowly formed. Consequently the induction time of the decomposition decreased such that only 30 minutes were left at 130 °C.

Thermal Safety of Chemical Processes: Risk Assessment and Process Design. Francis Stoessel

ISBN: 978-3-527-31712-7

Lessons drawn

- Even a very slow decomposition may impinge on the thermal stability of a substance decomposing by an autocatalytic mechanism. Thus, the time factor plays an important role.
- Thermal analysis should always be performed on samples that are representative of the substance to be processed.

12.1 Introduction

The first section of this chapter is an introduction of basic definitions, describing the behavior of autocatalytic reactions, their reaction mechanism, and a phenomenological study. The second section is devoted to their characterization and the last section gives some hints on mastering this category of reaction in the industrial environment.

12.2 Autocatalytic Decompositions

Autocatalytic decompositions are common in fine chemistry [1]. They are considered hazardous because they give rise to sudden heat evolution, often with unexpected initiation and unknown exogenic influences [2], consequently they are often perceived as unpredictable. The sudden heat evolution stems from the special nature of the reaction kinetics and results in a violent reaction often associated with important destructive power. For these reasons, it is worth dedicating a special chapter to autocatalytic reactions.

12.2.1 Definitions

12.2.1.1 Autocatalysis

There are several definitions of autocatalytic reactions:

A reaction is called autocatalytic, when a reaction product acts as a catalyst on the reaction course [3].

Autocatalytic reaction is a chemical reaction in which a product also functions as a catalyst. In such a reaction, the observed rate of the reaction is often found to increase with time from its initial value [4].

In fact, in most mechanisms, the reaction rate is proportional to the concentration of a reaction product. Thus, the term of self-accelerating reaction would be more appropriate. But for sake of simplicity, we maintain the term autocatalytic. Our use of the word “autocatalysis” does not imply any molecular mechanism.

12.2.1.2 **Induction Time**

The induction time is the time involved between the instant where the sample reaches its initial temperature and the instant where the reaction rate reaches its maximum. In practice, two types of induction times must be considered: the isothermal and the adiabatic. The isothermal induction time is the time a reaction takes to reach its maximum rate under isothermal conditions. It can typically be measured by DSC or DTA. This assumes that the heat release rate can be removed by an appropriate heat exchange system. Since the induction time is the result of a reaction producing the catalyst, the isothermal induction time is an exponential function of temperature. Thus, a plot of its natural logarithm, as a function of the inverse absolute temperature, delivers a straight line. The adiabatic induction time corresponds to the time to maximum rate under adiabatic conditions (TMR_{ad}). It can be measured by adiabatic calorimetry or calculated from kinetic data. This time is valid if the temperature is left increasing at the instantaneous heat release rate. In general, adiabatic induction time is shorter than isothermal induction time.

12.2.2
Behavior of Autocatalytic Reactions

Reactions often follow nth-order kinetic law. Under isothermal conditions, for example, under conditions where the sample temperature remains constant, the heat release rate decreases uniformly with time. In the case of autocatalytic decomposition, the behavior is different: an acceleration of the reaction with time is observed. The corresponding heat release rate passes through a maximum and then decreases again (Figure 12.1), giving a bell-shaped heat release rate curve or

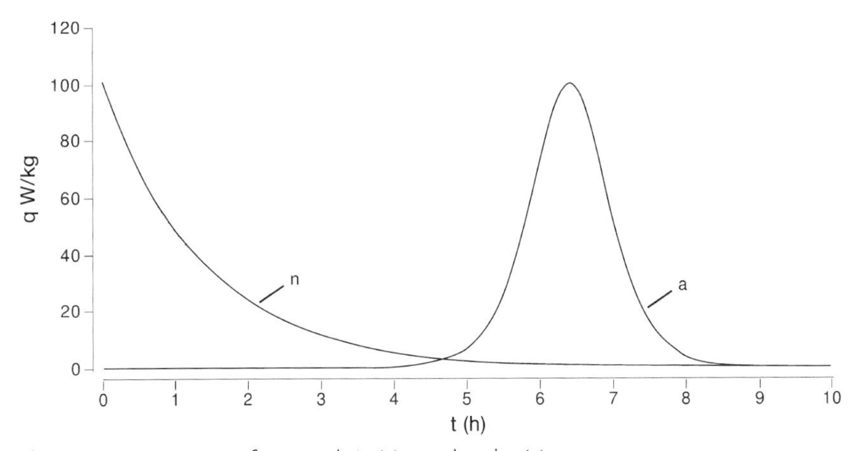

Figure 12.1 Comparison of autocatalytic (*a*) an nth-order (*n*) reactions in an isothermal DSC experiment performed at 200 °C. Both reactions present a maximum heat release rate of $100\,W\,kg^{-1}$ at 200 °C. The induction time of the autocatalytic reaction leads to a delay in the reaction course.

an S-shaped conversion curve. Hence, the acceleration period is often preceded by an induction period with no thermal signal and therefore no noticeable thermal conversion can be observed. An isothermal calorimetric experiment, for example, DSC, immediately shows to which type the nth-order or autocatalytic reaction belongs.

If one considers the case of adiabatic runaway, these two types of reaction lead to totally different temperature versus time curves. With nth-order reactions, the temperature increase starts immediately after cooling failure, while with autocatalytic reactions the temperature remains stable during the induction period and suddenly increases very sharply, as shown in Figure 12.2.

This is due to the fact that under isothermal conditions, the nth-order reaction presents its maximum heat release rate at the beginning of the exposure to initial temperature, whereas the autocatalytic reaction presents no heat release rate at this time. Thus, temperature increase is delayed and only detected later after an induction period, as the reaction rate becomes sufficiently fast. Hence acceleration, due to both product concentration and temperature increase, becomes very sharp.

This has essential consequences for the design of emergency measures. A technical measure to prevent a runaway could be a temperature alarm set at, for example, 10 K above the process temperature. This works well with nth-order reactions, where the alarm is activated at approximately half of the TMR_{ad}. However, autocatalytic reactions are not only accelerated by temperature, but also by time. This can lead to a sharp temperature increase. In the case shown in Figure 12.2, a temperature alarm is not effective, because there is no time left to take measures: in the example given, only a few minutes are left before runaway. Therefore, it is important to know if a decomposition reaction is of autocatalytic nature or not: that is, the safety measures must be adapted to this type of reaction.

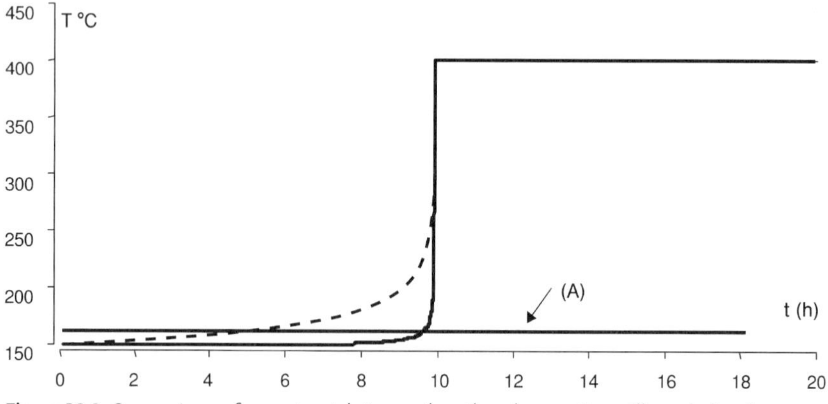

Figure 12.2 Comparison of an autocatalytic (solid line) and an nth-order reaction (dashed line) under adiabatic conditions starting from 150 °C. Both reactions have the same adiabatic induction time or TMR_{ad} of 10 hours. If an alarm level is set at 160 °C, the *n*th-order reaction will reach the alarm level after 4 hours 15 minutes, whereas the autocatalytic reaction reaches this level only after 9 hours 35 minutes. Temperature alarm level (A).

12.2.3 Rate Equations of Autocatalytic Reactions

Numerous models of autocatalytic reactions are described in the literature [2, 5–9]. Here we describe three of them: the Prout–Tompkins [8], the Benito–Perez [6] models, and a model stemming from the Berlin school [1, 10, 11]. These models describe the phenomenon in a simple way and are the most used in practice, especially with respect to process safety.

12.2.3.1 The Prout–Tompkins Model

The Prout–Tompkins model is the oldest described in the literature [8] and it is also the simplest, since it is based on only one reaction and one rate equation:

$$\mathrm{A}+\mathrm{B}\xrightarrow{k}2\mathrm{B}\quad \text{with}\quad -r_A=\frac{-dC_A}{dt}=k\cdot C_A\cdot C_B \tag{12.1}$$

Since the reaction rate is proportional to the concentration of the product, the rate increases when the product is formed, until the reactant concentration decreases. For this reason, even under isothermal conditions, the reaction first accelerates, passes a maximum, and then decreases (Figure 12.3). The rate equation may be expressed as a function of conversion:

$$\frac{dX}{dt}=\dot{X}=k\cdot X\cdot(1-X)\Rightarrow X=\frac{1}{1+e^{-kt}} \tag{12.2}$$

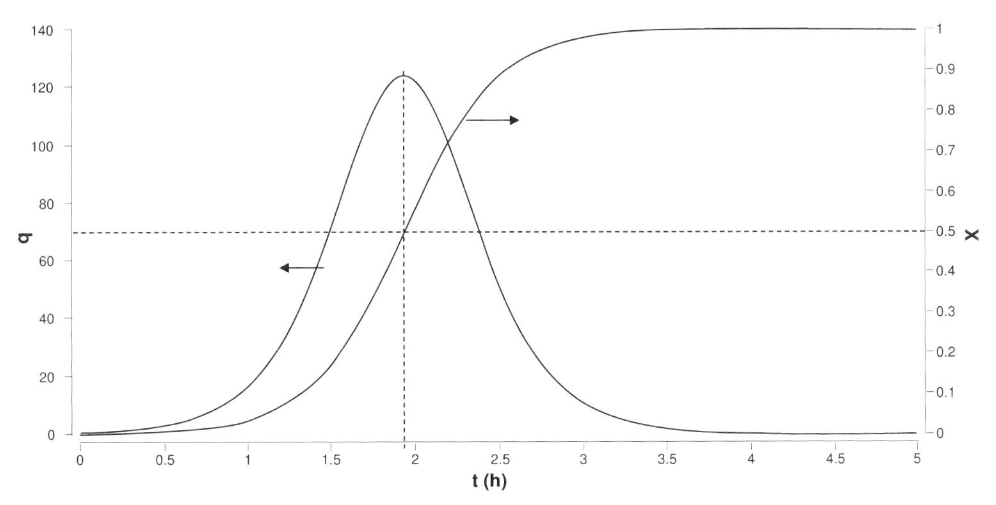

Figure 12.3 Heat release rate and conversion under isothermal condition for the Prout–Tompkins model.

This expression describes the characteristic S-shaped conversion curve as a function of time. Under isothermal conditions, the maximum reaction rate and consequently the maximum heat release rate is obtained for a conversion of 0.5:

$$\frac{d\dot{X}}{dX} = 1 - 2X = 0 \Rightarrow X = 0.5 \tag{12.3}$$

The kinetic constant can be calculated from the maximum heat release rate measured under isothermal conditions:

$$k = \frac{4 \cdot q'_{max}}{Q'} \tag{12.4}$$

This model gives a symmetrical peak with its maximum at half conversion. Hence the model is unable to describe non-symmetrical peaks as they are often observed in practice. Moreover, in order to obtain a reaction rate other than zero, some product B must be present in the reaction mass. Therefore, the initial concentration of B (C_{B0}) or the initial conversion (X_0) is a required parameter for describing the behavior of the reaction mass. This also means that the behavior of the reacting system depends on its "thermal history," that is, on the time of exposure to a given temperature. This simple model requires three parameters: the frequency factor, the activation energy, and the initial conversion that must be fitted to the measurement in order to predict the behavior of such a reaction under adiabatic conditions.

12.2.3.2 The Benito–Perez Model

The Benito–Perez model [6] also includes a first reaction, called the initiation reaction, producing some product, which then enters into an autocatalytic reaction similar to the schema described above (Section 12.2.3.1). It can be described by a formal kinetic model as

$$\begin{aligned} &\nu_1 \mathrm{A} \xrightarrow{k_1} \nu_1 \mathrm{B} \\ &\nu_2 \mathrm{A} + \nu_3 \mathrm{B} \xrightarrow{k_2} (\nu_3 + 1)\mathrm{B} \\ &-r_A = k_1 C_A^{a_1} + k_2 C_A^{a_2} C_B^{b} \end{aligned} \tag{12.5}$$

This model comprises eight parameters, that is, two frequency factors, two activation energies, three exponents for the reaction orders, and the initial conversion. It is often used in a simplified form, with all reaction orders equal to one:

$$\begin{aligned} &\mathrm{A} \xrightarrow{k_1} \mathrm{B} \\ &\mathrm{A} + \mathrm{B} \xrightarrow{k_2} 2\mathrm{B} \\ &-r_A = k_1 C_A + k_2 C_A C_B \end{aligned} \tag{12.6}$$

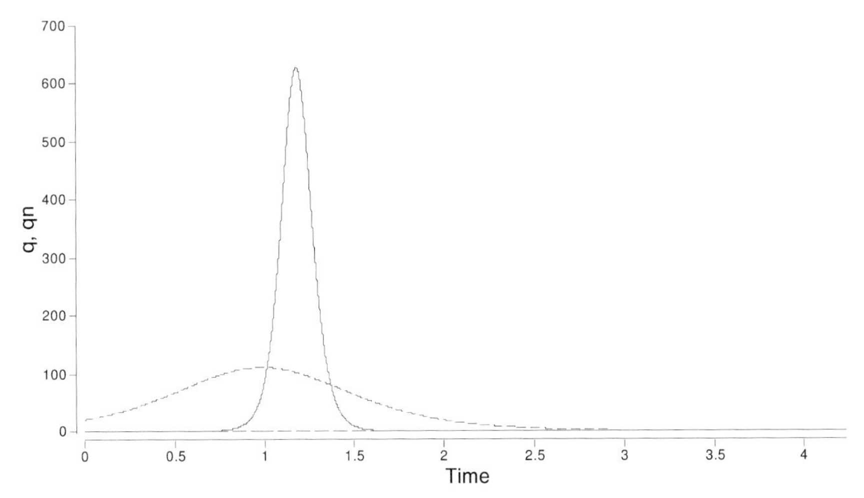

Figure 12.4 Comparison of a strongly autocatalytic reaction with a weak autocatalytic reaction (dashed line). Isothermal DSC experiment at 200 °C. The heat release rate for the strong autocatalytic reaction is zero at the beginning of the exposure.

The rate equation then becomes

$$\frac{dX}{dt} = k_1(1 - X) + k_2X(1 - X) \tag{12.7}$$

Thanks to its versatility, this model has proved to describe a great number of autocatalytic reaction systems [5]. Systems with a slow initiation reaction are called "strong autocatalytic." Because the rate of the initiation reaction is low, product is formed slowly, leading to a long induction time under isothermal conditions. For such systems, the initial heat release rate is low or practically zero. Consequently, the reaction may remain undetected for a relatively long period of time (Figure 12.4). When the reaction accelerates, such an acceleration appears suddenly and may lead to runaway. A strong autocatalytic reaction is formally equivalent to a Prout–Tompkins mechanism.

Systems with a faster initiation reaction provide an initial heat release rate which detects them earlier. These systems are called "weak autocatalytic."

12.2.3.3 **The Berlin Model**

Another rate equation is often referred to in the German literature [1, 11, 12]:

$$\frac{dX}{dt} = k(1 + PX)(1 - X) \tag{12.8}$$

The factor P is called the autocatalytic factor: for $P = 0$, the reaction becomes a single first-order reaction. With increasing P, the autocatalytic character becomes more important.

It can be shown that this model is equivalent to the Benito–Perez model by setting

$$P = \frac{k_2 C_{A0}}{k_1}$$

whereas it corresponds to the simplified Benito–Perez model with equal activation energies, and frequency factors for both steps and all reaction orders equal to 1. Since P is the ratio of two rate constants, both reactions do not present the same activation energy and P is an exponential function of temperature.

The simplicity and nevertheless versatility of the model makes it useful for studying phenomenological aspects of autocatalytic reaction.

12.2.4
Phenomenological Aspects of Autocatalytic Reactions

Different factors may strongly affect the behavior of autocatalytic reactions, especially if we consider the adiabatic temperature course that is used to predict the TMR_{ad}. Such effects are shown by numerical simulations using the Berlin model.

A higher degree of autocatalysis produces shorter TMR_{ad} (Figure 12.5). This has a practical consequence for process safety: For a non-autocatalytic reaction, the alarm set at 10 °C above the initial temperature would be triggered at approximately half of TMR_{ad}, leaving enough time to take counter measures. This time is reduced as the degree of autocatalysis increases, hence the alarm would be

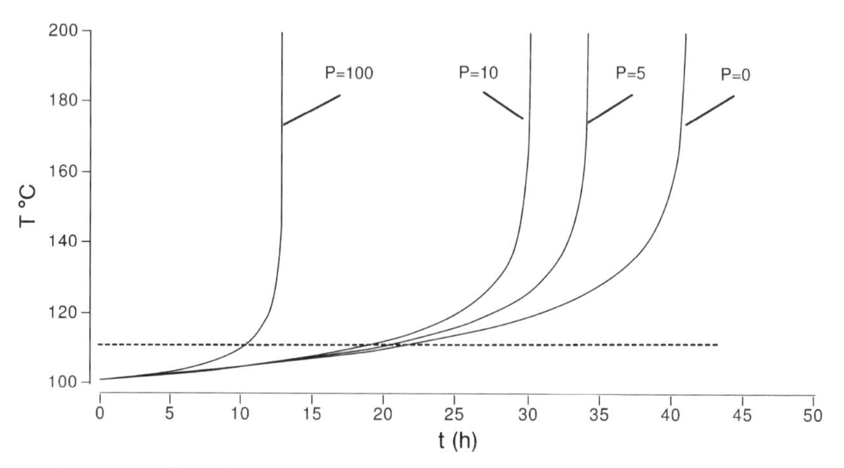

Figure 12.5 Effect of the degree of autocatalysis on the adiabatic temperature course using different values of the parameter P: 0, 5, 10, and 100. The initial temperature is 100 °C. The dashed line represents a temperature alarm level set at 110 °C.

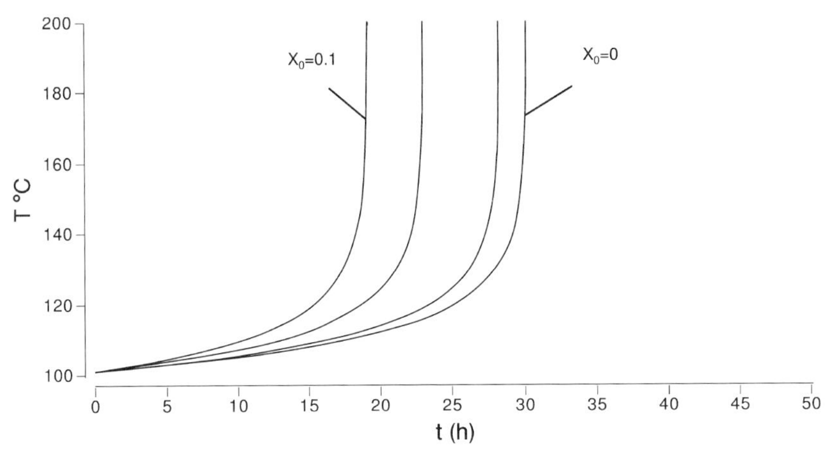

Figure 12.6 Effect of the initial conversion of the substance on the adiabatic temperature course. These numerical simulations were performed with a parameter *P* of 10. The initial conversion was $X_0 = 0$, 0.01, 0.05, and 0.1.

triggered too late. Therefore, autocatalytic reactions are very sensitive to catalytic effects, which may be due to impurities present in the substance. Thus, when dealing with autocatalytic reactions, one must be aware of the product quality or purity.

Since a reaction product catalyses the reaction, the initial concentration of product also has a strong effect on the TMR_{ad}. In the case illustrated in (Figure 12.6), an initial conversion of 10% leads to a reduction of the TMR_{ad} by a factor of 2. This also has direct implications for process safety: the "thermal history" of the substance, that is, its exposure to temperature for a certain time increases initial product concentration, leading to effects comparable to those illustrated in Figure 12.5. Hence it becomes obvious that substances showing an autocatalytic decomposition are very sensitive to external effects, such as contaminations and previous thermal treatments. This is important for industrial applications as well as during the experimental characterization of such decompositions: the sample chosen must be representative of the industrial situation, or several samples must be analysed.

12.3 Characterization of Autocatalytic Reactions

12.3.1 Chemical Characterization

Certain classes of compounds are known to decompose following an autocatalytic mechanism. Among them are:

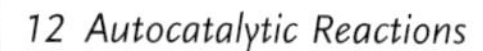

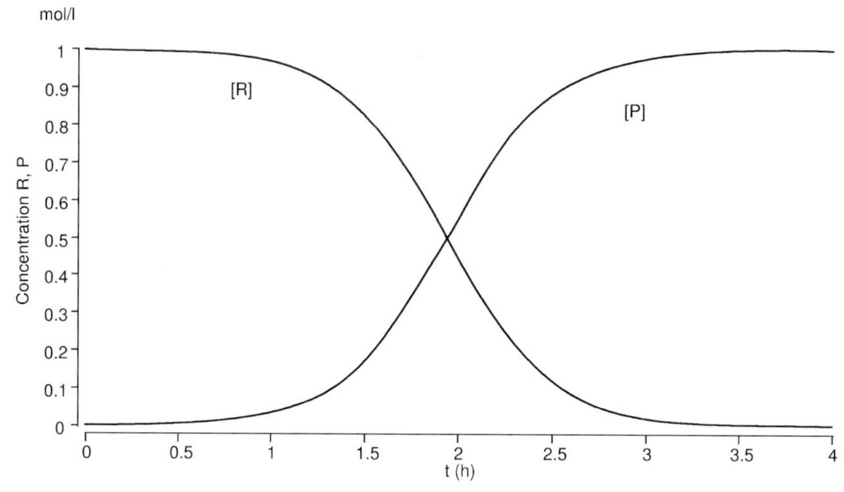

Figure 12.7 Variation of concentration as a function of time: *R* is reactant and *P* is product.

- Aromatic nitro compounds: the exact mechanism is unknown [1].
- Monomers: polymerizations show a strong self-accelerating reaction rate [13, 14].
- Chlorinated aromatic amines: polycondensation is catalyzed by hydrochloric acid that is produced by the reaction [15].
- Dimethylsulfoxide [16].
- Cyanuric chloride and its mono- and di-substituted derivatives: the decomposition is catalyzed by hydrochloric acid, which is also a reaction product [2].

An autocatalytic decomposition can be followed by isothermal aging and periodic sampling for a chemical analysis of the substance. The reactant concentration first remains constant and decreases after an induction period (Figure 12.7). This is characteristic for self-accelerating or autocatalytic behavior. The chemical analysis may also be replaced by a thermal analysis using dynamic DSC or other calorimetric methods, following the decrease of the thermal potential as a function of the aging time.

12.3.2
Characterization by Dynamic DSC

12.3.2.1 Peak Aspect in Dynamic DSC

The only reliable way to detect and characterize an autocatalytic decomposition is by isothermal or isoperibolic measurement (see Figure 12.1). Nevertheless, for experienced users, the results of thermal screening experiments in the dynamic mode can also give indications: the thermograms show narrow signals with a high heat release rate maximum and high energy potentials (Figure 12.8). Dynamic or

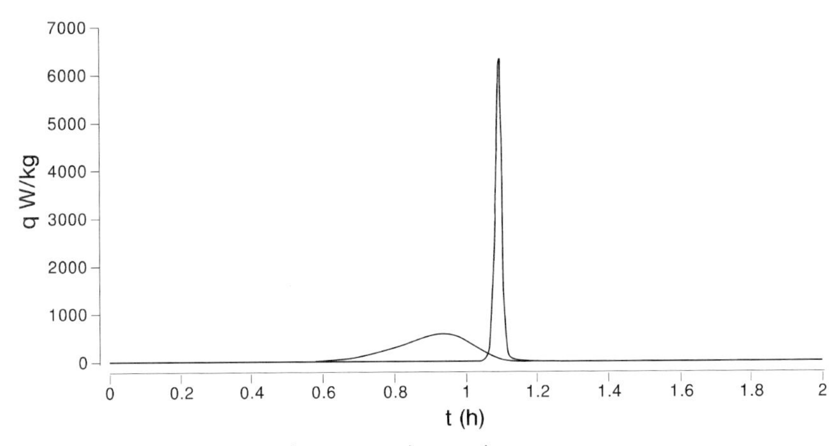

Figure 12.8 Dynamic DSC thermogram showing the difference in signal shape between autocatalytic (sharp peak) and *n*th-order reaction (flat peak).

temperature-programmed DSC or DTA measurements can only suggest the autocatalytic nature of the decomposition.

In Figure 12.8, both reactions the nth-order and the autocatalytic reactions have the same maximum heat release rate of $100\,\mathrm{W\,kg^{-1}}$ at 200 °C, the same activation energy of $100\,\mathrm{kJ\,mol^{-1}}$, and the same decomposition energy of $500\,\mathrm{J\,g^{-1}}$. This is the same system as used in Figures 12.1 and 12.2. One notices that the peak of the autocatalytic reaction is strongly shifted towards high temperatures. This could easily lead to a false interpretation that the sample corresponding to the autocatalytic reaction is more stable than the sample corresponding to the *n*th-order reaction, which is totally misleading. This is a further reason for detecting the autocatalytic nature of decomposition reactions. Moreover, the autocatalytic reaction presents a far higher maximum heat release rate than the nth-order reaction.

12.3.2.2 **Quantitative Characterization of the Peak Aspect**

The fact that the peak appears sharp and narrow with a high maximum heat release rate may be expressed in a quantitative way. The first idea is to measure the peak height and width and to use a ratio in order to determine if the reaction is autocatalytic or not. Even if this method looks simple, its drawback is that only a few points are used to describe the peak. Thus, the statistic significance of such an evaluation is poor.

Therefore, a more efficient and reliable method was developed at the Swiss Institute for the Promotion of Safety & Security [17], which has a higher statistic significance. In fact, the beginning of the reaction (or of the peak) determines the behavior of the reaction mass under adiabatic conditions. Therefore, a simple first-order model is fitted to the beginning of the peak by adjusting two parameters of the model (Figure 12.9): the activation energy and the frequency factor. The activation energy is an *apparent* activation energy and characterizes the steepness

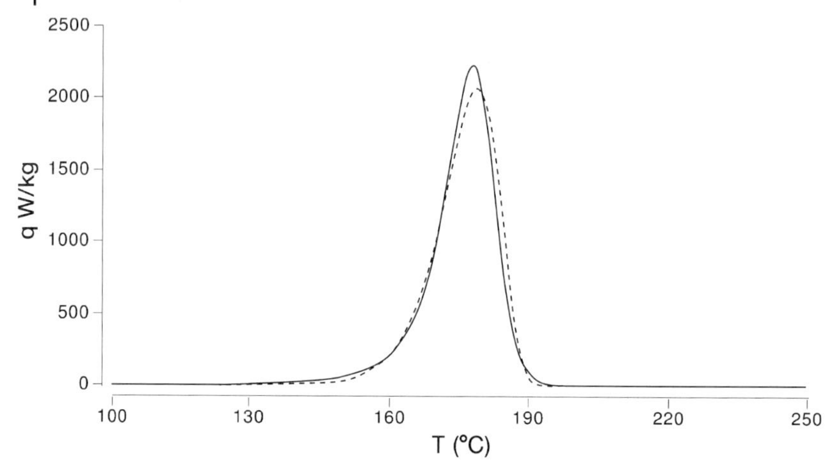

Figure 12.9 First-order model fitted to an autocatalytic peak in a dynamic experiment. The apparent activation energy is 280 kJ mol^{-1}, showing the autocatalytic nature of the reaction.

of the peak, that is, the higher the apparent activation, the greater the chance that the reaction is autocatalytic.

This method was validated with over 100 substances and compared with the results of the classical study by isothermal experiments. For apparent activation energies above a level of 220 kJ mol^{-1}, 100% of the samples showed an autocatalytic character in isothermal experiments. This method can be used as a screen to distinguish clearly autocatalytic reactions from others that should be studied by isothermal experiments. This reduces the number of isothermal experiments required.

Temperature-programmed DSC, or DTA measurements, can only suggest the autocatalytic nature of the decomposition. Neither the influence of the thermal history and contamination can be detected by them, nor can the kinetic parameters be determined from a single experiment.

12.3.2.3 **Characterization by Isothermal DSC**

This is a reliable way to detect and characterize an autocatalytic decomposition. Nevertheless, there are a certain number of precautions to take; especially the choice of the experiment temperature is crucial:

- At too low a temperature, the induction time may be longer than the experiment time, suggesting that there is no decomposition. This false interpretation can be avoided by comparing with the dynamic DSC experiment: the measured energy must be the same in both experiments.
- At too high a temperature, the induction time may be so short that only the decreasing part of the signal is detected, suggesting a non-autocatalytic decomposition. Here too, this false interpretation can be avoided by comparing with the dynamic DSC experiment: the measured energy must be the same in both experiments.

At a correct temperature level, the typical bell-shaped signal, as shown in Figure 12.3, can be identified: the reaction rate first increases, passes a maximum, and decreases again.

12.3.2.4 Characterization Using Zero-order Kinetics

This type of experiment can be repeated at other temperatures, determining the activation energy and the estimation of time to explosion. The concept of time to explosion or TMR_{ad} (Time to Maximum Rate under Adiabatic conditions) is extremely useful for that purpose [18]. This TMR_{ad} can be estimated by

$$TMR_{ad} = \frac{c'_P \cdot R \cdot T_0^2}{q'_0 \cdot E} \tag{12.9}$$

This expression was established for zero-order reactions, but can also be used for other reactions, if the influence of concentration on reaction rate can be neglected. This approximation is particularly valid for fast and exothermic reactions (see Section 2.4.3).

Worked Example 12.1: TMR_{ad} from Zero-order Approximation

The natural logarithms of maximum heat release rates determined on each thermogram are plotted as a function of the inverse temperature in an Arrhenius diagram. In Figure 12.10, the temperature axis is scaled using inverse temperature.

From this diagram, the activation energy and the heat release rate for every temperature can be calculated. As an example, in Figure 12.10 we take two points: $80\,W\,kg^{-1}$ at 120 °C and $16\,W\,kg^{-1}$ at 100 °C. With these values we calculate the activation energy as

$$E = \frac{R \cdot \ln\left(\frac{q_2}{q_1}\right)}{\frac{1}{T_1} - \frac{1}{T_2}} = \frac{8.314 \cdot \ln\left(\frac{80}{16}\right)}{\frac{1}{373} - \frac{1}{393}} \approx 98150\,J \cdot mol^{-1}$$

Using this energy of activation and heat capacity of $1.8\,kJ\,kg^{-1}\,K^{-1}$, the TMR_{ad} can be estimated according to Equation 12.9 for a temperature of 80 °C, where the heat release rate is $2.7\,W\,kg^{-1}$ (from Arrhenius Diagram in Figure 12.10):

$$TMR_{ad} = \frac{c'_P \cdot R \cdot T_0^2}{q'_0 \cdot E} = \frac{1800 \times 8.314 \times 353^2}{2.7 \times 98150} = 7037\,s \approx 2\,hrs$$

To obtain a TMR_{ad} longer than 8 hours, the temperature must not exceed 65 °C, and 50 °C for a TMR_{ad} longer than 24 hours.

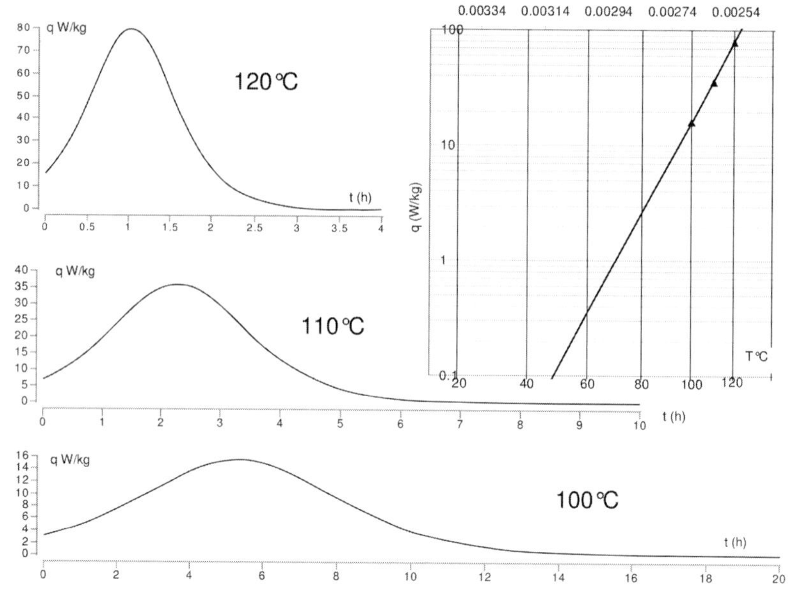

Figure 12.10 Isothermal DSC thermograms and Arrhenius Diagram.

Thus, it is possible to define the maximum allowable temperature by considering the acceptable induction time for the thermal explosion. This assessment of the probability of a runaway by using a zero-order reaction model may be too cautious.

The approximation for the TMR_{ad} is valid if conversion can be ignored. For autocatalytic reactions, the maximum heat release rate is reached at non-zero conversion, and thus conversion can no longer be ignored. Moreover, since the maximum of the heat release rate is often used to obtain an estimate of the TMR_{ad}, the obtained TMR_{ad} is definitely too short. The resulting risk assessment is therefore too conservative and may even endanger the development of a profitable process.

12.3.2.5 Characterization Using a Mechanistic Approach

To obtain a more realistic estimation of the behavior of an autocatalytic reaction under adiabatic conditions, it is possible to identify the kinetic parameters of the Benito–Perez model from a set of isothermal DSC measurements. In the example shown in Figure 12.11, the effect of neglecting the induction time assumes a zero-order reaction leading to a factor of over 15 during the time to explosion. Since this factor strongly depends on the initial conversion or concentration of "catalyst" initially present in the reaction mass, this method must be applied with extreme care. The sample must be truly representative of the substance used at industrial scale. For this reason, the method should be only be applied by specialists.

12.3.2.6 Characterization by Isoconversional Methods

These methods were presented in the previous chapter (Section 11.4.3.2). In their principles, isoconversional methods use a mathematical description of the conver-

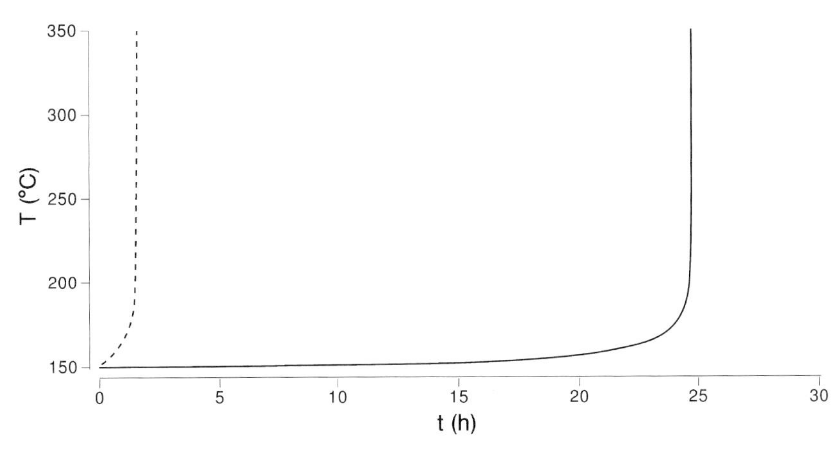

Figure 12.11 Adiabatic temperature course of an autocatalytic reaction (solid line) compared to the zero-order approximation (dashed line). Both reactions have a maximum heat release rate of $100\,W\,kg^{-1}$ at 200 °C and an energy of $500\,J\,g^{-1}$.

sion as a function of temperature, hence the autocatalytic nature of a reaction is implicitly accounted for (as far as the reaction presents this behavior). The great advantage of this method is that its results are based on a greater amount of data, since all measured points enter the evaluation procedure, giving a significantly larger database. This results in improved reliability and even more complex behaviors can be described accurately. Further, in the AKTS software, a feature calculates the confidence interval of the results [19]. In the example shown in Figure 12.15, the TMR_{ad} was calculated at 24 hours and the 95% confidence interval ranged from 19 to 30 hours. This again shows that the values of TMR_{ad} should be used with care, since under adiabatic conditions, errors are amplified.

12.3.2.7 **Characterization by Adiabatic Calorimetry**

Since autocatalytic reactions often show only a low initial heat release rate, the temperature rise under adiabatic condition will be difficult to detect. Therefore, the sensitivity of the adiabatic calorimeter must be carefully adjusted. A small deviation in temperature control may lead to large differences in the measured time to maximum rate. This method should only be applied by specialists and is often used to confirm results obtained by other methods.

12.4 Practical Safety Aspects for Autocatalytic Reactions

12.4.1 Specific Safety Aspects of Autocatalytic Reactions

Autocatalytic reactions are by definition catalyzed by their reaction products. Thus, fresh material can be contaminated when mixed with material that underwent thermal stress and thus contains some decomposition product. The TMR_{ad} of a substance decomposing by an autocatalytic mechanism depends therefore, not only on the temperature, but strongly on the thermal history [5]. This was exemplified by Dien [20] where 1,3-dinitrobenzene was exposed to 390 °C during a first period of time, then cooled to a lower temperature during 10 minutes and again heated at 390 °C in a DSC to measure the isothermal induction time (Figure 12.16) as the sum of the first and second heating period. Since it remains constant, the catalyst formed during the first heating period is stable and survives during the cooling period, shortening the induction time measured in the second heating period.

In addition, autocatalytic reactions may also be catalyzed by impurities, for example, by heavy metals or acids. As an example, it is known that the decomposition of dihydroxy-diphenylsulfone is catalyzed by iron [2]. In a diagram, as represented in Figure 12.17, the maximum allowed iron concentration can easily be defined. This gives a reliable way of establishing a critical limit for the process, which can also easily be checked before a batch is started.

For this reason, it is strongly recommended to check for thermal stability in case the raw material supplier is changed, or in case previous reaction steps did not follow specifications, and so on. Therefore, it is also recommended to check for the thermal history of the material, especially when abnormally long holding times of storage at higher temperature were used before handling the material again.

Worked Example 12.2: Kinetic Study of an Autocatalytic Decomposition

A reaction mass is to be concentrated by vacuum distillation in a 1600 liter stirred tank. Before distillation, the contents of the vessel are 1500 kg, containing 500 kg of product. The solvent should be totally removed from the solution at 120 °C, with a maximum wall temperature of 145 °C (5 bar steam). In order to evaluate the thermal stability of the concentrated product, a dynamic DSC experiment was performed (Figure 12.12).

Since this thermogram shows a steep peak, the autocatalytic nature of the decomposition is likely. Thus, two isothermal DSC experiments were performed at 240 and 250 °C, in order to confirm this hypothesis and to evaluate the probability of triggering the decomposition (Figure 12.13). The results can be summarized as follows: at 240 °C the initial heat release rate is $8.5\,W\,kg^{-1}$ and the maximum heat release rate $260\,W\,kg^{-1}$. At 250 °C, the measured heat release rates are 15 and $360\,W\,kg^{-1}$, respectively.

Questions:

How do you assess the severity of a runaway of this concentrate?

How do you assess the probability of triggering it?

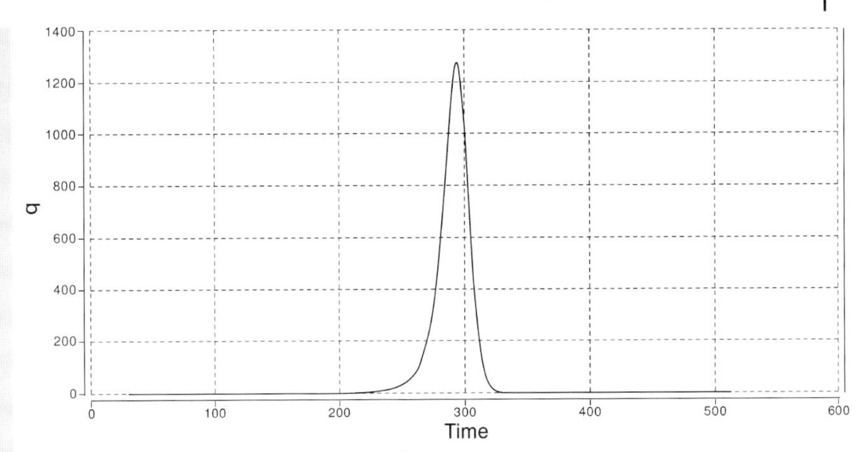

Figure 12.12 Dynamic DSC thermogram of 12.3 mg concentrate in a gold plated high pressure crucible. The scan rate is $4\,K\,min^{-1}$. The energy is $500\,J\,g^{-1}$. Temperature in °C, heat release rate in $W\,kg^{-1}$.

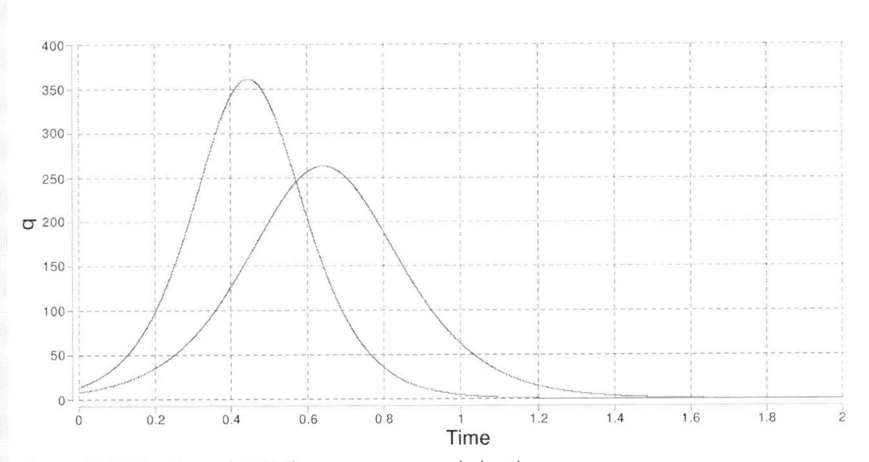

Figure 12.13 Isothermal DSC thermograms recorded with 25.2 mg sample at 240 °C and 23.4 mg at 250 °C in gold plated high pressure crucibles. For both experiments, the energy was close to $500\,J\,g^{-1}$. Time in hours and heat release rate in $W\,kg^{-1}$.

Potential: With a specific heat capacity of $1.7\,kJ\,kg^{-1}\,K^{-1}$, the adiabatic temperature rise is

$$\Delta T_{ad} = \frac{Q'}{c'_p} = \frac{500}{1.7} = 294\,K$$

Thus, the severity is "High" and the probability of triggering the decomposition must be assessed.

These differential equations can be integrated over time and give the temperature course under adiabatic conditions. The results are illustrated in Figure 12.14.

Conclusions of the worked example:

- The evaluation using the zero-order approximation neglects the induction period of the reaction and leads to a strong over-estimation of the risks: T_{D24} = 95 °C.
- The use of a kinetic model leads to more realistic prediction of the behavior of the distillation residue under adiabatic conditions: T_{D24} = 145 °C.
- In case of total failure of the system, the heat exchange system will become inactive and the vacuum pump also stops: the temperature of reactor contents will equilibrate with the reactor itself, thus the temperature would reach a level somewhat above 120 °C. The TMR_{ad} is longer than 24 hours.
- The failure of the vacuum pump stops the evaporation cooling and the temperature of the product may reach 145 °C or a TMR_{ad} of ca. 24 hours.
- In such a case, it is recommended to limit the heat carrier temperature and to install a trip between pressure (vacuum) and the heating system.
- It is also recommended to use a liquid heat carrier rather than steam, which ensures cooling when the temperature gradient is inversed.

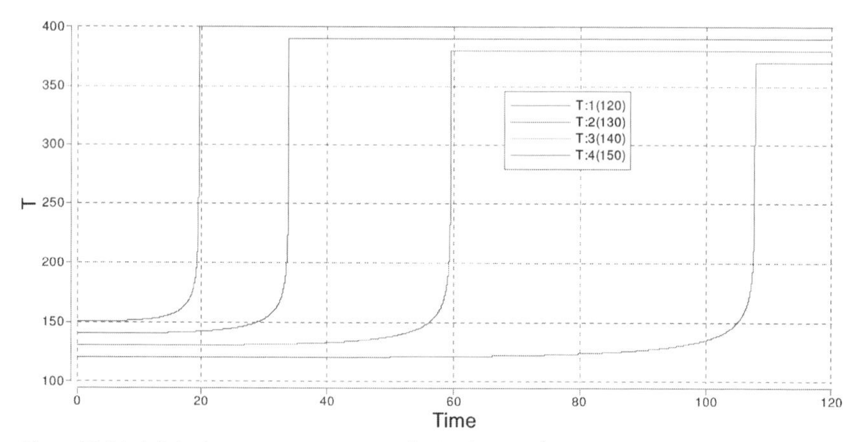

Figure 12.14 Adiabatic temperature course obtained using the kinetic data from the DSC thermograms from different starting temperatures: 120, 130, 140, and 150 °C. Temperature in °C, time in hours.

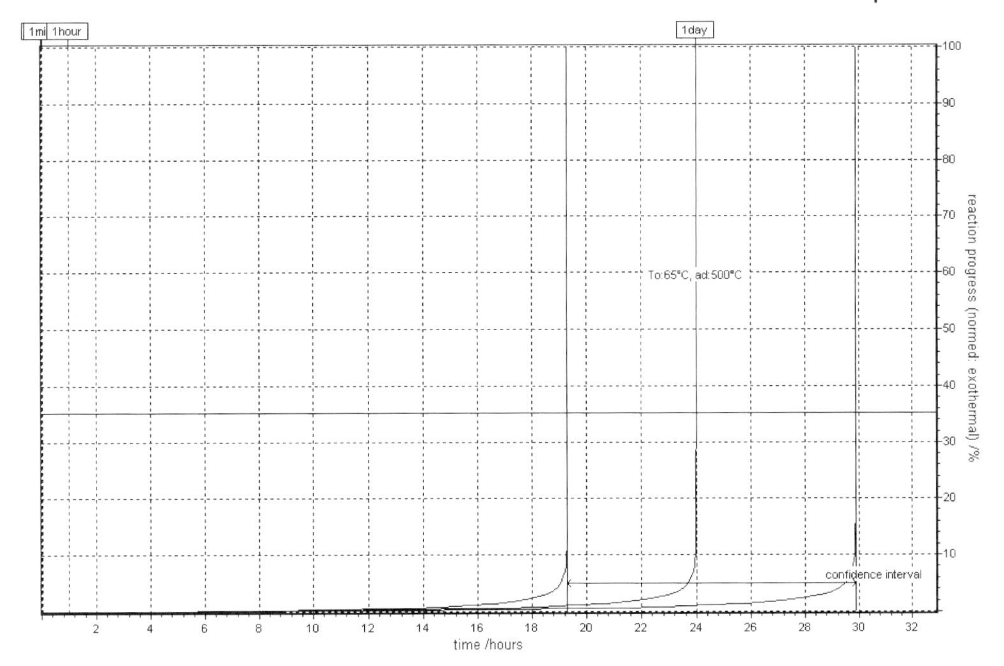

Figure 12.15 Adiabatic temperature course obtained from the AKTS software with the confidence interval for 10% relative error on the energy.

12.4.2 Assessment Procedure for Autocatalytic Decompositions

Once it is known with certainty that a decomposition follows an autocatalytic mechanism, it is recommended to proceed as follows:

Calculate the TMR_{ad} according to the conservative model mentioned above, using the results of isothermal measurements. If the resulting TMR_{ad} turns out to be critical, that is less than 24 hours (for a reaction, not for storage), the following points have to be clarified:

- Can contamination of the material or mixture in question, by acid or heavy metals, be excluded? Examples are sulfuric acid or iron in the form of rust.
- Do analytical specifications exist for the materials, mixture, or chemicals taking part in the mixture?
- Can mixing of thermally stressed material with fresh material be excluded?
- Do isothermal measurements from the past exist and if yes, does the observed isothermal induction time remain constant?

If the answer to all these questions is yes, then the autocatalytic model for the decomposition can be used with some reliability to predict the adiabatic behavior. This has to be done by specialists.

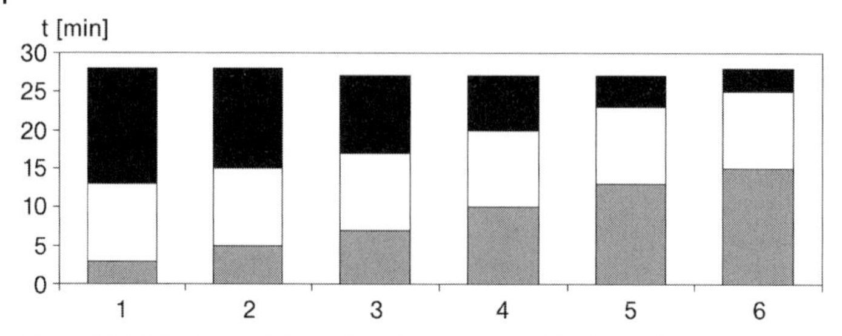

Figure 12.16 Isothermal induction time measured in 6 experiments with a first heating period at 390 °C with a variable duration, followed by a 10 minute cooling period and again heated at 390 °C.

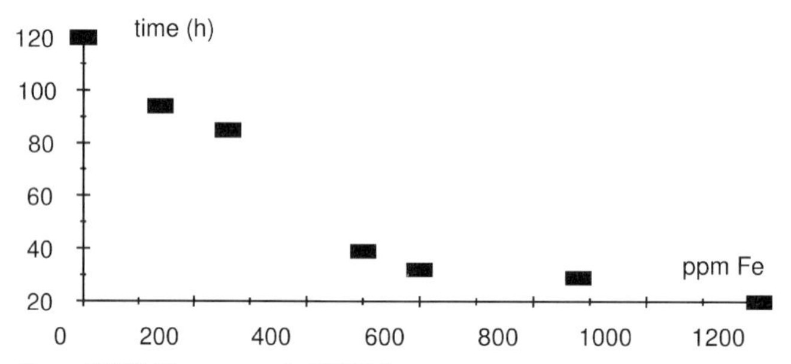

Figure 12.17 Time to reach 200 °C from process temperature in the synthesis of dihydroxydiphenylsulfone under adiabatic conditions as a function of the iron contents of the reaction mixture.

12.5 Exercises

▶ Exercise 12.1

Why is the zero-order approximation for the reaction kinetics especially conservative in the case of autocatalytic reactions?

▶ Exercise 12.2

What do you think about the statement: "For a strongly autocatalytic reaction, the isothermal induction time is close to TMR_{ad}?"

▶ Exercise 12.3

A product is to be purified by distillation: the desired product is obtained as a distillate and secondary products remain as a concentrated tar in the distillation

Table 12.2 Results of isothermal experiments.

Temperature (°C)	q'_{max} (W KG^{-1})	τ_{iso} (min)
90	51	566
100	124	234
110	287	102
120	640	45

vessel. It is known that this residue decomposes following an autocatalytic mechanism. In order to save time and handling operations, the plant manager decides to leave the residue in the distillation vessel and to charge the following batch over the residue of the previous one. The residue should be emptied only after every five batches. What do you think about this practice?

Exercise 12.4

Show that for the Prout–Tompkins model the activation energy calculated from the heat release rate and from the isothermal induction time is the same.

Exercise 12.5

A series of isothermal DSC experiments were performed on a sample. The samples were contained in pressure-resistant tight gold plated crucibles. The oven of the DSC was previously heated to the desired temperature with the reference in place. At time zero, the sample crucible was placed in the oven. The maximum heat release rate and the time at which it appeared were measured. The results are summarized in Table 12.2.

Calculate the activation energy from the maximum heat release rate and from the isothermal induction time. What conclusions could be drawn?

References

1 Grewer, T. (1994) *Thermal Hazards of Chemical Reactions*. Industrial safety series, Vol. 4, Elsevier, Amsterdam.

2 Brogli, F., Grimm, P., Meyer, M. and Zubler, H. (1980) Hazards of self-accelerating reactions, in *3rd International Symposium Loss Prevention and Safety Promotion in the Process Industries, Swiss Society of Chemical Industry*, Basel, 665–83.

3 Oswald, W. (1970) *Physikalische Organische Chemie*, Louis Plack, Hammett.

4 Gold, V., Loenig, K. and Sehmi, P. (1987) *Compendium of Chemical Terminology IUPAC Recommendations*, Blackwell Scientific Publications, Oxford.

5 Dien, J.M., Fierz, H., Stoessel, F. and Killé, G. (1994) The thermal risk of

autocatalytic decompositions: a kinetic study. *Chimia*, **48** (12), 542–50.

6 Maja-Perez, F. and Perez-Benito, J.F. (1987) The kinetic rate law for autocatalytic reactions. *Journal of Chemical Education*, **64** (11), 925–7.

7 Townsend, D.I. (1977) Hazard evaluation of self-accelerating reactions. *Chemical Engineering Progress*, **73**, 80–1.

8 Prout, E.G. and Tompkins, F.C. (1944) The thermal decomposition of potassium permanganate. *Transactions of the Faraday Society*, **40**, 488–98.

9 Chervin, S. and Bodman, G.T. (2002) Phenomenon of autocatalysis in decomposition of energetic chemicals. *Thermochimica Acta*, **392–393**, 371–83.

10 Hugo, P. (1992) Grundlagen der thermisch sicheren Auslegung von chemischen Reaktoren, in *Dechema Kurs Sicherheit chemischer Reaktoren*, Dechema, Frankfurt am Main.

11 Steinbach, J. (1999) *Safety Assessment for Chemical Processes*. VCH, Weinheim.

12 Hugo, P., Wagner, S. and Gnewikow, T. (1993) Determination of chemical kinetics by DSC measurements. Part 1: Theoretical foundations. *Thermochimica Acta*, **225** (2), 143–52.

13 Moritz, H.U. (ed.) (1995) Reaktionskalorimetrie und sicherheitstechnische Aspekte von Polyreaktionen, in *Praxis der Sicherheitstechnik; Sichere Handhabung chemischer Reaktionen*, Vol. 3 (eds G. Kreysa and O.-U. Langer), Frankfurt a.M., Dechema, pp. 115–73.

14 Moritz, H.U. (1989) Polymerisation calorimetry- a powerful tool for reactor control, in *Third Berlin International Workshop on Polymer Reaction Engineering*, VCH, Weinheim, Berlin.

15 Gygax, R. (1993) *Thermal Process Safety, Data Assessment, Criteria, Measures*, Vol. 8 (ed. ESCIS), ESCIS, Lucerne.

16 Hall, J. (1993) Hazards involved in the handling and use of dimethyl sulphoxide (DMSO). *Loss Prevention Bulletin*, **114** (12), 9–14.

17 Bou-Diab, L. and Fierz, H. (2002) Autocatalytic decomposition reactions, hazards and detection. *Journal of Hazardous Materials*, **93** (1), 137–46.

18 Townsend, D.I. and Tou, J.C. (1980) Thermal Hazard evaluation by an accelerating rate calorimeter. *Thermochimica Acta*, **37**, 1–30.

19 Roduit, B., Borgeta, C., Berger, B., Folly, P., Alonso, B., Aebischer, J.-N. and Stoessel, F. (2004) Advanced kinetic tools for the evaluation of decomposition reactions, in *ICTAC*, North American Thermal Analysis Society (NATAS), Sacramento.

20 Dien, J.M. (1995) *Contribution à l'étude de la sécurité des procédés chimiques conduisant à des décompositions autocatalytiques*. Université de Haute Alsace, Mulhouse.

13
Heat Confinement

Case History

A solid product is formulated by blending different solids, the blend being granulated and dried before packing into 25-kg bags. This process was run for several years in a relatively small plant unit, processing in one pass, that is, without an intermediate storage operation. The product becoming more in demand, it had to be processed in a larger unit, but the process required intermediate storage before being grinded. This storage was made in 3 m^3 mobile containers that were stored in the basement of the building before transporting them to a silo located on the upper floor. During the first warm weekend in May, in the night of Saturday to Sunday, one of the containers began emitting fumes. The fire brigade isolated the container, but it was the first that had been filled and was located close to the wall behind all the others. They eventually were able to bring it outside the building and flood it with water to stop the start of a runaway reaction. In the next night, of Sunday to Monday, a second container began emitting fumes. It was then decided to empty all the containers onto plastic tarpaulins on the floor outside the building to allow the contents to cool.

Lessons drawn

- Heat transfer in large volumes of solid is poor, thus a reactive solid may slowly raise the temperature to a level where runaway is unavoidable.
- Appreciating such heat confinement situations requires specialist knowledge.

13.1
Introduction

In this section, different typical heat accumulation situations encountered in the process industry are reviewed and analysed. The next section introduces different types of heat balance used in assessment of heat confinement situations.

Thermal Safety of Chemical Processes: Risk Assessment and Process Design. Francis Stoessel

ISBN: 978-3-527-31712-7

A special section then deals with the use of time-scales for the heat balance, which provides a simple to use assessment technique. The third section is devoted to the heat balance with purely conductive heat removal. The chapter closes on practical aspects for the assessment of industrial heat confinement situations.

13.2
Heat Accumulation Situations

In the chapters devoted to reactors, it was considered that a situation is thermally stable due to the relatively high heat removal capacity of reactors compensating for the high heat release rate of the reaction. We considered that in the case of a cooling failure, adiabatic conditions were a good approximation for the prediction of the temperature course of a reacting mass. This is true, in the sense that it represents the worst case scenario. Between these two extremes, the actively cooled reactor and adiabatic conditions, there are situations where a small heat removal rate may control the situation, when a slow reaction produces a small heat release rate. These situations with reduced heat removal, compared to active cooling, are called heat accumulation conditions or thermal confinement.

Thermal confinement situations are encountered as "nominal" conditions in storage and transport of reactive material. The may also happen in failure of production equipment, such as loss of agitation, pump failure, and so on.

In practice, truly adiabatic conditions are difficult to realize (see also Chapter 4), being seldom encountered and then only for short time periods. Therefore, considering a situation to be adiabatic may lead to a pointlessly severe assessment that may lead to abandoning a process, which in fact would have been possible to carry through safely, if a more realistic judgement had been made.

As an introduction, it is worth qualitatively analysing some common industrial situations. In an analogy to the two film theory (Section 9.3.1), we consider three contributions to the resistance against heat transport [1]:

1. Agitated jacketed vessel: the main resistance to heat transfer is located at the wall, where there is practically no resistance to heat transfer inside the reaction mass. Due to agitation, there is no temperature gradient in the reactor contents. Only the film near the wall presents a resistance. The same happens outside the reactor in its jacket, where the external film presents a resistance. The wall itself also presents a resistance. In summary, the resistance against heat transfer is located at the wall.
2. Unstirred storage tank containing a liquid: the main resistance to heat transfer is located at the outside of the wall: without agitation, natural convection will equalize the contents temperature. The wall itself, since it is not insulated, represents a weak resistance to heat transfer. The outside air film with natural convection represents a comparatively larger resistance.

Table 13.1 Effect of the container size on heat accumulation.

Heat release rate ($W\,kg^{-1}$)	T (°C)	Mass 0.5 kg	Mass 50 kg	Mass 5000 kg	Adiabatic
10	129	191 °C after 0.9 h	200 °C after 0.9 h	200 °C after 0.9 h	200 °C after 0.9 h
1	100	5.8 °C after 8 h	200 °C after 7.4 h	200 °C after 7.4 h	200 °C after 7.4 h
0.1	75	0.5 °C after 12 h	13.2 °C after 64 h	200 °C after 64 h	200 °C after 64 h
0.01	53		0.7 °C after 154 h	165 °C after 632 h	200 °C after 548 h

3. Storage silo containing a solid: the main resistance to heat transfer is located in the bulk of the product contained in the silo. The resistance of the wall and the external film are low compared to the conductive resistance of the solid.

In these examples, the degree of confinement increases from the first to last case. In general, when the heat transfer mechanism is conduction, it contributes significantly to the resistance. Thus, the nature of the reacting mass and its contents must be considered first in the assessment of a confinement situation.

Besides this, the thermal behavior of the reacting mass and the dimensions of its containment play a key role in the analysis. This is illustrated in the example given in Table 13.1. Here the ambient temperatures were chosen in such a way that the heat release rate differed by one order of magnitude in each line [1]. To simplify the calculations, the containers were considered spherical. The heat accumulation conditions increase with the size of the container, that is, from left to right.

It can be observed, in one line, that under severe heat accumulation conditions, there is no difference in the time-scale that corresponds to the time to maximum rate under adiabatic conditions (TMR_{ad}). Thus, severe heat accumulation conditions are close to adiabatic conditions. At the highest temperature, even the small container experienced a runaway situation. Even at this scale, only a small fraction of the heat release rate could be dissipated across the solid: the final temperature was only 191 °C instead of 200 °C. For small masses, the heat released is only partly dissipated to the surroundings, which leads to a stable temperature profile with time. Finally, it must be noted that for large volumes, the time-scale on which the heat balance must be considered is also large. This is especially critical during storage and transport.

13.3 Heat Balance

In this section, we reconsider the heat balance from the specific viewpoint of heat accumulation. Three different mechanisms of heat transfer are considered: forced convection, natural convection, and conduction. Before considering these mechanisms, we introduce the heat balance by time-scales.

13.3.1 Heat Balance Using Time Scale

A practical approach of heat balance, often used in assessment of heat accumulation situations, is the time-scale approach. The principle is as in any race: the fastest wins the race. For heat production, the time frame is obviously given by the time to maximum rate under adiabatic conditions. Then the removal is also characterized by a time that is dependent of the situation and this is defined in the next sections. If the TMR_{ad} is longer than the cooling time, the situation is stable, that is, the heat removal is faster. At the opposite, when the TMR_{ad} is shorter than the characteristic cooling time, the heat release rate is stronger than cooling and so runaway results.

13.3.2 Forced Convection, Semenov Model

In an agitated vessel, with heat transfer through the wall, the heat removal is given by

$$q_{ex} = U \cdot A(T_c - T) \tag{13.1}$$

Compared with the heat release rate by a reaction, which follows Arrhenius law, one obtains the Semenov diagram (Figure 2.6). From this diagram, we can calculate the critical temperature difference (Equations 2.32–2.34). But this also calculates the critical heat release rate as a function of q_0:

$$q_{crit} = q_0 \cdot e^{\frac{-E}{R}\left(\frac{1}{T_{crit}} - \frac{1}{T_0}\right)} \tag{13.2}$$

and placing into the heat balance, we find:

$$\rho \cdot V \cdot Q \cdot k_0 \cdot e^{\frac{-E}{R}\left(\frac{1}{T_{crit}} - \frac{1}{T_0}\right)} = U \cdot A \cdot \Delta T_{crit} \tag{13.3}$$

Since, from Equation 2.34, $\Delta T_{crit} = \dfrac{E}{RT_0^2}$, a simplification can be introduced:

$$\frac{-E}{R}\left(\frac{1}{T_{crit}} - \frac{1}{T_0}\right) \approx \frac{-E}{R}\left(\frac{T_0 - T_{crit}}{T_0^2}\right) = 1 \tag{13.4}$$

and the balance becomes

$$Q \cdot \rho \cdot V \cdot k_0 \cdot e = U \cdot A \cdot \frac{RT_0^2}{E} \tag{13.5}$$

Dividing by $\rho \cdot V \cdot c_P'$, this expression can be rearranged:

$$k_0 \cdot e \cdot \Delta T_{ad} = \frac{U \cdot A}{\rho \cdot V \cdot c_P'} \cdot \frac{RT_0^2}{E} \tag{13.6}$$

where we recognize the inverse of the thermal time constant

$$\tau = \frac{\rho \cdot V \cdot c_P'}{U \cdot A},$$

by introducing the thermal half-life

$$t_{1/2} = \ln(2) \cdot \tau$$

and noticing that

$$\frac{1}{k_0 \cdot \Delta T_{ad}} \cdot \frac{RT_0^2}{E} = TMR_{ad} \tag{13.7}$$

we obtain

$$TMR_{ad} = \frac{e}{\ln(2)} \cdot t_{1/2} = 3.92 \cdot t_{1/2} \tag{13.8}$$

Therefore a stable situation corresponds to the following condition:

$$TMR_{ad} > 3.92 \cdot t_{1/2} \tag{13.9}$$

This expression compares the characteristic time of runaway (TMR_{ad}) with the characteristic cooling time. Thus, knowing the mass, specific heat capacity, heat transfer coefficient, and heat exchange area allows the assessment. It is worth noting that, since the thermal time constant contains the ratio V/A, heat losses are proportional to the characteristic dimension of the container.

13.3.3 Natural Convection

When a liquid warms up, its density decreases, which results in buoyancy and an ascendant flow is induced. Thus, a reactive liquid will flow upwards in the center of a container and flow downwards at the walls, where it cools: this flow is called natural convection. Thus, at the wall, heat exchange may occur to a certain degree. This situation may correspond to a stirred tank reactor after loss of agitation. The exact mathematical description requires the simultaneous solution of heat and impulse transfer equations. Nevertheless, it is possible to use a simplified approach based on physical similitude. The mode of heat transfer within a fluid can be characterized by a dimensionless criterion, the Rayleigh number (*Ra*). As the Reynolds number does for forced convection, the Rayleigh number characterizes the flow regime in natural convection:

$$Ra = \frac{g \cdot \beta \cdot \rho^2 \cdot c'_P \cdot L^3 \cdot \Delta T}{\mu \cdot \lambda} \tag{13.10}$$

For convection along a vertical plate, $Ra > 10^9$ indicates that turbulent flow is established and heat transfer by convection dominates. For smaller values of the Rayleigh number, $Ra < 10^4$, the flow is laminar and conduction dominates. Thus, the Rayleigh number discriminates between conduction and convection [2].

For natural convection, a correlation was established between the Nusselt criterion, which compares convective and conductive resistances to heat transfer and the Rayleigh criterion, which compares buoyancy forces with viscous friction:

$$Nu = C^{te} \cdot Ra^m \tag{13.11}$$

with

$$Nu = \frac{h \cdot L}{\lambda}$$

The Rayleigh criterion can also be written as a function of the Grashof criterion, which compares convective with conductive heat transfer and the Prandtl criterion, which compares the momentum diffusivity (kinematic viscosity) with the thermal diffusivity:

$$Ra = Gr \cdot Pr$$

with

$$Pr = \frac{\nu}{a} = \frac{\mu \cdot c'_P}{\lambda} \tag{13.12}$$

$$Gr = \frac{g \cdot \beta \cdot L^3 \cdot \rho^2 \cdot \Delta T}{\mu^2}$$

For natural convection along a vertical surface, the following correlations can be used [3]:

$$\begin{array}{lll} Ra > 10^9 & \text{turbulent flow} & Nu = 0.13 \cdot Ra^{1/3} \\ 10^4 < Ra < 10^9 & \text{intermediate flow} & Nu = 0.59 \cdot Ra^{1/4} \\ Ra < 10^4 & \text{laminar flow} & Nu = 1.36 \cdot Ra^{1/6} \end{array} \quad (13.13)$$

In practice, the calculation of the Rayleigh criterion essentially serves to decide if a turbulent flow occurs in the film along the wall. If turbulent flow occurs, heat exchange by natural convection is likely to take place. With high containers, layering may occur, meaning that the upper zone of the container contents keep a higher temperature than in the bottom. Therefore, the length (L) used in the Rayleigh criterion must be chosen as rather short (typical value: 1 m). If the physical properties used for the calculation of the film heat transfer coefficient in agitated vessels (γ) are known, the Rayleigh criterion can be rearranged to make this coefficient appear as

$$h = 0.13 \cdot \gamma \cdot \sqrt[3]{\beta \cdot \Delta T}$$

with

$$\gamma = \sqrt[3]{\frac{\rho^2 \cdot \lambda^2 \cdot c_P' \cdot}{\mu}} \quad (13.14)$$

In general, and to give an order of magnitude, the film heat transfer coefficient with natural convection is approximately 10% of the heat transfer coefficient with agitation [4].

13.3.4 High Viscosity Liquids and Solids

Here we consider the case of a viscous or even solid reactive material contained in a vessel of known geometry. In this case, heat transfer takes place by pure conduction: there is no flow within the reactive material. The situation is stable, when the heat losses by conduction compensate for the heat release in the material. Thus, the following questions must be answered: "Under which conditions may a thermal explosion (runaway) be triggered? Under which conditions is the heat transfer by conduction sufficient to compensate for the heat release?"

Conductive heat transfer does not require any motion of atoms or molecules, as only the interactions between atoms or molecules transfers heat. The heat flux expressed in $W m^{-2}$ can be described using Fourier's law:

$$\vec{q} = -\lambda \cdot \vec{\nabla} T \quad (13.15)$$

This equation expresses the proportionality of the heat flux to the temperature gradient in the material considered. The transfer takes place in the opposite

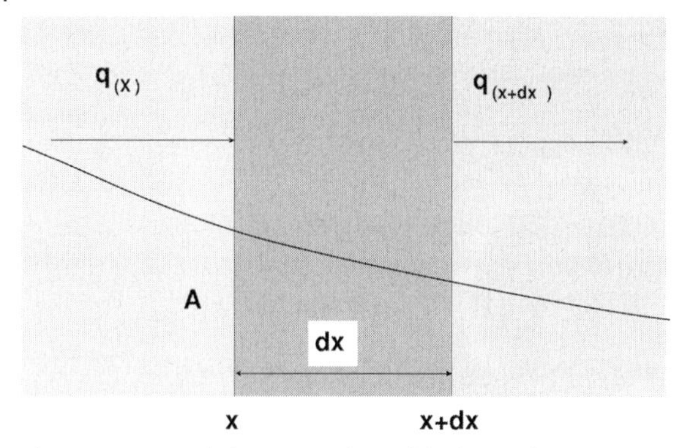

Figure 13.1 Heat balance in a slice of thickness *dx*.

direction to the temperature gradient and the proportionality constant is the thermal conductivity of the material, λ in $W\,m^{-1}\,K^{-1}$. If we consider the heat transfer along one axis (one-dimensional problem), the expression becomes

$$\vec{q} = -\lambda \cdot \frac{\vec{\partial} T}{\partial x} \tag{13.16}$$

If we consider the heat balance on a slice of thickness dx and section A, the heat accumulated in the slice is equal to the difference of the entering flux minus the leaving flux (Figure 13.1):

$$(\vec{q}_{x+dx} - \vec{q}_x) \cdot A \cdot dt = \frac{\partial \vec{q}_x}{\partial x} A \cdot dx \cdot dt = -\lambda \cdot \left[\frac{\left.\frac{\partial T}{\partial x}\right|_{x+dx} - \left.\frac{\partial T}{\partial x}\right|_{x}}{\partial x} \right] \cdot A \cdot dx \cdot dt \tag{13.17}$$

$$\text{Thus: } \frac{\partial q_x}{\partial x} = -\lambda \cdot \left.\frac{\partial^2 T}{\partial x^2}\right|_x$$

Applying the first principle of thermodynamics, we find that the rate of temperature change is

$$q = \rho \cdot c'_P \cdot \frac{\partial T}{\partial t} \cdot A \cdot dx \tag{13.18}$$

By combining Equations 13.17 with 13.18, and assuming a constant thermal conductivity, we obtain the second Fourier law:

$$\frac{\partial^2 T}{\partial x^2} = \frac{\rho \cdot c'_P}{\lambda} \cdot \frac{\partial T}{\partial t} = \frac{1}{a} \cdot \frac{\partial T}{\partial t} \tag{13.19}$$

Here

$$a = \frac{\lambda}{\rho \cdot c'_P}$$

and a represents the thermal diffusivity expressed in $m^2 s^{-1}$, which has the same dimensions as a diffusion coefficient used in Fick's law. The mathematical similarity of both laws is worth noting: the mathematical treatment is exactly the same.

A dimensionless criterion, the Biot number, is often used in transient heat transfer problems by comparing the heat transfer resistance within the body with the resistance at its surface:

$$Bi = \frac{h \cdot r_0}{\lambda} \tag{13.20}$$

A high Biot number means that the conductive transfer is small compared to convection and the situation is close to that considered by a Frank-Kamenetskii situation (Section 13.4.1). Inversely, a small Biot number, that is $Bi < 0.2$, means that the convective heat transfer dominates and the situation is close to a Semenov situation.

Conductive problems described by Equation 13.19 can be solved algebraically or graphically using nomograms based on dimensionless coordinates, where the dimensionless time is given by the Fourier number:

$$Fo = \frac{a \cdot t}{r^2} \tag{13.21}$$

The dimensionless temperature profile is described as a function of the Biot number and Fourier numbers [3, 5].

13.4 Heat Balance with Reactive Material

The problem of conductive heat transfer in an inert solid can be solved algebraically, when there is no heat source in the solid. Nevertheless, this problem is not within the scope of our considerations about thermal confinement, since we are interested in the thermal behavior of a reactive solid, that is, a solid comprising a heat source in itself, which requires specific mathematical treatment.

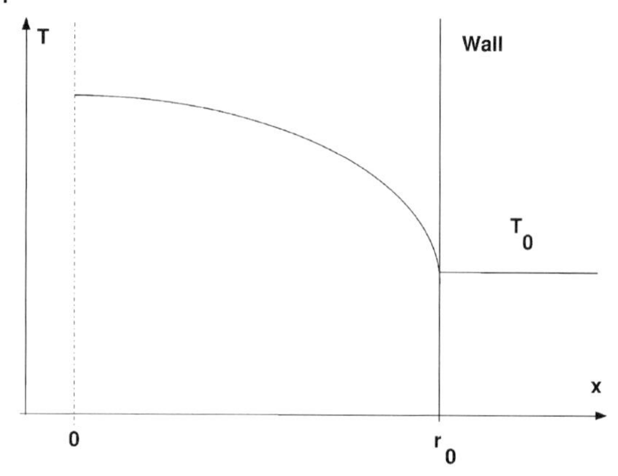

Figure 13.2 Temperature profile in a reactive solid.

13.4.1
Conduction in a Reactive Solid with a Heat Source, Frank-Kamenetskii Model

This problem was addressed and solved by Frank-Kamenetskii [6], who established the heat balance of a solid with a characteristic dimension *r*, an initial temperature T_0 equal to the surrounding temperature, and containing a uniform heat source with a heat release rate *q* expressed in $W \cdot m^{-3}$. The object is to determine under which conditions a steady state, that is, a constant temperature profile with time, can be established. We further assume that there is no resistance to heat transfer at the wall, that is, there is no temperature gradient at the wall. The second Fourier Law can be written as (Figure 13.2)

$$\lambda \frac{\partial^2 T}{dx^2} = q_{rx} \tag{13.22}$$

The border conditions are

$$T = T_0 \text{ at the wall } x = r_0$$

$$\frac{\partial T}{\partial x} = 0 \text{ at the center } x = 0 \text{ (symetry)} \tag{13.23}$$

The solution of this differential equation, when it exists, describes the temperature profile in the solid. In order to solve this equation, we must assume that the exothermal reaction taking place in the solid follows a zero-order rate law, that is, the reaction rate is independent of the conversion. Then the variables can be changed to dimensionless coordinates. Thus, generalizing the solutions:

Temperature:

$$\theta = \frac{E(T - T_0)}{RT_0^2} \tag{13.24}$$

Space coordinate:

$$z = \frac{x}{r_0} \tag{13.25}$$

The differential equation then becomes

$$\nabla_z^2 \theta = -\delta \cdot e^{\theta} \tag{13.26}$$

$$\delta = \frac{\rho_0 \cdot q_0}{\lambda} \cdot \frac{E}{RT_0^2} \cdot r_0^2 \tag{13.27}$$

The parameter δ is called the form factor or the Frank-Kamenetskii number. When a solution of Equation 13.26 exists, a stationary temperature profile can be established and the situation is stable. When there is no solution, no steady state can be established and the solid enters a runaway situation. The existence, or not, of a solution to the differential Equation 13.26, depends on the value of parameter δ, which therefore is a discriminator. The differential equation can be solved for simple shapes of the solid body, for which the Laplacian can be defined:

Slab of thickness $2r_0$ and infinite surface:

$$\frac{d^2\theta}{dz^2} = -\delta \cdot e^{\theta} \quad \delta_{crit} = 0.88 \quad \text{and} \quad \theta_{max,crit} = 1.19 \tag{13.28}$$

Cylinder of radius r_0 and infinite length:

$$\frac{d^2\theta}{dz^2} + \frac{1}{z}\frac{d\theta}{dz} = -\delta \cdot e^{\theta} \quad \delta_{crit} = 2.0 \quad \text{and} \quad \theta_{max,crit} = 1.39 \tag{13.29}$$

Sphere of radius r_0:

$$\frac{d^2\theta}{dz^2} + \frac{2}{z}\frac{d\theta}{dz} = -\delta \cdot e^{\theta} \quad \delta_{crit} = 3.32 \quad \text{and} \quad \theta_{max,crit} = 1.61 \tag{13.30}$$

Hence for a given container shape, Equation 13.27 can be solved for r. This means that the critical radius can be calculated using the geometric characteristics of the vessel, the physical properties of the contents, and the kinetic characteristics of the reaction taking place in the solid:

$$r_{crit} = \sqrt{\frac{\delta_{crit} \cdot \lambda \cdot RT_0^2}{\rho \cdot q_0' \cdot E}} \tag{13.31}$$

This expression is useful in practice, since it calculates the greatest dimension of the vessel, allowing a stable temperature profile for a given surrounding

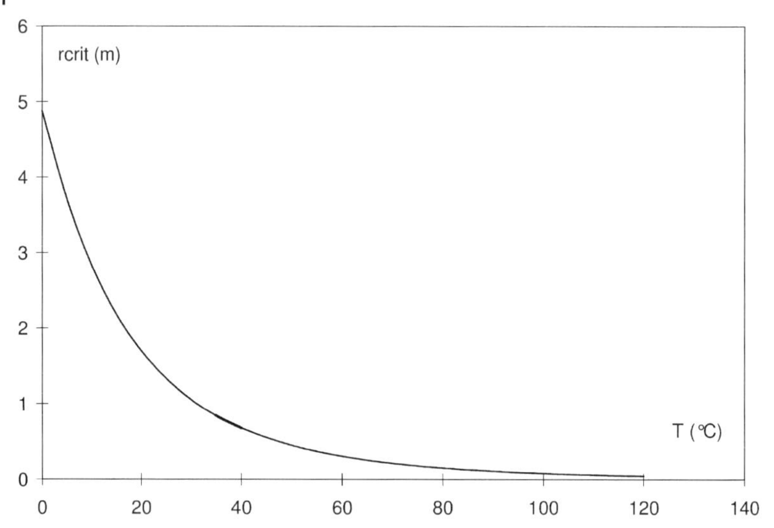

Figure 13.3 Critical radius as a function of temperature. This curve was calculated for a q'_{ref} of $10\,W\,kg^{-1}$ at 150 °C, $E = 75\,kJ\,mol^{-1}$, $c_P' = 1.8\,J\,kg^{-1}\,K^{-1}$, $\rho = 1000\,kg\,m^{-3}$, $\lambda = 0.1\,W\,m^{-1}\,K^{-1}$ and $\delta_{crit} = 2.37$.

temperature T_0 (Figure 13.3), or knowing the dimension of the vessel r, the highest surrounding temperature can be calculated. In the second case, an iterative solution is required since q_0' is an exponential function of temperature (Arrhenius). Due to the strong non-linearity of $q_0' = f(T)$ (Equation 13.31), the system is parametrically sensitive, that is, for small variations of one of the parameters the system may "switch" from stable to runaway. Consequently, here again we find the parametric sensitivity of the systems that is characteristic for thermal explosion phenomena: for each system, there is a limit beyond which the system becomes unstable and enters a runaway situation. In the example given in Figure 13.4, the container is a sphere of 0.2 m radius filled with a solid having a heat release rate of $10\,W\,kg^{-1}$ at 150 °C and an activation energy of $160\,kJ\,mol^{-1}$.

As for agitated systems, we may express the heat balance in terms of characteristic times, comparing the characteristic time of the reaction (TMR_{ad}) with the characteristic cooling time of the solid:

$$r_{crit}^2 = \frac{\delta_{crit} \cdot \lambda}{\rho \cdot c_P'} \cdot TMR_{ad} = \delta_{crit} \cdot a \cdot TMR_{ad} \tag{13.32}$$

For simple geometric shapes, the stability criteria can be expressed as

$$\overbrace{TMR_{ad} > \frac{0.3 \cdot r^2}{a}}^{sphere} \quad \overbrace{TMR_{ad} > \frac{0.5 \cdot r^2}{a}}^{infinite\ cylinder} \quad \overbrace{TMR_{ad} > \frac{1.14 \cdot r^2}{a}}^{slab} \tag{13.33}$$

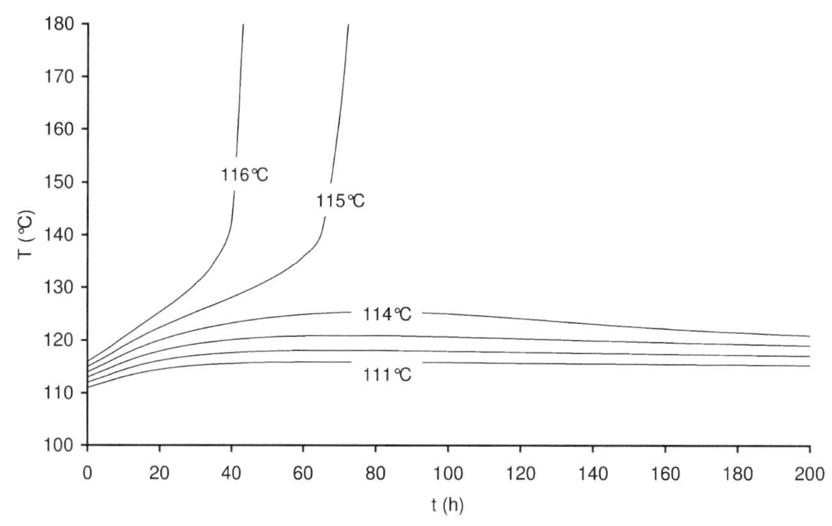

Figure 13.4 Temperature at the center of a solid submitted to heat confinement for 6 different surrounding temperatures: 111, 112, 113, 114, 115, and 116 °C.

It is worth noting that the dimension of the container (r) appears squared in the equations: in other words, the heat losses decrease with the square of the dimension. This is different from agitated vessels, where it increases proportional to r (see Sections 13.4.1 and 13.3.2). Further on, the temperature of the outermost element of the reactive solid, that is, the element closest to the wall, retains a temperature that is close to the surrounding temperature. This has a practical meaning: self-heating of the bulk material cannot be detected by measuring the wall temperature. Moreover, it is useless to cool the wall if a runaway is detected, since practically no heat can be removed from the reacting solid by this means. This is the reason for which, in the case history presented at the beginning of the chapter, the contents of the containers were spread on tarpaulins on the floor to allow cooling: this changed the characteristic dimension of the solid that became a slab with a small thickness presenting improved heat losses.

For the treatment of practical cases, it is often necessary to assess other shapes other than a slab, infinite cylinder, or sphere. In such a case, it is possible to calculate the Frank-Kamenetskii criterion for some commonly used shapes. For a cylinder of radius r and height h, the critical value of the Frank-Kamenetskii criterion is given by [7]

$$\delta_{crit} = 2.0 + 3.36\left[\frac{h}{r}(ad+1)\right]^{-2} \tag{13.34}$$

In this expression, the parameter ad is equal to 1, if the bottom is adiabatic, and zero in other cases. The radius of a sphere that is thermally equivalent to a cylinder is

$$r_{sph} = r_{cyl}\sqrt{\frac{3.32}{\delta_c}} \tag{13.35}$$

In the specific case of "chemical drums" with a height equal to three times the radius, the Frank-Kamenetskii criterion is $\delta_{crit} = 2.37$ [1]. A cube with a side length $2r_0$, can be converted to its thermally equivalent sphere. The Semenov number then becomes

$$\delta_{crit} = \frac{3.32}{\left(\frac{r_{sph}}{r_0}\right)^2} \tag{13.36}$$

The radius of the thermally equivalent sphere is r_{sph}. Different analogies used for this conversion are summarized in Table 13.1.

In Table 13.2, the best approximation of the cube is obtained with a sphere of radius $r_{sph} = 1.16 \cdot r_0$. The Frank-Kamenetskii number then is 2.5 for a cube with a side length 2 r_0.

Table 13.2 Approximation of a cube by a thermally equivalent sphere.

Analogy	Equivalent sphere radius	Justification
Inscribed cube	$R = r_0$	sub critical $V < V_{\text{sphere}}$
Circumscribed cube	$R = 1.73\ r_0$	over critical $V > V_{\text{sphere}}$
Cube with identical surface	$R = 1.38\ r_0$	over critical $V > V_{\text{sphere}}$
Cube with identical volume	$R = 1.24\ r_0$	over critical $S > S_{\text{sphere}}$
Approximation	$R = 1.16\ r_0$	cube with side length a = 2 r_0

13.4.2
Conduction in a Reactive Solid with Temperature Gradient at the Wall, Thomas Model

In the Frank-Kamenetskii model, the surroundings temperature is set equal to the temperature of the reacting solid. Thus, there is only a small temperature gradient between this element and the wall, so only a limited heat transfer to the surroundings. This simplification establishes the above described criteria, but it is not really representative of a certain number of industrial situations. In fact, there are numerous situations where the surrounding temperature different from the initial product temperature, for example, discharging of a hot product from a dryer to a container placed at ambient temperature, and so on. Therefore, Thomas [7] developed a model that accounts for heat transfer at the wall. He added a convective term to the heat balance:

$$\lambda \frac{dT}{dx} + h(T_s - T_0) = 0 \quad \text{at} \quad x = r_0 \tag{13.37}$$

In this equation, h is the convective heat transfer coefficient at the external side of the wall and T_s the temperature of the surroundings. In this equation, the heat capacity of the wall is ignored, which is justifiable in most industrial situations. The border conditions are the same as for the Frank-Kamenetskii model, that is, the problem is considered symmetrical:

$$\frac{dT}{dx} = 0 \quad \text{at} \quad x = 0 \tag{13.38}$$

Dimensionless variables are introduced:

$$z = \frac{x}{r_0}$$
$$\theta = \frac{E(T - T_0)}{RT_0^2} \tag{13.39}$$

For a zero-order kinetic law, we obtain:

$$\nabla_z^2 \theta = \frac{d^2\theta}{dz^2} + \frac{k}{z}\frac{d\theta}{dz} = \frac{d\theta}{\tau} - \delta e^{\theta} \tag{13.40}$$

The variable τ is the dimensionless thermal relaxation time, or Fourier number:

$$\tau = \frac{a \cdot t}{r_0^2} = \frac{\lambda \cdot t}{\rho \cdot c_P' \cdot r_0^2} \tag{13.41}$$

Thomas showed that this equation has solutions for values of δ below a limit δ_{crit}:

$$\delta_{crit} = \frac{1 + k}{e \cdot \left(\frac{1}{\beta_\infty} - \frac{1}{Bi} \right)} \tag{13.42}$$

The parameter β_∞ is called the effective Biot number and k is a shape coefficient defined as:

- $k = 0$ for the infinite slab of thickness $2r_0$ with $\beta_\infty = 2.39$
- $k = 1$ for the infinite cylinder of radius r_0 with $\beta_\infty = 2.72$
- $k = 2$ for the sphere of radius r_0 with $\beta_\infty = 3.01$

The Frank-Kamenetskii number, or parameter δ, characterizing the reaction is

$$\delta = \frac{\rho_0 \cdot q_0'}{\lambda} \cdot \frac{E}{RT_0^2} \cdot r_0^2 \tag{13.43}$$

A situation can be evaluated by calculating both parameters δ and δ_{crit}: if $\delta > \delta_{crit}$, runaway takes place. Inversely, for $\delta < \delta_{crit}$, a stable temperature profile will be established. It is also possible to define the highest temperature allowing a stable situation in a given container, by searching for at which temperature T_0, both parameters are equal. This equation must be solved by iterations since it is transcendental: q_0' is an exponential function of the temperature.

13.4.3
Conduction in a Reactive Solid with Formal Kinetics, Finite Elements Model

The solutions of conductivity problems shown in the previous sections were obtained for zero-order kinetics. When the approximation by zero-order kinetics is not justified, which is the case, especially for autocatalytic reactions, a numerical solution is required. Here the use of finite elements is particularly efficient. The geometry of the container is described by a mesh of cells and the heat balance is established for each of these cells (Figure 13.5). The problem is then solved by iterations. As an example, a sphere can be described by a succession of concentric shells (like onion skins). In each cell, a mass and a heat balance are established. This gives access to the temperature profile if one considers the temperature of the different cells, or the temperature and conversion may be obtained as a function of time.

Moreover, the conversion in the different cells, which can be an important parameter for assessing the "quality loss" of the product, is also obtained. An example of such a calculation is given in Figure 13.6. In the left part of the figure, starting from an initial temperature of 124.5 °C, self heating of the reactive substance is observed with a maximum temperature of ca. 160 °C. This would mean a loss of quality, but the temperature stabilizes again. In the right part of the figure, starting from an initial temperature of 124.75 °C, that is, only 0.25 °C above the previous temperature, leads to a thermal explosion, which may have serious con-

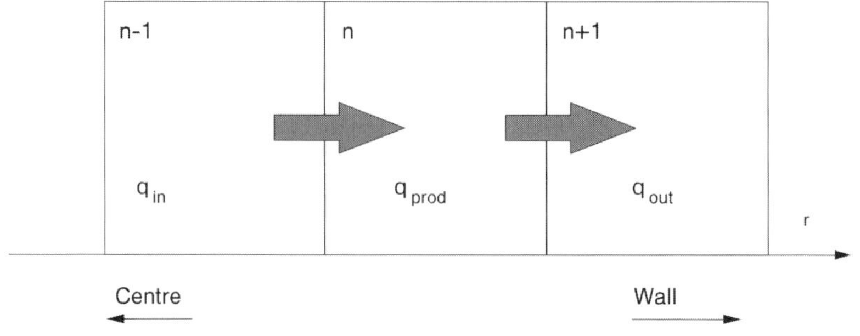

Figure 13.5 Finite elements with heat balance.

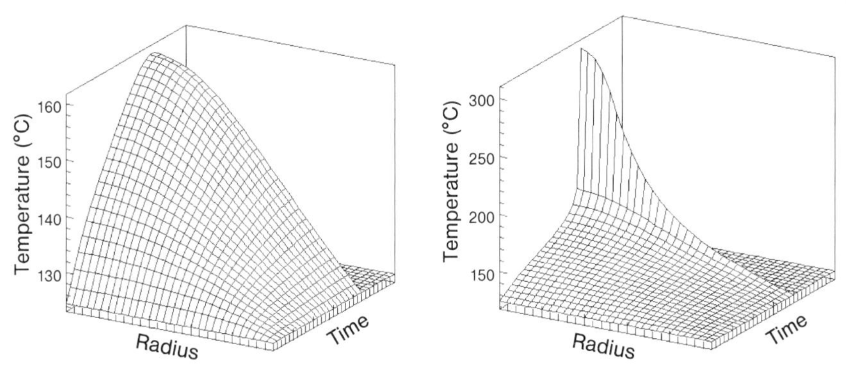

Figure 13.6 Temperature profiles obtained by the finite elements method. The left-hand figure is obtained with an initial temperature of 124.5 °C, the right-hand figure with an initial temperature of 124.75 °C.

sequences. Here again the parametric sensitivity is enhanced. In this figure, it is also worth noting that the elements close to the container wall do not show any significant temperature increase: an ongoing thermal explosion would not be detected by monitoring the wall temperature. Such problems may be solved using advanced kinetic methods as developped by Roduit [8].

13.5 Assessing Heat Accumulation Conditions

As in every safety study, the solution of heat accumulation problems can be undertaken by applying the two commonly used principles of simplification and worst-case approach, as described in Section 3.4.1. A typical procedure following these principles is illustrated with the example of a decision tree for the assessment of the risks linked with thermal confinement (Figure 13.7).

The first step is to assume adiabatic conditions. This is obviously the most penalizing assumption since it is assumed that no heat loss occurs. Under these conditions, the confinement time is compared with the characteristic time of a runaway reaction under adiabatic conditions, that is, the TMR_{ad}. If the TMR_{ad} is significantly longer than the confinement time, for example, the intended storage time, the situation is stable and the analysis can be stopped at this stage. The only data required for this assessment step is the TMR_{ad} as a function of temperature. In such a case, heat losses improve the situation. Since the heat losses were ignored, they are not required to ensure safe conditions. If a situation is found to be safe at this stage, depending on the energy potential of the reactive substance, temperature monitoring is recommended. If the condition is not fulfilled, that is, the TMR_{ad} is too close or even shorter than the confinement time; a more detailed data set is required.

In the second step, the question is to check the stability of an agitated system by comparing the TMR_{ad} with the characteristic cooling time of an agitated system.

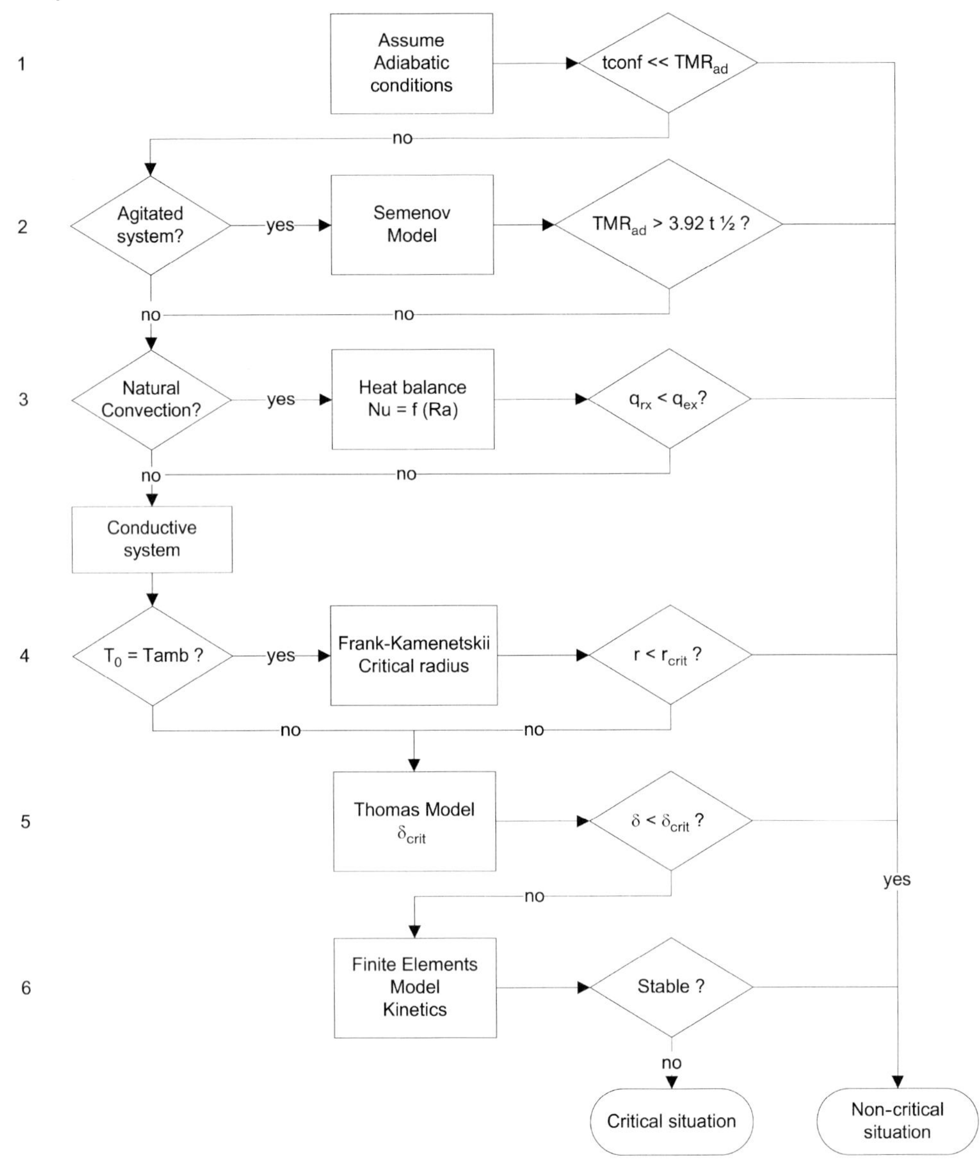

Figure 13.7 Decision tree for the assessment of thermal confinement situations.

Obviously, this comparison is only meaningful when the system is agitated, or at least it can be stated that an agitation is required to ensure safe conditions. In addition to the TMR_{ad} as a function of temperature, the data set also comprises the heat exchange data, that is, overall heat transfer coefficient and heat exchange area. When there is no agitation, or no agitation can be installed as for solids, a more detailed data set is required.

The third step checks if natural convection is sufficient to maintain heat losses, to provide a sufficient cooling capacity. The additional data required in this step comprises the variation of density as a function of temperature (β), the viscosity, and the thermal conductivity. This again is only meaningful as long as the reacting mass has a low viscosity allowing for buoyancy. If this data set is not sufficient or the natural convection cannot be established, for example, as for solids, the system must be considered as purely conductive.

In the fourth step, we consider a purely conductive system with an ambient temperature equal to the initial temperature of the reacting mass, corresponding to the Franck-Kamenetskii conditions. Besides the knowledge of the TMR_{ad} as a function of temperature, the additional data are the density of the system, its thermal conductivity, and the geometry of the vessel containing the reacting mass, that is, the form factor δ and the dimension of the vessel. If the situation is found to be stable under these relatively severe confinement conditions, the procedure can be stopped at this stage. In case the situation is assessed to be critical, the next step is required.

In the fifth step, heat exchange with the surroundings is also considered: the ambient temperature is different from the initial temperature of the reactive mass. This assessment also requires the heat transfer coefficient from the wall to the surroundings and uses the Thomas model. If the situation is assessed to be critical under these conditions, real kinetics can be used in order to give a more precise assessment.

In the sixth and last step, the system is still considered to be purely conductive, with heat exchange at the wall to the surroundings and the zero-order approximation of the kinetics is replaced by a more realistic kinetic model. This technique is very powerful in autocatalytic reaction, since a zero-order approximation leads to the very conservative assumption that the maximum heat release rate is realized at the beginning of the exposure and maintained at this level, respectively increasing with temperature, during the whole time period. In reality, the maximum heat release rate is delayed, and only achieved later on. Thus, heat losses may lead to a decreasing temperature during the induction time of the autocatalytic reaction.

By proceeding successively with these six steps, it is guaranteed that all required data are used, but without wasting time and experimental effort in the determination of useless data.

Worked Example 13.1: Storage Tank

An intermediate product with a melting point of 50 °C should be stored during two months in a cylindrical tank at 60 °C. The tank is a cylinder with vertical axis equipped with a jacket on the vertical wall allowing hot water circulation, but no agitator. The bottom and the lid are not heated. The volume is $4\,m^3$, the height 1.8 m, and the diameter 1.2 m. The corresponding shape factor is $\delta_{crit} = 2.37$. The overall heat transfer coefficient of the jacket is $50\,W m^{-2} K^{-1}$.

The physical properties of the product are

$$c'_P = 1800\,\mathrm{J \cdot kg^{-1} \cdot K^{-1}}$$

$$\rho = 1000\,\mathrm{kg \cdot m^3}$$

$$\lambda = 0.1\,\mathrm{W \cdot m^{-1} \cdot K^{-1}}$$

$$\mu = 100\,\mathrm{mPa \cdot s}$$

A slow exothermal decomposition with a high energy potential of 400 kJ kg^{-1} takes place in the product. The decomposition kinetics are characterized by the TMR_{ad}: 3500 days at 20 °C, 940 days at 30 °C, 280 days at 40 °C, 92 days at 50 °C, 32 days at 60 °C, and 12 days at 70 °C.

Question:
Assess the thermal safety of the intended storage. Propose technical solutions to improve the safety.

Solution:
The problem can be treated using the decision tree in Figure 13.7.

Thus, the first step is to assume adiabatic conditions: at 60 °C, the TMR_{ad} is 32 days, which is significantly shorter than the intended storage time of 2 months, thus assuming adiabatic conditions lead to a runaway situation.

Step 2 consists of assessing the heat exchange in an agitated system and since the tank is not equipped with a stirrer, this solution is not practicable.

Step 3 assumes natural convection. This may be difficult, because at 10 °C above the melting point, the viscosity may be too high for an efficient natural convection. This can be checked by calculating the Rayleigh criterion. The height of the tank is 1.8 m, but since it is not always full, a height of 0.9 m can be taken. With a dynamic viscosity of

$$\mu = 100\,\mathrm{mPa \cdot s}$$

and a thermal volume expansion of

$$\beta = 10^{-3}\,\mathrm{K^{-1}},$$

one obtains

$$Ra = \frac{g \cdot \beta \cdot L^3 \cdot \rho^2 \cdot C'_p \cdot \Delta T}{\mu \cdot \lambda} = 1.3 \cdot 10^9$$

This Rayleigh number corresponds to a turbulent film along the wall, but its value is close to the lower end of turbulent flow. If, for example, the viscosity increases to 1000 mPa·s, the Rayleigh criterion is only 10^8. Moreover, it is not certain that the convection will be established on the total height of the cylinder, that is, layering may occur. Thus, heat transfer by natural convection cannot be guaranteed.

Step 4: The system can be considered purely conductive following the Frank-Kamenetskii model. The thermal diffusivity is

$$a=\frac{\lambda}{\rho\cdot c_P'}=\frac{0.1}{1000\times 1800}=5.56\cdot 10^{-8}\,\mathrm{m^2\cdot s^{-1}}$$

This leads to a characteristic cooling time of

$$t_{cool}=\frac{0.5\cdot r^2}{a}=\frac{0.5\times 0.6^2}{5.56\cdot 10^{-8}}=3.24\cdot 10^6\,\mathrm{s}\simeq 900\,\mathrm{h}$$

This time corresponds to 37.5 days and is too short. Alternatively, the situation can be assessed using the critical radius:

$$r_{crit}=\sqrt{\frac{\delta_{crit}\cdot\lambda\cdot RT_0^2}{\rho\cdot q_0'\cdot E}}=\sqrt{\frac{2.37\cdot 0.1\cdot 8.314\cdot 333^2}{1000\cdot 0.006\cdot 100000}}=0.603\,\mathrm{m}$$

This radius is only slightly larger than the radius of the tank so no stable temperature profile will be established and a runaway develops.

Step 5: The heat exchange at the wall with the overall heat transfer coefficient of $50\,\mathrm{W m^{-2}\,K^{-1}}$ can also be considered, following the Thomas model. Thus, the Frank-Kamenetskii criterion (δ) is to be compared with the Thomas criteria (δ_{crit}):

The Biot criterion is

$$Bi=\frac{h\cdot r_0}{\lambda}=\frac{50\times 0.6}{0.1}=300$$

For a cylinder

$$\delta_{crit}=\frac{1+k}{e\cdot\left(\frac{1}{\beta_\infty}-\frac{1}{\mathrm{Bi}}\right)}=\frac{1+1}{2.71818\times\left(\frac{1}{2.72}-\frac{1}{300}\right)}=1.983$$

the reaction is

$$\delta=\frac{\rho_0\cdot q_0'}{\lambda}\cdot\frac{E}{RT_0^2}\cdot r_0^2=\frac{1000\times 0.006}{0.1}\times\frac{100000}{8.314\times 333^2}\times 0.6^2=2.34$$

Thus, since $\delta>\delta_{crit}$, the situation is instable and runaway will develop. Both criteria would have the same value at 55 °C, showing that the storage is close to the stability limit.

As a first conclusion, the thermal risk linked to the storage, as it is intended, appears to be high: the severity is high and the probability of occurrence is high too. Thus, the situation must be improved.

A first attempt could be to decrease the storage temperature, but this would mean decreasing the temperature to 50 °C, which corresponds to the melting point and so is not feasible.

A second attempt could be to install a stirrer in the tank.

With a stirrer, the overall heat transfer coefficient could be increased to 200 W m^{-2} K^{-1}. Since the heat transfer area for the filled tank is 2.26 m^2, the thermal time constant is

$$\tau = \frac{\rho \cdot V \cdot C_p'}{U \cdot A} = \frac{1000 \times 4 \times 1800}{200 \times 2.26} = 1.6 \cdot 10^4\ \mathrm{s} \simeq 4.4\ \mathrm{h}$$

and the half life is

$$t_{1/2} = \ln(2) \cdot \tau = 0.693 \times 4.4 = 3.1\ \mathrm{h}$$

Since the TMR_{ad} must be longer than 3.92 times the half-life, a TMR_{ad} of 12.1 hours is required. This value is reached at 105 °C. Thus, the heat release rate of the decomposition can easily be removed at 60 °C. Nevertheless, an additional heat input by the stirrer must also be accounted for. This solution is feasible. An alternative could be to circulate the tank contents through an external heat exchanger.

A third attempt could be to improve the natural convection. This could be achieved by a higher storage temperature that would also lead to a lower viscosity increasing the Rayleigh criterion. But this would also lead to an exponential increase of the heat release rate. Hence the situation cannot be improved in this way.

A fourth attempt could be to use a smaller tank, such as a tank with a diameter of only 1 m that would lead to a stable situation in the frame of the Frank-Kamenetskii model: the radius of 0.5 m is smaller than the critical radius of 0.603 m. But this solution means building a new tank.

As a fifth attempt, an increase of the heat transfer at the wall in the Thomas model is not practicable and would not be efficient, since the major part of the resistance to heat transfer is the conductivity in the product itself, as shown by the high value of the Biot criterion, 300, which is closer to Frank-Kamenetskii conditions than to Semenov conditions.

Thus, the last solution is to use a finite elements model to assess the situation. Such a calculation was performed with 40 concentric elements. The results are represented graphically in Figure 13.8. This shows that the situation is less critical than assumed, using the more conservative models. The temperature passes a maximum at 68 °C after approximately 1700 hours, that is, 70 days. It is worth noticing the long time-scale. Nevertheless, another point must also be considered: after 60 days of storage, the conversion is about 12%, meaning a significant loss in quality.

As a final recommendation, with respect to the high energy potential, the temperature should be monitored and provided with an alarm that would start a stirrer or a loop passing through an external heat exchanger. Since the phenomena are slow, the cooling system could be started manually. Moreover, the temperature should be monitored at the center of the tank as close as possible to the upper liquid surface, which is problematic in cases where the filling level of the tank changes. Thus, several temperature probes at different levels should be installed.

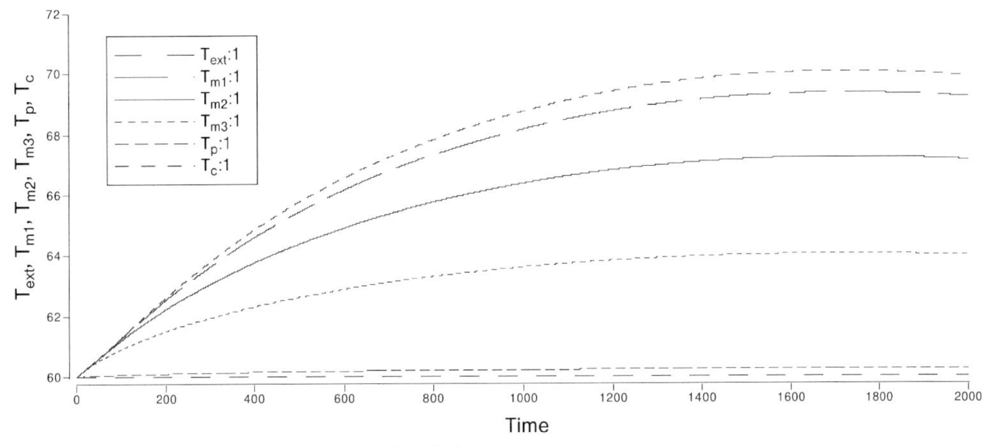

Figure 13.8 Temperature profiles in the tank at different positions: T_c at the center, T_{m1} at 25% of the radius, T_{m2} at 50% of the radius, T_{m3} at 75% of the radius, and T_p at the wall. Text is the surroundings temperature. The heat transfer at the wall is $50\,W\,m^{-2}\,K^{-1}$.

13.6 Exercises

► Exercise 13.1

A tubular reactor is to be designed in such a way that it can be stopped safely. The reaction mass is thermally instable and a decomposition reaction with a high energetic potential may be triggered if heat accumulation conditions occur. The time to maximum rate under adiabatic conditions of the decomposition is 24 hours at 200 °C. The activation energy of the decomposition is $100\,kJ\,mol^{-1}$. The operating temperature of the reactor is 120 °C. Determine the maximum diameter of the reactor tubes, resulting in a stable temperature profile, in case the reactor is suddenly stopped at 120 °C.

Physical properties of the reaction mass:

$$\rho = 800\,\mathrm{kg\,m^{-3}}$$

$$\lambda = 0.12\,\mathrm{W\,m^{-1}\,K^{-1}}$$

$$c_P' = 1800\,\mathrm{J\,kg^{-1}\,K^{-1}}$$

Hint: Consider that the heat transfer is purely conductive when the flow is stopped and the reactor wall remains at 120 °C.

Answer: $d \leq 30\,\mathrm{mm}$.

▶ Exercise 13.2

A reactive resin is to be discharged from a reactor to drums (radius 0.3 m, height 0.9 m). In order to obtain a low viscosity allowing a practicable transfer time, the discharge temperature should be above 75 °C. It is known that the heat release rate of the resin is $10\,\mathrm{W\,kg^{-1}}$ at 180 °C and the activation energy of the reaction is $80\,\mathrm{kJ\,mol^{-1}}$.

Question:
Is this operation thermally safe?

Data:

$$\rho = 1100\,\mathrm{kg\,m^{-3}}$$

$$\lambda = 0.1\,\mathrm{W\,m^{-1}\,K^{-1}}$$

$$c_P' = 2100\,\mathrm{J\,kg^{-1}\,K^{-1}}$$

$$\delta_c = 2.37$$

Answer: yes.

▶ Exercise 13.3

A liquid is stored in a $10\,\mathrm{m^3}$ tank (cylinder with vertical axis, diameter 2 m). The lower part of the tank is equipped with a jacket (height 1 m). At the storage temperature of 30 °C, the liquid shows a heat release rate of $15\,\mathrm{mW\,kg^{-1}}$. The tank is stirred with a propeller type agitator.

Question:
Would natural convection be sufficient to ensure a stable temperature in case of failure of the stirrer?

Data:

$$\rho = 1000\,\mathrm{kg\,m^{-3}}$$

$$\lambda = 0.1\,\mathrm{W\,m^{-1}\,K^{-1}}$$

$$c'_P = 2000\,\mathrm{J\,kg^{-1}\,K^{-1}}$$

$$\mu = 10\,\mathrm{mPa\,s}$$

$$\beta = 10^{-3}\,\mathrm{K^{-1}}$$

Answer: yes, even assuming a ΔT of 1 K.

References

1 Gygax, R. (1993) *Thermal Process Safety, Assessment, Criteria, Measures*, Vol. 8 (ed. ESCIS), ESCIS, Lucerne.

2 Taine, J. and Petit, J.P. (1995) *Cours et données de base transferts thermiques mécanique des fluides anisothermes*, 2nd edn. Dunod, Paris.

3 Perry, R. and Green, D. (eds) (1998) *Perry's Chemical Engineer's Handbook*, 7th edn. McGraw-Hill, New York.

4 Bourne, J.R., Brogli, F., Hoch, F. and Regenass, W. (1987) Heat transfer from exothermally reacting fluid in vertical unstirred vessels-2 free convection heat transfer correlations and reactor safety. *Chemical Engineering Science*, **42** (9), 2193–6.

5 VDI (1984) *VDI-Wärmeatlas Berechnungsblätter für den Wärmeübergang*, VDI-Verlag, Düsseldorf.

6 Frank-Kamenetskii, D.A. (1969) Diffusion and heat transfer in chemical kinetics, in *The Theory of Thermal Explosion* (ed. J.P. Appleton), Plenum Press, New York.

7 Gray, P. and Lee, P.R. (eds) (1967) Thermal explosion theory, in *Oxidation and Combustion Reviews*, Vol. 2 (ed. C.F.H. Tipper), Elsevier, Amsterdam.

8 Roduit, B., Borgeat, C., Berger, B., Folly, P., Alonso, B. and Aebischer, J.N. (2005) The prediction of thermal stability of self-reactive chemicals, from milligrams to tons. *Journal of Thermal Analysis and Calorimetry*, **80**, 91–102.

14 Symbols

Symbol	Name	Practical unit	SI unit
a	Thermal diffusivity	$m^2 s^{-1}$	$m^2 s^{-1}$
A	Heat exchange area	m^2	m^2
A, b, c	Polynomial coefficients	—	—
A,B,C,P,R . .	Chemical compounds	—	—
B	Reaction number	—	—
C	Concentration	$mol\,l^{-1}$	$mol\,m^{-3}$
C'	Concentration (in mass units)	$mol\,kg^{-1}$	$mol\,kg^{-1}$
c_P	Heat capacity	$J\,K^{-1}$	$J\,K^{-1}$
c'_P	Specific heat capacity	$kJ\,kg^{-1}\,K^{-1}$	$J\,g^{-1}\,K^{-1}$
d	Diameter or thickness	m	m
E	Activation energy	$J\,mol^{-1}$	$J\,mol^{-1}$
F	Molar flow rate	$mol\,h^{-1}$	$mol\,s^{-1}$
g	Acceleration of gravity	$m\,s^{-2}$	$m\,s^{-2}$
G	Controller gain	—	—
h	Film heat transfer coefficient	$W\,m^{-2}\,K^{-1}$	$W\,m^{-2}\,K^{-1}$
H	Height	m	m
ΔH	Molar enthalpy	$kJ\,mol^{-1}$	$J\,mol^{-1}$
$\Delta H'$	Specific enthalpy	$kJ\,kg^{-1}$	$J\,g^{-1}$
K	Constant, esp. equilibrium constant	—	—
k	Form coefficient	—	—
k	Kinetic rate constant	function of rate law	function of rate law
k_0	Frequency factor	function of rate law	function of rate law
l	Length	m	m
$\dot{m}$	Mass flow rate	$kg\,h^{-1}$	$g\,s^{-1}$
M	Mass	kg	g
M	Molar ratio	—	—
M_w	Mol weight	$g\,mol^{-1}$	$g\,mol^{-1}$
n	Order of reaction	—	—
n	Revolution frequency	rpm	s^{-1}
N	Number of moles	—	—
P	Pressure	bar	Pa
q	Heat release rate	W	W

Thermal Safety of Chemical Processes: Risk Assessment and Process Design. Francis Stoessel

ISBN: 978-3-527-31712-7

Symbol	Name	Practical unit	SI unit
q'	Specific heat release rate	$W\,kg^{-1}$	$W\,g^{-1}$
Q	Thermal energy (heat)	kJ	J
Q'	Specific energy	$kJ\,kg^{-1}$	$J\,g^{-1}$
r	Reaction rate	$mol\,m^{-3}\,h^{-1}$	$mol\,m^{-3}\,s^{-1}$
r	Radius	m	m
R	Universal gas constant	$J\,mol^{-1}\,K^{-1}$ $l \cdot mbar \cdot mol^{-1}K^{-1}$	$J\,mol^{-1}\,K^{-1}$
S	Cross-section	m^2	m^2
t	Time	h	s
T	Temperature	°C	K
ΔT_{ad}	Adiabatic temperature rise	°C	K
u	Linear velocity	$m\,s^{-1}$	$m\,s^{-1}$
U	Overall heat transfer coefficient	$W\,m^{-2}\,K^{-1}$	$W\,m^{-2}\,K^{-1}$
$\dot{v}$	Volume flow rate	$m^3\,h^{-1}$	$m^3\,s^{-1}$
V	Volume	m^3	m^3
X	Conversion	—	—
z	Length coordinate	m	m
z	Equipment constant (stirred tank)	—	—

Subscripts

Subscript	Meaning	Example	
0	Initial value	T_0	Initial temperature
A,B,P,R,S	Reference to chemical compounds	C_A	Concentration of A
ac	Accumulation	X_{ac}	Degree of accumulation
ad	Adiabatic	ΔT_{ad}	Adiabatic temperature rise
amb	Ambient	T_{amb}	Ambient temperature
b	Boiling	T_b	Boiling point
c	Coolant	T_c	Temperature of coolant
$cell$	Calorimetric cell	M_{cell}	Mass of calorimetric cell
cf	Cooling failure	T_{cf}	Temperature after cooling failure
$cond$	Condensation	q_{cond}	Cooling capacity of condenser
$crit$	Critical	T_{crit}	Critical temperature
cx	Convective	q_{cx}	Convective heat flow
d	Decomposition	ΔH_d	Enthalpy of decomposition
$D24$	TMR_{ad} = 24 hours	T_{D24}	Temperature at which TMR_{ad} = 24 hours
ex	Exchange (heat exchange)	q_{ex}	Heat dissipation rate
ex	Explosion	V_{ex}	Volume of explosive atmosphere
f	Final	T_f	Final temperature
fd	Feed	T_{fd}	Feed temperature
g	Gas	ρ_G	Gas density
H	Hydraulic	d_h	Hydraulic diameter

Subscript	Meaning	Example	
i	Index	C_i	Concentration of ith compound
l	Liquid	ρ_l	Liquid density
$loss$	Loss	q_{loss}	Rate of heat loss
max	Maximum	T_{max}	Maximum temperature
mes	Measure	T_{mes}	Measurement temperature
p	Process	T_P	Process temperature
r	Reactor, reaction mass	d_r	Diameter of reactor
ref	Reference	T_{ref}	Reference temperature
rx	Reaction	q_{rx}	Heat release rate of reaction
s	Stirrer	d_s	Stirrer diameter
st	Stoichiometric	$M_{R,st}$	Reaction mass at stoichiometry
tox	Toxic	V_{tox}	Volume of toxic atmosphere
v	Vapor	ΔH_v	Latent enthalpy of evaporation
w	Wall	T_w	Wall temperature

Greek

Symbol	Name	Practical unit	SI unit
α	Relative volume increase	—	—
α	Heat loss coefficient	$W \cdot K^{-1}$	$W\,K^{-1}$
β	Effective Biot number	—	—
β	Volumic expansion coefficient	K^{-1}	K^{-1}
γ	Material constant for heat transfer	$W\,m^{-2}\,K^{-1}$	$W\,m^{-2}\,K^{-1}$
δ	Form factor, Frank–Kamenetskii number	—	—
Δ	Difference (used as prefix)	—	—
ε	Expansion coefficient	—	—
φ	Heat tranfer coefficient of equipment	$W\,m^{-2}\,K^{-1}$	$W\,m^{-2}\,K^{-1}$
Φ	Adiabacity coefficient	—	—
λ	Thermal conductivity	$W\,m^{-1}\,K^{-1}$	$W\,m^{-1} \cdot K^{-1}$
μ	Dynamic viscosity	cP = mPa.s	Pa.s
ν	Kinematic viscosity	$m^2\,s^{-1}$	$m^2\,s^{-1}$
ν_A	Stoichiometric coefficient of A	—	—
π	3.141 59 ...		
θ	Dimensionless temperature	—	—
θ	Dimensionless time	—	—
ρ	Specific weight	$kg\,m^{-3}$	$g\,m^{-3}$
σ	Surface tension	$N\,m^{-1}$ = 103 dyne $\cdot$ cm^{-1}	$N\,m^{-1}$ (= $kg\,s^{-2}$)
σ	Variance (square root of ...)	—	—
τ	Time constant	h	s
τ	Space time	h	s
ψ	Semenov criterion	—	—

Acronyms

BR	Batch reactor
CSTR	Continuous stirred tank reactor
DSC	Differential scanning calorimetry
DTA	Differential thermal analysis
IDLH	Immediately dangerous to life and health
LEL	Lower explosion limit
MTSR	Maximum temperature of synthesis reaction
MTT	Maximum temperature for technical reasons
PFR	Plug flow reactor
SBR	Semi-batch reactor
TMR_{ad}	Time to maximum rate under adiabatic conditions
TNR	Time of no return

Dimensionless Groups

Symbol	Name	Expression	Parameter	
B	Thermal reaction number	$B=\frac{\Delta T_{ad}\cdot E}{R\cdot T^2}$	ΔT_{ad}	Adiabatic temperature rise
			E	Activation energy
			R	Universal gas constant
			T	Temperature
Bi	Biot	$Bi=\frac{h\cdot r_0}{\lambda}$	h	Film heat transfer coefficient
			r	Radius
			λ	Thermal conductivity
Da	Damköhler	$Da=\frac{r_0\cdot t}{C_0}$	r_0	Reaction rate
			t	Reaction time
			C_0	Initial concentration
Fo	Fourier	$Fo=\frac{at}{r^2}$	a	Thermal diffusivity
			t	Time
			r	Characteristic dimension
Gr	Grashof	$Gr=\frac{g\cdot\beta\cdot L^3\cdot\rho^2\cdot\Delta T}{\mu^2}$	g	Gravity constant
			β	Volumetric expansion
			ρ	Specific weight
			L	Characteristic length
			ΔT	Temperature difference
			μ	Dynamic viscosity
			λ	Thermal conductivity
Ne	Newton (power number)	$Ne=\frac{P}{\rho\cdot n_S^3\cdot d_S^5}$	P	Power of stirrer
			ρ	Specific weight of fluid
			n_S	Revolution speed
			d_S	Stirrer diameter
Nu	Nusselt	$Nu=\frac{h\cdot d}{\lambda}$	h	Film heat transfer coefficient
			d	Characteristic length
			λ	Thermal conductivity

Symbol	Name	Expression	Parameter	
Pr	Prandtl	$Pr = \frac{\mu \cdot c_P'}{\lambda}$	μ	Dynamic viscosity
			cp	Specific heat capacity
			λ	Thermal conductivity
Ra	Rayleigh	$Ra = \frac{g \cdot \beta \cdot \rho^2 \cdot c_P' \cdot L^3 \cdot \Delta T}{\mu \cdot \lambda}$	g	Gravity constant
			β	Volumetric expansion coefficient
			ρ	Specific weight
			cp	Specific heat capacity
			L	Characteristic length
			ΔT	Temperature difference
			μ	Dynamic viscosity
			λ	Thermal conductivity
Re	Reynolds (tube)	$Re = \frac{u \cdot d \cdot \rho}{\mu}$	u	Flow velocity
			d	Diameter
			ρ	Specific weight
			μ	Dynamic viscosity
Re	Reynolds (stirred tank)	$Re = \frac{n \cdot d^2 \cdot \rho}{\mu}$	n	Stirrer frequency
			d	Diameter of agitator
			ρ	Specific weight
			μ	Dynamic viscosity
St	Stanton (modified)	$St = \frac{U \cdot A \cdot t}{\rho \cdot V \cdot c_p'}$	U	Heat transfer coefficient
			A	Heat transfer area
			t	Characteristic time
			ρ	Specific weight
			V	Volume
			c_p'	Specific heat capacity

Index

a
accelerating rate calorimetry (ARC) 85, 89, 128, 298
acceleration 313f.
accidents 179
– chemical industry 5
– statistics 5
accumulation 62f., 72f., 85, 96, 104, 110ff., 122, 139, 153, 156ff., 163ff., 167ff., 174, 183, 186, 198ff., 233, 244ff., 255, 264f., 268, 272
– controlled feed 173ff.
activation energy 41f., 49, 56, 106, 248, 271, 286f., 289ff., 297, 316ff.
addition 149
– by portions 167
adiabacity coefficient 193
adiabatic 31ff., 48f., 60, 66, 83ff., 112, 123, 127ff., 133, 142f., 159, 166, 182ff., 193, 215, 255, 294ff., 313, 319, 324ff., 336ff., 347, 351f., 381
adiabatic calorimetry 330
adiabatic temperature 72
adiabatic temperature course 50
adiabatic temperature rise 37, 48f., 62ff., 105f., 108, 111f., 127ff., 141, 160, 165, 183, 244, 257, 269, 283, 311, 325
agitation 220, 352
agitator 46, 221
agitator constant 222
alarm 251f., 273, 318
alarm systems 251
Arrhenius 40, 261, 286,289, 290, 295
as low as reasonably practicable (ALARP) 14, 273
assessment 73
– thermal risk 72
assessment criteria 12f., 65ff., 104, 260
attenuation 244
autocatalytic 288, 294, 311f., 314ff., 324f., 327, 330f., 350, 353
autoclave 119ff.
autothermal 185
average temperature difference 222

b
batch 299, 302, 331
batch distillation 311
batch process 23
batch reaction 43, 61, 105, 215
batch reactor 111, 119ff., 267
Biot 343, 349, 355f.
boiling 103, 226, 230f.
boiling point 39f., 228f., 231f., 259, 268, 271
boiling rate 227
brainstorming 23
brine 207ff.
buoyancy 340, 353
bursting disk 255

c
calibration 88, 91
calorimeter 201, 298
– differential 83
calorimetry 37, 42ff., 62f., 72, 82, 133, 288, 313
– adiabatic 329
Calvet calorimetry 138, 244, 247, 261, 296ff.
carcinogenic 18
cascade controller 169, 219
cascade reactor 198
cascade temperature controller 135, 159
catalyst 313
catalytic 302f., 319
catalytic effect 285
charge 128, 141f., 162

Thermal Safety of Chemical Processes: Risk Assessment and Process Design. Francis Stoessel

ISBN: 978-3-527-31712-7

charging errors 120, 302
check list 21
chemical properties 17
circulation systems 209
closed loop 207
coils 208
combustion energy 286
combustion index 19
competitive reactions 41
concentration 141, 149f., 153f., 156f., 160ff., 170, 179ff., 200
condensation 103
condenser 228
conduction 43, 337f., 348ff.
conductive 336f., 353
– heat transfer 220
confidence interval 329
confinement 209, 336
– heat 335
– thermal 94
consecutive reactions 41
containment 256, 267
contamination 322, 332
continuous processes 244
continuous reactor 46
continuous stirred tank reactor (CSTR) 43, 179
control of feed 245
control parameter 214
controllability 132, 260, 263, 266
controlled depressurization 70
convection 43, 46f., 355
convective cooling 163
conversion 40, 43, 62, 75, 83, 87, 96, 132, 135, 138ff., 150f., 167, 182, 186ff., 198, 250, 261, 297, 315f., 319, 324, 328, 344, 350
– thermal 158, 167
cooling, evaporative 125, 229, 265f.
cooling capacity 44, 50, 53, 62, 73, 107, 124ff., 149, 152, 159, 162ff., 174f., 193f., 200, 204, 208f., 215, 226f., 246, 261f., 353
cooling curve 226
cooling experiment 222ff., 234
cooling failure 61ff., 104, 111f., 127, 136, 140, 143, 155f., 171, 186ff., 194, 246, 255, 270
cooling failure scenario 64, 67, 126, 162f., 243
cooling rate 107
cooling system 50f., 114, 123, 129, 132ff., 147, 164, 222
cooling time 218, 338f., 351
corrosion 179
criteria 14
critical limits 10
critical radius 195, 345f., 355f.
critical temperature 51ff.
criticality 67ff., 163, 257f., 264ff.
crucible 86, 90ff., 300
cycle time 217ff.
cyclone 256

d

damage 12
Damköhler 109, 150, 165, 172
Damköhler criterion 109
decision table 25
decomposition 3, 31, 37f., 59ff., 69f., 92ff., 120, 159, 163, 169, 172f., 180, 227f., 258, 268ff., 282ff., 291, 298f., 303f., 311, 314, 321f., 325, 331, 354
– energies 35
– self-sustaining 19
decomposition reaction 54, 113
deductive method 11, 26
degree of filling 233, 263
depressurization 248ff., 265f.
design 158
design institute of emergency relief systems (DIERS) 71, 254
design intention 23
detection limit 286, 290ff.
deviations 104, 169f.
– search 10
Dewar 88f., 128
differential calorimetry 83, 92
differential scanning calorimetry (DSC) 82, 85, 90, 128, 138, 141, 161, 244, 261, 284ff., 313f., 320ff., 328
– thermogram 31f.
dimensionless criterion 109
dimensionless numbers 220
direct steam 206
direct cooling 208
direct heating 208
direct steam injection 206
distance rule 294
domino effect 245
dumping 70, 93, 248
dust explosion 39
dynamic 125ff., 169, 199, 228, 234
dynamic simulation 255
dynamic stability 114f., 130
dynamics 204, 215ff., 233, 252

e
ecotoxic properties 17
ecotoxicity 19
effluent treatment 256
electrical heating 207
electrostatic charges 19
eliminating measures 243f.
emergency 241, 247f.
emergency cooling 207f., 227, 246, 251f., 281
emergency cooling system 55, 261
emergency measures 66, 141, 243, 253, 267
emergency plans 25
emergency pressure relief 69ff., 253ff., 263
emergency response planning guidelines (EPRG) 18, 258
endothermal 42, 92
energy 34, 82f.
– activation, *see* activation energy
– combustion 286
– loss 9
– mechanical 38
– of decomposition 35
energy balance 181
energy potential 72, 87, 92, 120, 159, 162, 351, 354
energy release 244, 285f.
enthalpy 34, 37, 284
– evaporation 39f.
– formation 35
– latent 39
– of formation 34
– specific 127
environment 4, 7, 12, 15
estimation 284f.
ethylene glycol 207
evaporation
– enthalpy 39f.
– rate 228
evaporative cooling 69f., 125, 226ff., 248, 270ff.
evaporator 207
event tree analysis 25
excess 157
exothermal 42ff., 61, 64, 81, 90f.
exothermal reaction 89, 92, 107, 228
explosion 39, 59, 65, 119, 200, 228, 253, 258ff., 283, 305
– dust 39
– thermal 31ff.
– vapor 39
explosion limits 19
external film 220ff., 234

f
failure mode and effect analysis 22
fatal accident rate (FAR) 5
fault tree 27
fault tree analysis 26
feed 46, 148, 154ff., 159, 181, 185, 188
feed by portions 245
feed control 158, 167, 173ff., 245
feed rate 149ff., 164ff., 245, 264, 267
feed time 158, 163ff.
filling level 357
finite elements 350
flashpoint 19
flood 335
flooding 71, 227ff., 247ff., 272
forced convection 337
fouling 52, 220, 227
Fourier 341ff.
Frank-Kamenetskii 195, 200, 343ff.
friction 20

g
gain of the cascade 219
gas production 259
gas release 39, 95, 153, 203, 254, 258, 262, 268ff.
gas release rate 73, 141, 154, 158, 262, 271
gas velocity 262ff.
golden rules 170
Grashof 340

h
half-coils 208
half-life 217, 339
half-welded coils 224
hazard and operability study (HAZOP) 23f., 28
hazard catalog 28
hazard identification 3f.
hazards 7
– catalog 11
– search 11, 20
health safety and environment 7
heat 93
– confinement 335
– decomposition 35
– latent, *see* latent heat
– mixing 247
– overall 43, 47
– reaction 85, 92
– sensible 183
heat accumulation 45, 69, 152, 215, 264, 335ff., 351

heat balance 42, 48, 51, 83ff., 92, 104ff., 115, 122, 134, 151ff., 164ff., 180ff., 190f., 212, 222f., 247, 254, 283, 328, 335ff., 343, 349
heat capacity 45f., 87f., 130, 140, 151, 160, 183, 191ff., 198ff., 223, 234, 268, 287, 291ff., 323, 339
– specific 35ff., 127
heat carrier 205ff.
– required temperature 218
heat confinement 336, 347
heat exchange 45, 59, 85, 123, 127ff., 143, 152, 175, 188, 191f., 215, 219, 234
heat exchange area 44, 193
heat exchange capacities 214
heat exchange coefficient 105, 152, 169
heat exchange system 121
heat exchanger 185, 206, 209
heat flow 85
heat loss 47, 88f., 234, 347, 353
heat of reaction 34, 140, 166, 183, 268
– specific 37
heat production 43, 142, 155, 188
heat production rate 149
heat release 120, 198
heat release rate 40ff., 51, 62, 73, 83, 94ff., 107, 114f., 123ff., 140, 143, 151, 154, 158ff., 171, 174, 183, 227ff., 248, 251, 255, 261ff., 271, 283, 287ff., 315f., 320ff., 336ff., 353
– specific 291
heat removal 43, 115, 142, 151, 155, 162, 192, 195, 283, 336
heat removal capacity 160
heat transfer 199, 221f., 227, 233, 336, 343, 349, 355ff.
heat transfer area 52
heat transfer coefficient 45, 52, 195, 199, 208, 217ff., 234, 246, 339ff., 349ff.
heat transport 207
heating experiment 224
heating rate 136ff.
heating time 218
hot spot 19
hot water heating 206
human error 5ff., 23
hysteresis 186

i

ice 207
ideal reactors 196
IEC 61511 273
ignition 111, 186
ignition source 19
immediately dangerous to life and health (IDLH) 18, 258f.
impurities 303, 331
incident 9, 59, 147, 179, 203
independent protection levels (IPL) 273ff.
indirect heating 208
induction time 313, 322, 331f.
inductive methods 11, 25
inherent safety 16
inherently safe processes 15, 110, 244f., 252
inhibitor 241, 248, 304
initiation 123, 129ff., 141, 235, 317
instable operating point 51
instruments, design 23
integrated process development 244
intensification 244
interactions 20
interlock 16, 169, 246, 252, 265, 268
intermediates 301f.
internal film 220f., 227
intuitive methods 11
isoconversional 296ff., 324
isoperibolic 84f., 91, 123, 131ff., 159, 163, 166, 212ff., 236
isothermal 83ff., 89ff., 123ff., 142f., 151ff., 187f., 194, 200ff., 212ff., 244, 261, 288, 313, 316, 320, 331f.
isothermal reaction 194

j

jacket 208, 212, 227, 235

k

Kamenetskii, *see* Frank-Kamenetskii
kinetic 42, 83, 95, 109, 119, 124, 138, 235, 261, 290, 296ff., 326, 345, 349
kinetic data 113
kinetic excess 157
kinetics 87, 183, 201, 233, 250, 263, 286, 312
– n^{th}-order 40

l

latent enthalpy 39
latent heat 205ff., 230, 250, 259f.
laws 8
loss of control 125, 261

m

maintenance 7, 15, 22, 26
mass balance 121, 181ff., 188f., 328
material balance 105, 150

material safety data sheet (MSDS) 5, 17
maximum allowed work place concentration (MAC) 18
maximum temperature for technical reasons (MTT) 68ff., 245, 250, 258, 261, 264ff., 271
maximum temperature of the synthesis reaction (MTSR) 61ff., 104, 112, 125ff., 133ff., 160ff., 186, 245, 250, 258ff., 264ff.
measures 21
– eliminating 15
– mitigation 15
– organizational 16
– preventive 15
– risk-reducing 28
– technical 16
mechanical energy 38, 46
melting point 103
micro reactors 199, 244
minimum ignition energy 20
mixing 93, 151, 180, 233, 246, 332
mixing enthalpy 151
mixing heat 247
model 234f., 321
moderator 28
multi-purpose 5, 255, 273
multi-purpose plant 13, 17, 21, 27, 189, 233

n
natural convection 227, 246, 336ff., 353ff.
Newtonian fluids 220
nitro 120, 147, 281, 320
n^{th}-order reaction 124
numerical simulation 235
Nusselt 220, 340

o
omission 303
on-line measurement 175
onset 286, 294f.
operating conditions, normal 104
operating mode 21
optimization 161, 169f.
oscillations 184
overcharge 119
overheating 141
overshoot 130, 136, 209, 236
oxidation 285

p
parallel reaction 175, 235
parametric sensitivity 52, 104ff., 115, 130ff., 164ff., 185, 192, 351
passive measure 257
passive safety 206
performance equation 122, 181, 190
physical properties 17
physico-chemical properties 152
plant operation 95
plug flow 180, 189f., 198f.
polymerization 285, 320
polymers 220, 227
polytropic 123, 128f., 131, 166, 192, 209, 215
portions 159
potential 37
Prandtl 220
pre-exponential factor 297
pressure 38, 65, 87ff., 103, 111, 119, 125, 128, 141, 193, 242ff., 251ff., 298, 301
pressure control 205
pressure increase rate 261
pressure relief 229, 259, 267
pressure relief system 103, 119, 147f., 242
preventive measures 15, 243ff.
probability 7, 13, 27, 60, 66f., 71f., 252, 260ff., 283, 296, 303, 324, 356
probability of occurrence 12, 24
process 252
– automation 15
– batch 23
– continuous 23
– control 8, 235
– design 23, 245
– development 95, 244
– deviations 9, 302
– equipment 23
– intensification 199
– interruption 300
product safety 4
productivity 171, 233, 244, 247
protection system 27
public attention 4

q
QFS (Quick onset, Fair conversion and Smooth temperature profile) 110f., 164
quantification 26
quenching 70, 93, 247f., 273

r
radiation 43, 47
rate equation 113, 124, 149, 315
rate of evaporation 228

Rayleigh 340f., 354
reactant addition 149
reaction 37, 93
– n^{th}-order 43
– energy 64, 141ff.
– enthalpy 35, 43, 165
– heat 85, 92
– kinetics 56, 180
reaction calorimeter 95, 139, 158, 161, 167, 221, 224, 234ff., 261
reaction number 105f., 114, 133
reaction rate 40, 48, 106, 109, 121f., 125, 133, 139, 150, 153ff., 161, 254
reactor safety 120
reactor stability 52, 104
reactor wall 222
recycle reactor 198
reflux 95, 159, 214, 227
reflux system 71
reliability 13ff., 27, 242f., 273f., 329
relief capacity 255
residual risk 16
resistance 220
responsibility 6f., 16
Reynolds 220f., 340
Reynolds number 45
risk 7
– accepted 8, 14ff.
– analysis 4, 8
– analysis team 16, 28
– assessment 8ff., 21, 60, 273
– diagram 14
– elimination 15
– identified 16
– improvement 9
– matrix 14
– mitigation 15
– perception 5–f
– policy 13f.
– prevention 15
– profile 9, 13ff., 273
– reducing measures 14, 26, 69, 243
– residual 11, 16, 21
– thermal 60
– unidentified 16
runaway 15, 32, 37, 40, 50ff., 60ff., 84ff., 92, 107ff., 125, 160, 164, 170ff., 203, 209, 228f., 241ff., 256f., 263, 267, 273, 283, 294, 314, 317, 324f., 335ff., 355
runaway reaction 243
rupture disk 3

s

safe batch reaction 141
safe design 138
safe operation 141
safety 8
– margins 72ff., 294
– measures 112, 153
– process 112
safety assessment 46, 128, 160
safety barrier 70f.
safety data 10, 17
safety instrumented system 243, 273
safety integrated level (SIL) 273f.
safety trips 16
safety valves 255
sample 299
scale-down 136, 234ff.
scale-up 44, 47, 52, 95, 109, 139, 167, 217, 233f.
scan rate 286, 296f.
scenario 7, 11ff., 25, 60f., 68ff., 125, 139, 228, 254f., 261, 264, 273, 336
screening 300
secondary circulation 213
secondary circulation loop 211
secondary circulation system 207
secondary reaction 63f., 111f., 125, 128, 139
security 8
selectivity 153, 159, 167, 170, 233ff.
self-accelerating 312, 320
self-heating 49, 350
self-sustaining decomposition 19
Semenov 50ff., 108, 114, 338, 343, 352
Semenov criterion 107
semi-batch 147, 245, 264, 267, 281, 299ff.
semi-batch reactor 46, 110f., 244
sensible heat 46, 183
sensitive heat 88
sensitivity 105, 111, 133, 164, 201, 330
– parametric 132
separator 256
set point 219
severity 7, 12, 24, 37ff., 60ff., 92, 111, 162, 244, 257ff., 283, 325f., 356
shape coefficient 349
shock 20
simplification 245
single circuit 210
solid 341ff., 352
solvents 303
specific enthalpy 127

specific heat 34
specific heat capacity 35ff., 64, 127
specific heat of reaction 37
specific heat release rate 291
spillage 17ff., 253
stability 94, 138, 161, 182ff., 351
– criteria 104
– thermal 171
stability diagram 107f.
stable equilibrium point 51
stable operating point 51
Stanton-criterion 109
steady state 181
steam 103
steam heating 205, 208
stirred tank 221
stirred tank reactor 224
stirrer 45f., 88, 121, 147f., 170f., 179, 246, 354ff.
stirrer power 234
stirrer speed 221
stoichiometric point 153, 158
stoichiometric ratio 150
stoichiometry 154ff., 172f.
storage 31f., 54, 94, 335ff., 351, 356
sulfonation reaction 3
surroundings temperature 347f.
swelling 228ff., 262f.
synthesis reaction 113

t
T_{D24} 68ff., 73, 266, 271, 290ff., 328
team leader 28
technical failures 10
technical limits 67
technical measures 242, 245, 257
– organizational 15
– procedural 15
temperature alarm 170
temperature control 114, 120ff., 134, 141, 158f., 205, 211f.
temperature control system 169, 204, 233
temperature profile 195, 357
temperature sensitivity 106
tempered system 259
tempering 71
thermal activity monitor 94, 296
thermal analysis 72
thermal conductivity 342, 353
thermal confinement 94, 352
thermal conversion 158, 167
thermal diffusivity 340, 343
thermal energy 46
thermal explosion 31ff., 49f., 60, 65, 125, 180, 324, 350f.
thermal history 316ff., 331
thermal inertia 86f., 159, 200, 233
thermal potential 63, 127
thermal risk 60
thermal safety 64, 74
thermal safety assessment 71
thermal stability 40, 63, 72, 86, 93, 134, 171, 179, 228, 248f., 261, 282, 299, 302f., 311, 325, 331
thermal time constant 48, 108f., 183, 199, 215ff., 223ff., 339
thermochemical data 138
time constant 133, 217, 233, 242
– thermal 108f.
time factor 252
time of no return (TnoR) 55, 246
time scale 338
time to explosion 50
time to maximum rate 163, 286, 298
time-cycle 128, 171
time-scale 66, 336f.
TMR_{ad} (Time to Maximum Rate under adiabatic conditions) 54ff., 64, 67, 72ff., 83, 92, 252, 261ff., 283, 290ff., 313f., 318f., 323ff., 337ff., 351ff.
toxicity 17f., 39, 258ff., 270
trajectory 123, 130ff., 251
transfer coefficient 43, 47
tubular reactor 43, 189
tuning 214, 219
two-film theory 219f., 336
two-phase flow 120, 148, 231, 242, 251

v
vacuum 125, 214, 226
van't Hoff 33, 41, 290ff.
vapor 267
vapor explosion 39
vapor flow rate 40, 264
vapor mass flow rate 262
vapor pressure 38f., 112, 254, 258f., 270
vapor release 130, 258f., 264f., 269f.
vapor release rate 250, 262
vapor velocity 71, 229ff., 251, 261ff., 272
viscosity 43, 46, 62, 170, 220, 246, 340f., 353f.

volatile 92, 284
volume expansion 151
volume increase 231f.

w
wall breakthrough 211
warning 111
welded half coil 227
Wilson plot 221, 224
worst-case 72f., 92, 351

z
zero-order 50, 290, 295, 323f., 353
– reaction 49, 53f., 287